T0329565

Communication Systems Principles Using MATLAB®

Communication Systems Principles Using MATLAB®

John W. Leis

University of Southern Queensland

This edition first published 2018
© 2018 John Wiley & Sons, Inc.

Registered Office
John Wiley & Sons, Inc., 111 River Street, Hoboken, NJ 07030, USA

Editorial Office
111 River Street, Hoboken, NJ 07030, USA

For details of our global editorial offices, customer services, and for information about Wiley products visit us at www.wiley.com.

Wiley also publishes its books in a variety of electronic formats and by print-on-demand. Some content that appears in standard print versions of this book may not be available in other formats.

Library of Congress Cataloging-in-Publication Data

Names: Leis, John, 1966– author.
Title: Communication systems principles using MATLAB / by John W. Leis.
Description: Hoboken, NJ : John Wiley & Sons, 2018. | Includes bibliographical references and index. |
Identifiers: LCCN 2018008692 (print) | LCCN 2018021192 (ebook) | ISBN 9781119470687 (pdf) |
    ISBN 9781119470755 (epub) | ISBN 9781119470670 (cloth)
Subjects: LCSH: MATLAB. | Data transmission systems–Computer simulation. |
    Telecommunication systems–Computer simulation.
Classification: LCC TK5105 (ebook) | LCC TK5105 .L45 2018 (print) | DDC 621.3820285/53–dc23
LC record available at https://lccn.loc.gov/2018008692

Cover Design: Wiley
Cover Image: © Nongkran_ch/getty images

Set in 10/12pt Warnock by SPi Global, Pondicherry, India

10 9 8 7 6 5 4 3 2 1

To my father, Harold, and my mother, Rosalie.
For all you have done for us, and teaching that
mountains are there to be climbed.

To Debbie, Amy, and Kate.
For enriching my life beyond imagination.

Contents

Preface *xiii*
Acknowledgments *xv*
Introduction *xvii*
About the Companion Website *xxi*

1   **Signals and Systems** *1*
1.1   Chapter Objectives *1*
1.2   Introduction *1*
1.3   Signals and Phase Shift *2*
1.4   System Building Blocks *3*
1.4.1   Basic Building Blocks *3*
1.4.2   Phase Shifting Blocks *4*
1.4.3   Linear and Nonlinear Blocks *5*
1.4.4   Filtering Blocks *8*
1.5   Integration and Differentiation of a Waveform *10*
1.6   Generating Signals *16*
1.7   Measuring and Transferring Power *19*
1.7.1   Root Mean Square *19*
1.7.2   The Decibel *23*
1.7.3   Maximum Power Transfer *25*
1.8   System Noise *29*
1.9   Chapter Summary *32*
         Problems *32*

2   **Wired, Wireless, and Optical Systems** *37*
2.1   Chapter Objectives *37*
2.2   Introduction *37*
2.3   Useful Preliminaries *38*
2.3.1   Frequency Components When a Signal Waveform Is Known *38*
2.3.2   Frequency Spectrum When a Signal Is Measured *42*
2.3.3   Measuring the Frequency Spectrum in Practice *44*

2.4   Wired Communications   *50*
2.4.1   Cabling Considerations   *50*
2.4.2   Pulse Shaping   *52*
2.4.3   Line Codes and Synchronization   *62*
2.4.4   Scrambling and Synchronization   *66*
2.4.5   Pulse Reflection   *73*
2.4.6   Characteristic Impedance of a Transmission Line   *80*
2.4.7   Wave Equation for a Transmission Line   *83*
2.4.8   Standing Waves   *84*
2.5   Radio and Wireless   *92*
2.5.1   Radio-frequency Spectrum   *92*
2.5.2   Radio Propagation   *92*
2.5.3   Line-of-sight Considerations   *96*
2.5.4   Radio Reflection   *97*
2.5.5   Radio Wave Diffraction   *99*
2.5.6   Radio Waves with a Moving Sender or Receiver   *103*
2.5.7   Sending and Capturing a Radio Signal   *105*
2.5.8   Processing a Wireless Signal   *119*
2.5.9   Intermodulation   *128*
2.5.10   External Noise   *131*
2.6   Optical Transmission   *132*
2.6.1   Principles of Optical Transmission   *132*
2.6.2   Optical Sources   *134*
2.6.3   Optical Fiber   *139*
2.6.4   Optical Fiber Losses   *145*
2.6.5   Optical Transmission Measurements   *147*
2.7   Chapter Summary   *150*
           Problems   *151*

3   Modulation and Demodulation   *155*
3.1   Chapter Objectives   *155*
3.2   Introduction   *155*
3.3   Useful Preliminaries   *156*
3.3.1   Trigonometry   *157*
3.3.2   Complex Numbers   *159*
3.4   The Need for Modulation   *162*
3.5   Amplitude Modulation   *164*
3.5.1   Frequency Components   *167*
3.5.2   Power Analysis   *170*
3.5.3   AM Demodulation   *171*
3.5.4   Variations on AM   *173*
3.6   Frequency and Phase Modulation   *180*
3.6.1   FM and PM Concepts   *181*

3.6.2   FM and PM Analysis   *183*
3.6.3   Generation of FM and PM Signals   *185*
3.6.4   The Spectrum of Frequency Modulation   *186*
3.6.5   Why Do the Bessel Coefficients Give the Spectrum of FM?   *195*
3.6.6   FM Demodulation   *200*
3.7   Phase Tracking and Synchronization   *204*
3.8   Demodulation Using *IQ* Methods   *215*
3.8.1   Demodulation of AM Using *IQ* Signals   *216*
3.8.2   Demodulation of PM Using *IQ* Signals   *219*
3.8.3   Demodulation of FM Using *IQ* Signals   *222*
3.9   Modulation for Digital Transmission   *225*
3.9.1   Digital Modulation   *226*
3.9.2   Recovering Digital Signals   *228*
3.9.3   Orthogonal Signals   *237*
3.9.4   Quadrature Amplitude Modulation   *239*
3.9.5   Frequency Division Multiplexing   *242*
3.9.6   Orthogonal Frequency Division Multiplexing   *244*
3.9.7   Implementing OFDM: The FFT   *247*
3.9.8   Spread Spectrum   *254*
3.10   Chapter Summary   *261*
         Problems   *261*

4   **Internet Protocols and Packet Delivery Algorithms**   *269*
4.1   Chapter Objectives   *269*
4.2   Introduction   *269*
4.3   Useful Preliminaries   *270*
4.3.1   Packet Switching   *270*
4.3.2   Binary Operations   *272*
4.3.3   Data Structures and Dereferencing Data   *272*
4.4   Packets, Protocol Layers, and the Protocol Stack   *277*
4.5   Local Area Networks   *281*
4.5.1   Wired LANs   *282*
4.5.2   Wireless LANs   *284*
4.6   Device Packet Delivery: Internet Protocol   *286*
4.6.1   The Original IPv4   *286*
4.6.2   Extension to IPv6   *286*
4.6.3   IP Checksum   *290*
4.6.4   IP Addressing   *294*
4.6.5   Subnetworks   *296*
4.6.6   Network Address Translation   *298*
4.7   Network Access Configuration   *300*
4.7.1   Mapping MAC to IP: ARP   *301*
4.7.2   IP Configuration: DHCP   *302*

4.7.3   Domain Name System (DNS)   *302*
4.8   Application Packet Delivery: TCP and UDP   *303*
4.9   TCP: Reliable Delivery and Network Fairness   *309*
4.9.1   Connection Establishment and Teardown   *311*
4.9.2   Congestion Control   *311*
4.9.3   TCP Timeouts   *319*
4.10   Packet Routing   *321*
4.10.1   Routing Example   *322*
4.10.2   Mechanics of Packet Forwarding   *323*
4.10.3   Routing Tasks   *325*
4.10.4   Forwarding Table Using Supernetting   *326*
4.10.5   Route Path Lookup   *330*
4.10.6   Routing Tables Based on Neighbor Discovery: Distance Vector   *343*
4.10.7   Routing Tables Based on Network Topology: Link State   *348*
4.11   Chapter Summary   *359*
         Problems   *359*

5   Quantization and Coding   *363*
5.1   Chapter Objectives   *363*
5.2   Introduction   *363*
5.3   Useful Preliminaries   *364*
5.3.1   Probability Functions   *364*
5.3.2   Difference Equations and the z Transform   *366*
5.4   Digital Channel Capacity   *369*
5.5   Quantization   *372*
5.5.1   Scalar Quantization   *373*
5.5.2   Companding   *379*
5.5.3   Unequal Step Size Quantization   *382*
5.5.4   Adaptive Scalar Quantization   *383*
5.5.5   Vector Quantization   *385*
5.6   Source Coding   *389*
5.6.1   Lossless Codes   *390*
5.6.1.1 Entropy and Codewords   *390*
5.6.1.2 The Huffman Code   *392*
5.6.1.3 Adapting the Probability Table   *404*
5.6.2   Block-based Lossless Encoders   *405*
5.6.2.1 Sliding-Window Lossless Encoders   *405*
5.6.2.2 Dictionary-based Lossless Encoders   *407*
5.6.3   Differential PCM   *409*
5.6.3.1 Sample-by-sample Prediction   *410*
5.6.3.2 Adaptive Prediction   *417*
5.7   Image Coding   *420*
5.7.1   Block Truncation Algorithm   *422*

5.7.2   Discrete Cosine Transform   *425*

5.7.3   Quadtree Decomposition   *430*

5.7.4   Color Representation   *431*

5.8   Speech and Audio Coding   *433*

5.8.1   Linear Prediction for Speech Coding   *434*

5.8.2   Analysis by Synthesis   *439*

5.8.3   Spectral Response and Noise Weighting   *440*

5.8.4   Audio Coding   *442*

5.9   Chapter Summary   *447*

         Problems   *447*

6   **Data Transmission and Integrity**   *453*

6.1   Chapter Objectives   *453*

6.2   Introduction   *453*

6.3   Useful Preliminaries   *454*

6.3.1   Probability Error Functions   *454*

6.3.2   Integer Arithmetic   *458*

6.4   Bit Errors in Digital Systems   *461*

6.4.1   Basic Concepts   *461*

6.4.2   Analyzing Bit Errors   *463*

6.5   Approaches to Block Error Detection   *470*

6.5.1   Hamming Codes   *472*

6.5.2   Checksums   *478*

6.5.3   Cyclic Redundancy Checks   *482*

6.5.4   Convolutional Coding for Error Correction   *489*

6.6   Encryption and Security   *507*

6.6.1   Cipher Algorithms   *508*

6.6.2   Simple Encipherment Systems   *509*

6.6.3   Key Exchange   *512*

6.6.4   Digital Signatures and Hash Functions   *519*

6.6.5   Public-key Encryption   *520*

6.6.6   Public-key Authentication   *522*

6.6.7   Mathematics Underpinning Public-key Encryption   *522*

6.7   Chapter Summary   *526*

         Problems   *526*

References   *531*

Index   *541*

Preface

History has probably never witnessed such a dramatic rise in technical sophistication, coupled with blanket penetration into everyday life, as has occurred in recent times with telecommunications. The combination of electronic systems, together with readily available programmable devices, provides endless possibilities for interconnecting what were previously separated and isolated means of communicating, both across the street and across the globe.

How, then, is the college- or university-level student to come to grips with all this sophistication in just a few semesters of study? Human learning has not changed substantially, but the means to acquire knowledge and shape understanding certainly has. This is through the ability to experiment, craft, code, and create systems of our own making. This book recognizes that a valuable approach is that of *learn-by-doing*, experimenting, making mistakes, and altering our mental models as a result. Whilst there are many excellent reference texts on the subject available, they can be opaque and impenetrable to the newcomer.

This book is not designed to simply offer a recipe for each current and emerging technology. Rather, the underpinning theories and ideas are explained in order to motivate the *why does it work in this way?* questions rather than *how does technology X work?*.

With these observations as a background, this book was designed to cover many fundamental topics in telecommunications but without the need to master a large body of theory whose relevance may not immediately be apparent. It is suitable for several one-semester courses focusing on one or more topics in radio and wireless modulation, reception and transmission, wired networks, and fiber-optic communications. This is then extended to packet networks and TCP/IP and then to digital source and channel coding and the basics of data encryption. The emphasis is on understanding, rather than regurgitating facts. Digital communications is addressed with the coverage of packet-switched networks, with many fundamental concepts such as routing via shortest path introduced with simple, concrete, and intuitive examples.

The treatment of advanced telecommunication topics extends to OFDM for wireless modulation and public-key exchange algorithms for data encryption.

The reader is urged to try the examples as they are given. MATLAB® was chosen as the vehicle for demonstrating many of the basic ideas, with code examples in every chapter as an integral part of the text, rather than an afterthought. Since MATLAB® is widely used by telecommunication engineers, many useful take-home skills may be developed in parallel with the study of each aspect of telecommunications.

In addition to the coding and experimentation approach, many real-world examples are given where appropriate. Underpinning theory is given where necessary, and a Useful Preliminaries section at the start of each chapter serves to remind students of useful background theory, which may be required in order to understand the theoretical and conceptual developments presented within the chapter.

Although an enormous effort, it has been an ongoing source of satisfaction in writing the book over several years and developing the "learn-by-doing" concept in a field that presents so many challenges in formulating lucid explanations. I hope that you will find it equally stimulating to your own endeavors and that it helps to understand the power and potential of modern communication systems.

I will consider that my aims have been achieved if reading and studying the book is not a chore to you, but, rather, a source of motivation and inspiration to learn more.

John W. Leis

Acknowledgments

As with any work of this magnitude, a great many people helped contribute, directly and indirectly, along the journey. Some may not even have realized it.

I wish to thank Professor Derek Wilson for his unwavering enthusiasm for any project I put forward. Many discussions formed the core approaches used in this book, which were a little nebulous to start with, but became the self-directed learn-by-doing style, which I hope will aid many others.

I thank Professor Athanassios (Thanos) Skodras for his kind and positive comments at critical points in the manuscript preparation and to friend and colleague from afar, Professor Tadeusz Wysocki, for his interest in all things about telecommunications and support over the years. His unassuming style belies his knowledge and achievements.

To my earlier mentors, including Bruce Varnes, whose technical competency was, and remains, a source of inspiration.

To my students in signal processing, communications, and control courses, who have often provided the critical insight as to why a *thing* ought to be explained in a particular way, you have helped far more than you imagined. Their names are too numerous to mention, but their critical insight and questions helped sharpen my focus and writing.

I am grateful to Brett Kurzman, Editor, Professional Learning at Wiley, who helped to bring the manuscript to the light of day ahead of time; he had a genuine interest in the project and never failed to give assistance when needed.

Finally, to those who indirectly shaped the work you see here, my parents Harold and Rosalie, nothing was ever too much for them to sacrifice, and they instilled the desire to learn and achieve as much as your talents will permit. It does not matter which side of the street you come from, it's what you do and how you treat others that matter. Dedication and hard work overcome any obstacles, real or imagined.

John W. Leis
December 2017

Introduction

Telecommunications encompasses a vast range of subdisciplines, and any treatment must strike a balance between the breadth of treatment and depth in specific areas. This book aims to give an overview with sufficient technical detail to enable coverage from the physical layer (how the electrical or wireless or optical signal is encoded) through to the representation of real-world information (images, sounds) and then to the movement of that data from one point to another and finally how to encode information and ensure its secure transmission.

Apart from the first chapter, most chapters may be studied as stand-alone entities or chosen for specific courses. Each chapter includes a Useful Preliminaries section at the start, which reviews some important concepts that may have been studied previously, and places them in the context of telecommunications.

Chapter 1, "Signals and Systems," introduces and reviews some basic ideas about signals that convey information. The emphasis is on operations that can be performed on signals, which are important to create telecommunication subsystems such as modulators. The idea of block diagrams, and synthesizing complex systems from simpler functional blocks, is also introduced.

Chapter 2, "Wired, Wireless, and Optical Systems," covers the means of physical transmission of telecommunication signals – either through wired systems such as copper or coaxial cable, wireless or radio systems, or fiber optics. Each is treated separately, with common threads such as signal attenuation covered for all. The emphasis is on understanding the ideas behind each method and their shortcomings in terms of cost, complexity, interference, transmission, and throughput. The section on radio transmission covers transmission and reception, antennas, and related issues such as propagation and diffraction. Visualizing the propagation of a radio signal is shown through MATLAB® code, which students can try for themselves.

Chapter 3, "Modulation and Demodulation," explains how a signal is encoded or modulated. It starts from very basic signal types such as Amplitude Modulation (AM) and proceeds to develop the theory for other types of modulation, toward newer techniques such as Orthogonal Frequency Division Multiplexing (OFDM), and the concept of spread spectrum. Digital line codes are also covered. Synchronization is also introduced in this chapter, including the phase-locked loop and the Costas loop. The notion of IQ modulation and demodulation is explained, as it underpins so much digital modulation theory. MATLAB® examples are employed throughout, including the use of the Fourier transform in OFDM to cater for advanced-level students.

Chapter 4, "Internet Protocols and Packet Delivery Algorithms," builds upon the assumption that the physical signal is sent and received, but that a useful system needs higher-level functionality, which is provided by packet-switched networks. Some of the important principles of the Internet are covered, including packet routing, TCP/IP, congestion control, error checking, and routing of packets from source to destination. Algorithms for packet routing and shortest-path determination are explained, with MATLAB® examples using object-oriented principles employed to elucidate the concepts.

Chapter 5, "Quantization and Coding," moves to more advanced treatment of signal representation. The idea of quantization (both scalar and vector) is explained, as well as the theory of entropy and data encoding. Lossless codes are explained using object-oriented structures in MATLAB® to illustrate the design of Huffman coding trees. Algorithms for digital encoding that are explained include the Discrete Cosine Transform (DCT) for image encoding and the Linear Predictive Coding (LPC) approach to speech encoding.

Chapter 6, "Data Transmission and Integrity," extends the previous chapter to address the important topic of data integrity, encryption, and security. Classic algorithms for error checking such as the checksum and cyclic redundancy check (CRC) for error detection are introduced, as well as the Hamming code for error correction. For data security, the keydistribution and public-key approaches are explained with numerical examples. The mathematics behind encryption is explained, and its computational limitations are investigated using code examples. Once again, reference is made to MATLAB® examples where appropriate.

It is one thing to read about a topic, but quite another to really *understand* it. For this reason, end-of-chapter problems for each chapter serve to reinforce the concepts covered. They variously require explanation of understanding, algebraic derivations, or code writing. A solutions manual is available to instructors, which includes fully worked solutions together with MATLAB® code solutions where requested. Additionally, both lecture presentations and MATLAB® code from the text are available to instructors.

To gain maximum benefit from this book, it is recommended that the examples using MATLAB® be studied as they are presented. MATLAB® is a registered trademark of The MathWorks, Inc.

For MATLAB® product information, please contact:

The MathWorks, Inc.
3 Apple Hill Drive
Natick, MA, 01760-2098 USA
Tel: 508-647-7000
Fax: 508-647-7101
E-mail: info@mathworks.com
Web: mathworks.com
How to buy: www.mathworks.com/store

Although several additional toolboxes are available for separate purchase, only the core MATLAB® product is required for the examples in this book. All code examples in the book were developed and tested using MATLAB® version R2017a/R2017b.

About the Companion Website

This book is accompanied by a companion website:

www.wiley.com/go/Leis/communications-principles-using-matlab

BCS Instructor Website contains:

- Teaching slides for instructors
- Solutions for the problems given in the chapters.
- Matlab codes

1

Signals and Systems

1.1 Chapter Objectives

On completion of this chapter, the reader should:

1) Be able to apply mathematical principles to waveforms.
2) Be conversant with some important terms and definitions used in telecommunications, such as root-mean-square for voltage measurements and decibels for power.
3) Understand the relationship between the time- and frequency-domain descriptions of a signal and have a basic understanding of the operation of frequency-selective filters.
4) Be able to name several common building blocks for creating more complex systems.
5) Understand the reasons why impedances need to be matched, to maximize power transfer.
6) Understand the significance of noise in telecommunication system design and be able to calculate the effect of noise on a system.

1.2 Introduction

A signal is essentially just a time-varying quantity. It is often an electrical voltage, but it could be some other quantity, which can be changed or *modulated* easily, such as radio-frequency power or optical (light) power. It is used to carry information from one end of a communications channel (the sender or transmitter) to the receiving end. Various operations can be performed on a signal, and in designing a telecommunications transmitter or receiver, many basic operations are employed in order to achieve the desired, more complex operation. For example, modulating a voice signal so that it may be transmitted through free space or encoding data bits on a wire all entail some sort of processing of the signal.

Communication Systems Principles Using MATLAB®, First Edition. John W. Leis.
© 2018 John Wiley & Sons, Inc. Published 2018 by John Wiley & Sons, Inc.
Companion website: www.wiley.com/go/Leis/communications-principles-using-matlab

A voltage that changes in some known fashion over time is termed a *waveform*, and that waveform carries information as a function of time. In the following sections, several operations on waveforms are introduced.

1.3   Signals and Phase Shift

In many communication systems, it is necessary to delay a signal by a certain amount. If this delay is relative to the frequency of the signal, it is a constant proportion of the total cycle time of the signal. In that case, it is convenient to write the delay not as time, but as a phase angle relative to 360° or 2π rad (radians). As with delay, it is useful to be able to advance a signal, so that it occurs earlier with respect to a reference waveform. This may run a little counter to intuition, since after all, it is not possible to know the value of a signal at some point in the future. However, considering that a signal repetitive goes on forever (or at least, for as long as we wish to observe it), then an advance of say one-quarter of a cycle or 90° is equivalent to a delay of $90 - 360 = -270°$.

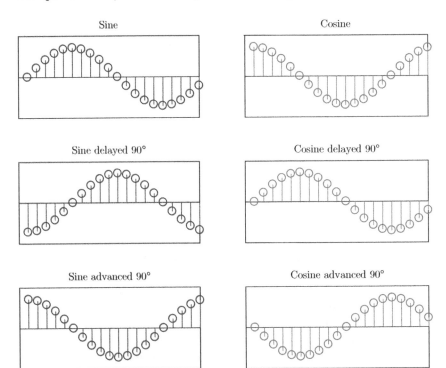

Figure 1.1 Sine and cosine, phase advance, and phase retard. Each plot shows amplitude $x(t)$ versus time t.

To see the effect of phase advance and phase delay, consider Figure 1.1, which shows these operations on both sine and cosine signals. The left panels show a sine wave, a delayed signal (moved later in time), and an advanced signal (moved earlier). The corresponding equations are

$$x(t) = \sin \omega t$$
$$x(t) = \sin \left(\omega t - \frac{\pi}{2} \right)$$
$$x(t) = \sin \left(\omega t + \frac{\pi}{2} \right)$$

Starting with a cosine signal, Figure 1.1 shows on the right the original, delayed (or retarded), and advanced signals, respectively, with equations

$$x(t) = \cos \omega t$$
$$x(t) = \cos \left(\omega t - \frac{\pi}{2} \right)$$
$$x(t) = \cos \left(\omega t + \frac{\pi}{2} \right)$$

1.4 System Building Blocks

Telecommunication systems can be understood and analyzed in terms of some basic building blocks. More complicated systems may be "built up" from simpler blocks. Each of the simpler blocks performs a specific function. This section looks initially at some simple system blocks and then at some more complex arrangements.

1.4.1 Basic Building Blocks

There are many types of blocks that can be specified according to need, but some common ones to start with are shown in Figure 1.2. The generic *input/output block* shows an input $x(t)$ and an output $y(t)$, with the input signal waveform being altered in some way on passing through. The alteration of the signal may be simple, such as multiplying the waveform by a constant A to give $y(t) = Ax(t)$. Alternatively, the operation may be more complex, such as introducing a phase delay. The *signal source* is used to show the source of a waveform – in this case, a sinusoidal wave of a certain frequency ω_0. The addition (or subtraction) block acts on two input signals to produce a single output signal, so that $y(t) = x_1(t) \pm x_2(t)$ for each time instant t. Similarly, a multiplier block produces at its output the product $y(t) = x_1(t) \times x_2(t)$.

These basic blocks are used to encapsulate common functions and may be combined to build up more complicated systems. Figure 1.3 shows two system blocks in cascade. Suppose each block is a simple multiplier – that is, the output

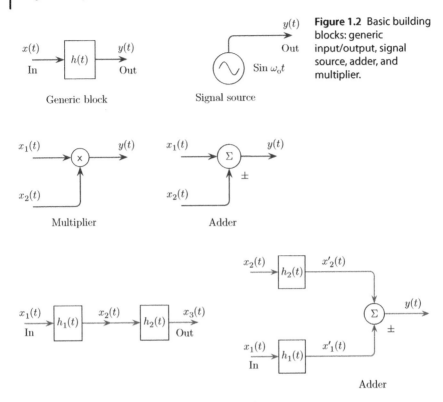

Figure 1.2 Basic building blocks: generic input/output, signal source, adder, and multiplier.

Figure 1.3 Cascading blocks in series (left) and adding them in parallel (right).

is simply the input multiplied by a gain factor. Let the gain of the $h_1(t)$ block be G_1 and that of the $h_2(t)$ block be G_2. Then, the overall gain from input to output would be just $G = G_1 G_2$.

To see how it might be possible to build up a more complicated system from the basic blocks, consider the system shown on the right in Figure 1.3. In this case, the boxes are simply gain multipliers such that $h_2(t) = G_1$ and $h_2(t) = G_2$, and so the overall output is $y(t) = G_1 x_1(t) + G_2 x_2(t)$.

1.4.2 Phase Shifting Blocks

In Section 1.3, the concept of phase shift of a waveform was discussed. It is possible to develop circuits or design algorithms to alter the phase of a waveform, and it is very useful in telecommunication systems to be able to do this. Consequently, the use of a phase-shifting block is very convenient. Most commonly, a phase shift of $\pm 90°$ is required. Of course, $\pi/2$ radians in the phase angle is equivalent to $90°$. As illustrated in the block diagrams of

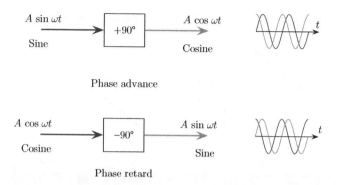

Figure 1.4 Phase shifting blocks. Note the input and output equations.

Figure 1.4, we use +90° to mean a *phase advance* of 90° and, similarly, −90° to mean a *phase delay* of 90°.

1.4.3 Linear and Nonlinear Blocks

Let us examine more closely what happens when a signal is passed through a system. Suppose for the moment that it is just a simple DC voltage. Figure 1.5 shows a transfer characteristic, which maps the input voltage to a corresponding output voltage. Two input values separated by δx, with corresponding outputs separated by δy, allow determination of the *change* in output as a function of the *change* in input. This is referred to as the *gain* of the system.

Suppose such a linear transfer characteristic with zero offset (that is, it passes through $x = 0, y = 0$) is subjected to a sinusoidal input. The output $y(t)$ is a linear function of input $x(t)$, which we denote as a constant α. Then,

$$y(t) = \alpha \, x(t) \tag{1.1}$$

With input $x(t) = A \sin \omega t$, the output will be

$$y(t) = \alpha \, A \sin \omega t \tag{1.2}$$

Thus the change in output is simply in proportion to the input, as expected.

This linear case is somewhat idealistic. Usually, toward the maximum and minimum range of voltages which an electronic system can handle, a characteristic that is not purely linear is found. Typically, the output has a limiting or saturation characteristic – as the input increases, the output does not increase directly in proportion at higher amplitudes. This simple type of nonlinear behavior is illustrated in Figure 1.6. In this case, the relationship between the input and output is not a simple constant of proportionality – though note that if the input is kept within a defined range, the characteristic may well be approximately linear.

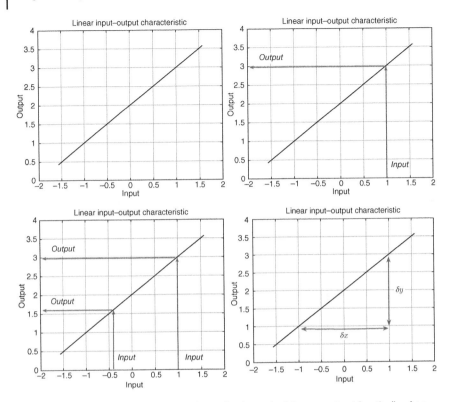

Figure 1.5 The process of mapping an input (horizontal axis) to an output (vertical), when the block has a linear characteristic. The constant or DC offset may be zero, or nonzero as illustrated.

To fix ideas more concretely, suppose the characteristic may be represented by a quadratic form, with both a linear constant multiplier α and a small amount of signal introduced that is proportional to the square of the input, via constant β. If the input $x(t)$ is again a sinusoidal function, the output may then be written as

$$y(t) = \alpha\, x(t) + \beta x^2(t)$$
$$= \alpha\, A \sin \omega t + \beta\, A^2 \sin^2 \omega t \qquad (1.3)$$

This is straightforward, but what does the sinusoidal squared term represent? Using the trigonometric identities

$$\cos(a + b) = \cos a \cos b - \sin a \sin b \qquad (1.4)$$

$$\cos(a - b) = \cos a \cos b + \sin a \sin b \qquad (1.5)$$

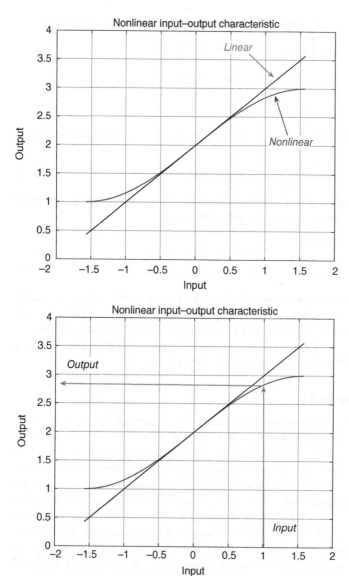

Figure 1.6 Example of mapping an input (horizontal axis) to an output (vertical), when the block has a nonlinear characteristic. Other types of nonlinearity are possible, of course.

we have by subtracting the first from the second, and then putting $b = a$,

$$\sin a \sin b = \frac{1}{2}[\cos(a - b) - \cos(a + b)]$$

$$\therefore \quad \sin^2 a = \frac{1}{2}[\cos(a - a) - \cos(a + a)]$$

$$= \frac{1}{2}(1 - \cos 2a) \tag{1.6}$$

After application of this relation, and simplification, the output may be written as

$$y(t) = \alpha A \sin \omega t + \frac{1}{2}\beta A^2(1 - \cos 2\omega t) \tag{1.7}$$

This can be broken down into a constant or DC term, a term at the input frequency, and a term at twice the input frequency:

$$y(t) = \overbrace{\alpha A \sin \omega t}^{\text{Linear term}} + \overbrace{\frac{1}{2}\beta A^2}^{\text{Constant term}} - \overbrace{\frac{1}{2}\beta A^2 \cos 2\omega t}^{\text{Double-frequency term}} \tag{1.8}$$

This is an important conclusion: the introduction of nonlinearity to a system may affect the frequency components present at the output. A linear system always has frequency components at the output of the exact same frequency as the input. A nonlinear system, as we have demonstrated, may produce harmonically related components at other frequencies.

1.4.4 Filtering Blocks

A more complicated building block is the frequency-selective filter, almost always just called a *filter*. Typically, a number of filters are used in a telecommunication system for various purposes. The common types are shown in Figure 1.7. The sine waves (with and without cross-outs) shown in the middle of each box are used to denote the operation of the filter in terms of frequency selectivity. For example, the lowpass filter shows two sine waves, with the lower one in the vertical stack indicating the lower frequency. The higher frequency is crossed out, thus leaving only lower frequency components. Representative input and output waveforms are shown for each filter type. Consider, for example, the bandpass filter. Lower frequencies are attenuated (reduced in amplitude) when going from input to output. Intermediate frequencies are passed through with the same amplitude, while high frequencies are attenuated. Thus, the term *bandpass filter* is used. Filters defining highpass and bandstop operation may be designated in a similar fashion, and their operation is also indicated in the figure.

When it comes to more precisely defining the operation of a filter, one or more cutoff frequencies have to be specified. For a lowpass filter, it is not sufficient to say merely that "lower" frequencies are passed through unaltered. It

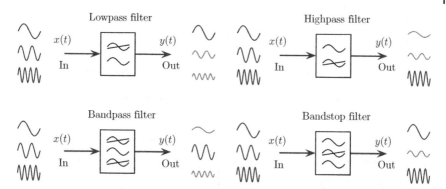

Figure 1.7 Some important filter blocks and indicative time responses. The waveforms and crossed-out waveforms in the boxes, arranged high to low in order, represent high to low frequencies. Input/ouput waveform pairs represent low, medium, and high frequencies, and the amplitude of each waveform at the output is shown accordingly.

is necessary to specify a boundary or cutoff frequency ω_c. Input waveforms whose frequency is below ω_c are passed through, but (in the ideal case) frequencies above ω_c are removed completely. In mathematical terms, the lower frequencies are passed through with a gain of one, whereas higher frequencies are multiplied by a gain of zero.

The operation of common filters may be depicted in the frequency domain as shown in the diagrams of Figure 1.8. First, consider the lowpass filter. This type of filter would, ideally, pass all frequencies from zero (DC) up to a specified cutoff frequency. Ideally, the gain in the *passband* would be unity, and the gain in the *stopband* would be zero. In reality, several types of imperfections mean that this situation is not always realized. The *order* of the filter determines how rapidly the response changes from one gain level to another. The order of a filter determines the number of components required for electronic filters or the number of computations required for a digitallyprocessed filter.

A low-order filter, as shown on the left, has a slower transition than a high-order filter (right). In any given design, a tradeoff must be made between a lower-cost, low-order filter (giving less rapid passband-to-stopband transitions) and a more expensive high-order filter.

Lowpass filters are often used to remove noise components from a signal. Of course, if the noise exists across a large frequency band, a filter can only remove or attenuate those components in its stopband. If the frequency range of the signal of interest also contains noise, then a simple filter cannot differentiate the desired signal from the undesired one.

In a similar fashion, a highpass filter may be depicted as also shown in Figure 1.8. As we would expect, this type of filter passes frequencies that are higher than some desired cutoff. A hybrid characteristic leads to a bandpass filter or bandstop filter. These types of filters are used in telecommunication

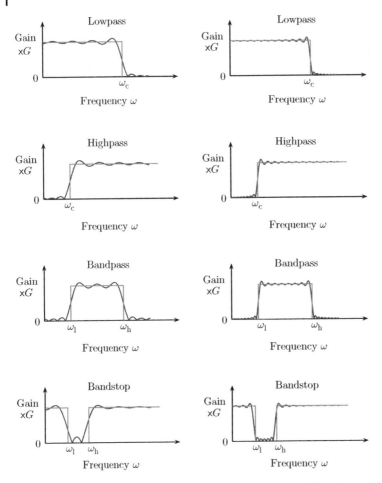

Figure 1.8 Primary filter types: lowpass, highpass, bandpass, and bandstop, with a low-order filter shown on the left and higher-order on the right. Ideally, the passband has a fixed and finite signal gain, whereas the stopband has zero gain.

systems for special purposes. For example, the bandstop filter may be used to remove interference at a particular frequency, and a bandpass filter may be used to pass only a particular defined range of frequencies (a channel or set of channels, for example).

1.5 Integration and Differentiation of a Waveform

This section details two signal operations that are related to fundamental mathematical operations. First, there is *integration*, which in terms of signals means

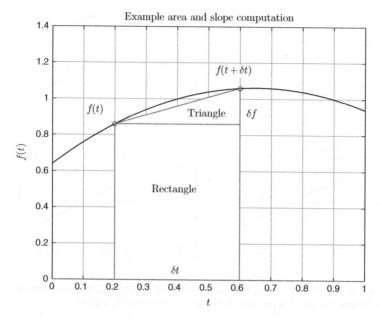

Figure 1.9 Calculating the area over a small time increment δt using a rectangle and the slope of the curve using a triangle.

the cumulative or sum total of a waveform over time. The opposite operation, *differentiation*, essentially means the rate of change of the voltage waveform over time. These are really just the two fundamental operations of calculus: Integration and differentiation. These are the inverse of each other, as will be explained. This intuition is useful in understanding the signal processing for communication systems presented in later chapters. The functions are presented in terms of time t, as this is the most useful formulation when dealing with time-varying signals.

Figure 1.9 shows the calculation of the area (integral) and slope (derivative) for two adjacent points. At a specific time t, the function value is $f(t)$, and at a small time increment δt later, the function value is $f(t + \delta t)$. The area (or actually, a small increment of area) may be approximated by the area of the rectangle of width δt and height $f(t)$. This small increment of area δA is

$$\delta A \approx f(t)\,\delta t \tag{1.9}$$

It could be argued that this approximation would be more accurate if the area of the small triangle as indicated were taken into account. This additional area would be the area of the triangle or $(1/2)(\delta t\,\delta f)$, which would diminish rapidly as the time increment gets smaller ($\delta t \to 0$). This is because it is not one small quantity δt, but the product of two small quantities $\delta t\,\delta f$.

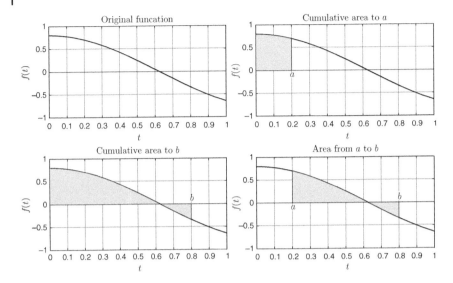

Figure 1.10 A function $f(t)$, calculating its cumulative area to a and b, and the area between $t = a$ and $t = b$. Note the negative portions of the "area" below the $f(t) = 0$ line.

Similarly, the slope at point $(t, f(t))$ is $\delta f / \delta t$. This is the instantaneous slope or derivative, which of course varies with t, since $f(t)$ varies. This slope may be approximated as the slope of the triangle, which changes from $f(t)$ to $f(t + \delta t)$ over a range δt. So the slope is

$$\frac{\delta f}{\delta t} \approx \frac{f(t + \delta t) - f(t)}{\delta t} \qquad (1.10)$$

The calculation of the derivative or slope of a tangent to a curve is a point-by-point operation, since the slope will change with $f(t)$ and hence the t value (the exception being a constant rate of change of value over time, which has a constant slope). The integral or area, though, depends on the range of t values over which we calculate the area. Since the integral is a continuous function, it extends from the left from as far back as we wish to the right as far as we decide. Figure 1.10 shows a function and its integral from the origin to some point $t = a$ (note that we have started this curve at $t = 0$, but that does not have to be the case). In the lower-left panel, we extend the area to some point $t = b$. This is essentially the same concept, except that the area below the horizontal $f(t) = 0$ line is in fact negative. While the concept of "negative area" might not be found in reality, it is a useful concept. In this case, the negative area simply subtracts from the positive area to form the net area. Finally, the lower-right panel illustrates the fact that the area from $t = a$ to $t = b$ is simply

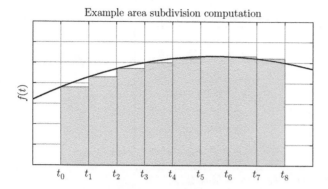

Figure 1.11 Calculating area using a succession of small strips of width δt.

the area to $t = b$, less the area to $t = a$. Mathematically, this is written as

$$\int_a^b f(t) \, \mathrm{d}t = F(b) - F(a) \tag{1.11}$$

where $F(\cdot)$ represents the cumulative area to that point. This is called the definite integral – an integration or area calculation with definite or known start and end boundaries.[1]

The area may be approximated by creating successive small strips of width δt as before, and joining enough of them together to make the desired area. This is illustrated in Figure 1.11, for just a few subdivisions. Using the idea of $F(t)$ as the cumulative area function under the curve $f(t)$, consider the area under the curve from t to $t + \delta t$, where δt is some small step of time. The *change* in area over that increment is

$$\delta A = F(t + \delta t) - F(t) \tag{1.12}$$

Also, the change in area is *approximated* by the rectangle of height $f(t)$ and width δt, so

$$\delta A = f(t) \, \delta t \tag{1.13}$$

Equating this change of area δA,

$$f(t) \, \delta t = F(t + \delta t) - F(t) \tag{1.14}$$

$$f(t) = \frac{F(t + \delta t) - F(t)}{\delta t} \tag{1.15}$$

---

1 The $\int$ symbol comes from the "long s" of the 1700s, so you can see the connection with the idea of "summation."

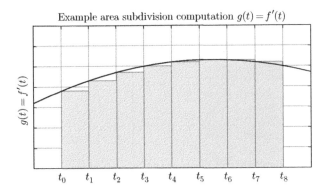

Figure 1.12 The area under a curve $g(t)$, but the curve happens to be the derivative of $f(t)$.

This is the same form of equation we had earlier for the definition of slope. Now, it is showing that the slope of some function $F(t)$, which happens to be the integral or area under $f(t)$, is actually equal to $f(t)$. That is, **the derivative of the integral equals the original function**. This is our first important conclusion.

Next, consider how to calculate the cumulative area by subdividing a curve $f(t)$ into successive small strips. However, instead of the plain function $f(t)$, suppose we plot its derivative, $f'(t)$ instead. This is illustrated in Figure 1.12, for just a few strips of area from t_0 at the start to an ending value t_8.

The *cumulative* area (call it $A(t)$) under this curve $f'(t)$ – which we defined to be the derivative of $f(t)$ – is the summation of all the individual rectangles, which is

$$A(t) = \delta t\, f'(t_0) + \delta t\, f'(t_1) + \cdots + \delta t\, f'(t_{n-1}) \tag{1.16}$$

$$= \delta t\, [f'(t_0) + f'(t_1) + \cdots + f'(t_{n-1})] \tag{1.17}$$

Now we can use the same concept for slope as developed before, where we had the approximation to the derivative

$$f'(t) = \frac{f(t + \delta t) - f(t)}{\delta t} \tag{1.18}$$

Substituting this for all the derivative terms, we have

$$A(t) = \delta t\, \left\{ \left[\frac{f(t_0 + \delta t) - f(t_0)}{\delta t} \right] + \left[\frac{f(t_1 + \delta t) - f(t_1)}{\delta t} \right] + \cdots \right.$$
$$\left. + \left[\frac{f(t_{n-1} + \delta t) - f(t_{n-1})}{\delta t} \right] \right\} \tag{1.19}$$

Canceling the δt and using the fact that each $t_k + \delta t$ is actually the next point t_{k+1} (for example, $t_1 = t_0 + \delta t$, $t_2 = t_1 + \delta t$), we can simplify things to

$$A(t) = \{[f(t_1) - f(t_0)] + [f(t_2) - f(t_1)] + \cdots + [f(t_n) - f(t_{n-1})]\} \qquad (1.20)$$

Looking carefully, we can see terms that will cancel, such as $f(t_1)$ in the first square brackets, minus the same term in the second square brackets. All these will cancel, except for the very first $-f(t_0)$ and the very last $f(t_n)$ to leave us with

$$A(t) = f(t_n) - f(t_0) \qquad (1.21)$$

So this time, we have found that the area under some curve $f'(t)$ (which happens to be the derivative or slope of $f(t)$) is actually equal to the original $f(t)$. That is, the *area under the slope curve* equals *the original function evaluated at the end* (right-hand side), less any start area. The subtraction of the start area seems reasonable, since it is "cumulative area to b less cumulative area to a," as we had previously. Thus, our second important result is that **the integral of a derivative equals the original function.**

We can see the relationship between differentiation and integration at a glance in the following figures. Figure 1.13 shows taking a function (top) and integrating it (lower); if we then take this integrated (area) function as shown

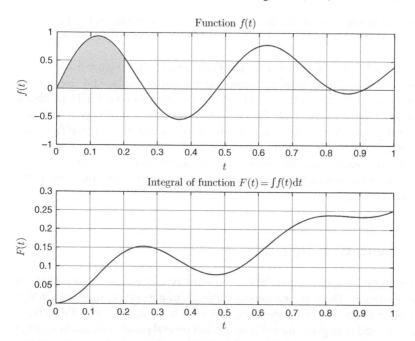

Figure 1.13 The cumulative area under $f(t)$. Each point on $F(t)$ represents the area up to the right-hand side of the shaded portion at some value of t (here $t = 0.2$ for the shaded portion). Note that when $f(t)$ becomes negative, the area reduces.

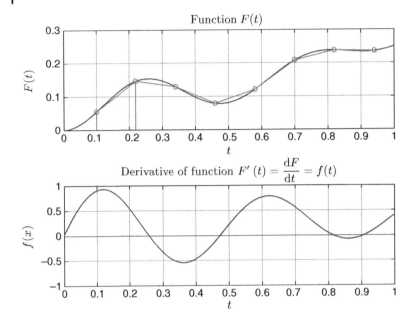

Figure 1.14 The derivative of f(t) as a function. It may be approximated by the slopes of the lines as indicated, though the spacing is exaggerated for the purpose of illustration.

in Figure 1.14 (top) and then take the derivative of that (Figure 1.14, lower), *we end up with the original function* that we started with. And the process is invertible: Take the derivative of the top function in Figure 1.14 to obtain the lower plot of Figure 1.14. Transferring this to the top of Figure 1.13, and then integrating it, we again end up where we started: the original function. So it is reasonable to say that integration and differentiation are the inverse operations of each other. We just have to be careful with the integration, since it is cumulative area, and that may or may not have started from zero at the leftmost starting point.

1.6 Generating Signals

Communication systems invariably need some type of waveform generation in their operation. There are numerous methods of generating sinusoids, which have been devised over many years, and each has advantages and disadvantages. The ability to generate not just one, but several possible frequencies (that is, to tune the frequency), is a desirable attribute. So too is the spectral purity of the waveform: How close it is to an ideal sine function. One method, which is relatively simple, has a tunable frequency, and can generate a wide range of

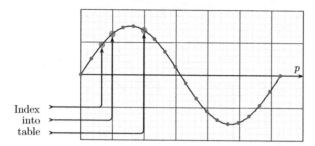

Figure 1.15 Generating a sinusoid using an index *p* into a table. The value at each index specifies the required amplitude at that instant.

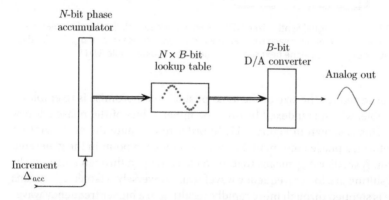

Figure 1.16 Using a lookup table to generate a waveform. Successive digital (binary-valued) steps are used to index the table. The digital-to-analog (D/A) converter transforms the sample value into a voltage.

possible frequencies, is the Direct Digital Synthesizer (DDS), whose working principle was originally introduced in Tierney et al. (1971).

Computing the actual samples of a sine function is often not feasible in real time for high frequencies. However, precomputing the values and storing in a table – a Lookup Table or LUT – is possible. Stored-table sampling with indexing is illustrated in Figure 1.15. Effectively, the index of each point in the table is the phase value, and each point's value represents the amplitude at that particular phase. All that is then required is to step through the table with a digital counter as shown in Figure 1.16.

The number of points on the waveform determines the accuracy and also the resolution of frequency tuning steps. This resolution is the clock frequency f_{clk} divided by the number of points 2^N, where N is the number of bits in the address counter. However, this also requires a table of size 2^N. It follows that for finer frequency tuning steps, N should be as large as possible.

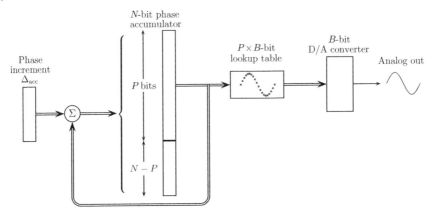

Figure 1.17 A Direct Digital Synthesizer (DDS) using a reduced lookup table. Samples are produced at a rate of f_s and for each new sample produced, a phase step of Δ_{acc} is added to the current index p to locate the next sample value in the Lookup Table (LUT).

In order to reduce the size of the lookup table, a compromise is to employ a smaller table, which is indexed by only the upper P bits of the phase address counter. This is shown in Figure 1.17. In order to compute the next point on the waveform, a phase increment Δ_{acc} is added for each point in the generated waveform. A smaller Δ_{acc} means that the table is stepped through more slowly, hence resulting in a lower frequency waveform. Conversely, a larger Δ_{acc} means the table is stepped through more rapidly, resulting in a higher frequency waveform. The tradeoff in using a smaller LUT means that the preciseness of the waveform is reduced, which is shown in Figure 1.18 for a small table size.

An interesting problem then arises. If the phase accumulator step Δ_{acc} is a power of 2, then at the end of the LUT, the counter will wrap back to the same relative starting position. The only problem with the output frequency spectrum will be the harmonics generated by stepping through at a faster rate, and these harmonics will not vary over time. However, if the step is such that, upon reaching the end of the table, the addition of the step takes the pointer back to a different start position, the next cycle of the waveform will be generated from a slightly different starting point. This means that there will be some jitter in the output waveform, and the frequency spectrum will contain additional phase noise components, as shown in Figure 1.19.

The DDS structure is able to generate multiple waveforms by using multiple index pointers. For example, sine and cosine may be generated by offsetting one pointer by the equivalent of a quarter of a cycle in the table. The phase and frequency are also easily changed by changing the relative position of the index pointer, and this is useful for generating modulated signals (discussed in Chapter 3).

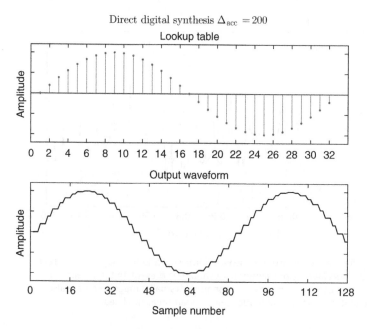

Figure 1.18 A lookup table (top) with $2^P = 32$ entries, requiring $P = 5$ bits. One possible waveform generated by stepping through at increments of $\Delta_{acc} = 200$ is shown below, when the total phase accumulator has $N = 14$ bits.

1.7 Measuring and Transferring Power

This section discusses the concept of the power transferred by a signal and the related concept of impedance of a circuit. The notion of power is important in telecommunications, since how much power is used to send a signal is clearly important, how far can a signal travel and how much power is enough are relevant questions. The impedance of a circuit appears a great deal in discussions about power and information transfer. It basically describes how much a current flow is "impeded" along its way.

1.7.1 Root Mean Square

Sinusoidal signals have their amplitude determined directly by the factor A in the equation of a sinusoid, $x(t) = A \sin(\omega t + \varphi)$. However, not all signals are pure sinusoids. It is useful to have a definition of power, which is not dependent on the wave shape of the underlying signal.

One of the most commonly used is the RMS, or Root Mean Square. This means that first, we square the signal and then take the mean or average of that result. This is necessary so as to measure power over a normalized time interval.

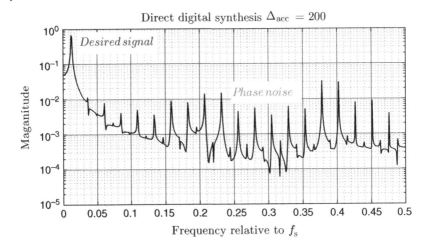

Figure 1.19 The frequency spectrum of the waveform, showing the magnitude of each signal component. Ideally, only one component should be present, but the stepping approach means that other unwanted components with smaller magnitudes are also produced. Note that the vertical amplitude scale is logarithmic, not linear.

Finally, to "undo" the squaring operation, we take the square root. Graphically, Figure 1.20 illustrates this operation for a sine wave. The first step is to square the waveform, which means that negative values are converted into positive, since squaring a negative value results in a positive result.

The second step after squaring the waveform is to add up all the squared values, as illustrated in Figure 1.21. This diagram shows individual bars or samples of the waveform in order to illustrate the point – in reality, the signal has no discontinuities. Next, we divide by the time we have averaged over. In the illustration, this is exactly one cycle of the wave. If need not we do say 2 or 100 cycles, then the summation would be correspondingly larger, and dividing by the number of samples (in the discrete-bar case) or the total time (for the continuous wave) would normalize things out. Finally, we take the square root of this quantity, and we have the RMS value.

We can calculate this mathematically for known signals. A simple and commonly used case is the pure sine wave, and to work this out let the period be

$$\tau = \frac{2\pi}{\omega_0} \tag{1.22}$$

where ω_0 is the radian frequency (rad s^{-1}). To convert from Hertz frequency f to radian frequency ω, the formula $\omega = 2\pi f$ is used, where f is in Hertz, or cycles per second, and ω is in radians per second. The equation of the sine wave is

$$x(t) = A \sin \omega t \tag{1.23}$$

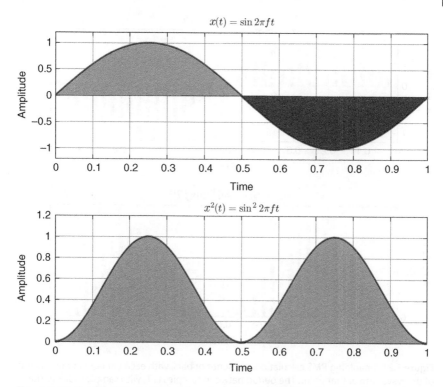

Figure 1.20 Graphical illustration of the calculation of RMS value. Squaring the waveform at the top results in the lower waveform.

Squaring gives

$$x^2(t) = A^2\sin^2\omega t \tag{1.24}$$

In order to calculate the mean square over one period τ, we need to integrate the squared waveform

$$\overline{x^2} = \frac{1}{\tau} \int_0^\tau A^2\sin^2\omega t \, dt \tag{1.25}$$

Evaluating this integral, we find that the mean-square value of a sine wave is

$$\overline{x^2} = \frac{A^2}{2} \tag{1.26}$$

The RMS is just the square root of this, or

$$\text{RMS } \{x(t)\} = \frac{A}{\sqrt{2}} \tag{1.27}$$

This is a very common result. It tells us that the RMS value of a sine wave is the peak divided by $\sqrt{2}$, or approximately 1.4. Equivalently, the peak is multiplied

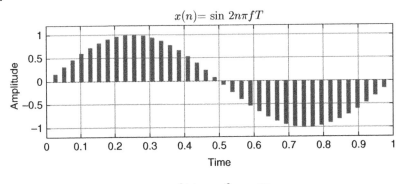

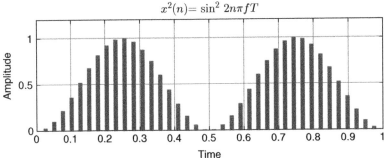

Figure 1.21 Imagining RMS calculation as a series of bars, with each bar equal to the height of the waveform at that point. The period between samples is *T*, with sample index *n*. The substitution required is then $t = nT$.

by $1/\sqrt{2} \approx 0.7$ to obtain the RMS value. Alternatively, if we know the RMS value, we multiply it by $\sqrt{2} \approx 1.4$ to find the peak value. The following MAT-LAB code shows how to generate a sine wave and calculate the RMS value from the peak.

```
% waveform parameters
dt = 0.01;
tmax = 2;
t = 0:dt:tmax;
f = 2;

% generate the signal
x = 1*sin(2*pi*f*t);
plot(t,x);

% calculate the signal's RMS value
sqrt((sum(x.*x)*dt)/tmax)
ans =
```

```
    0.7071

% it is a known factor
1/sqrt(2)
ans =
    0.7071
```

So, what use is the RMS value? Even though we calculated a mathematical expression relating the amplitude of a sine wave to its RMS value, the concept is applicable to *any* waveform. It gives us a measure of the power that the signal can deliver. If, for example, we simply averaged the waveform, then a sine wave would yield a figure of zero (since it is symmetrical about the time axis). This is not a very useful result. In the next section, it is demonstrated that the RMS value may be related to another quantity, termed the *decibel*, which is commonly used in telecommunication systems.

1.7.2 The Decibel

Another quantity that is frequently encountered in telecommunications is the *decibel* (dB). It is used in different contexts. One is to show the power of a signal, and in that way it might be regarded as similar to the RMS value mentioned above. Another context in which the decibel is used is to measure the gain or loss of a communication processing block, such as an amplifier.

The first use is to denote power, or more precisely, power relative to some reference value. For a power P, the relative *power in decibels* is calculated as

$$P_{dB} = 10\log_{10}\left(\frac{P}{P_{ref}}\right) \tag{1.28}$$

where P_{ref} is the reference power. There are several important points to note about this formula. First, it does not measure absolute power as such, but rather power relative to a defined reference power level. Secondly, we use the logarithm to base 10 in the computation of the decibel. The relative power is usually a standard amount, in which case standard symbols are used to denote this. For example, dBW is used when the reference power P_{ref} is 1W (Watt) and dBm when the reference power P_{ref} is 1mW (milliwatt), or 1×10^{-3} W.

The concept of power is meaningless in a practical sense unless it is applied to a load. The load must have a certain impedance. Suppose we had a purely resistive load of 50Ω. Power is $P = IV$ and Ohm's law is $V = IR$, and so power is V^2/R. Thus for a power of 1mW, we have

$$P = \frac{V^2}{R}$$
$$\therefore V = \sqrt{P \times R}$$

$$= \sqrt{1\text{mW} \times 50\Omega}$$

$$= \sqrt{0.05}$$

$$= 0.2236\text{V} \approx 223\text{mV} \tag{1.29}$$

This is the voltage needed across the load resistance to develop the given amount of power. Note that the voltage is RMS, not peak amplitude.

The second common use of the decibel is in measuring the gain of a system. That is to say, given an input power P_{in}, and a corresponding output power P_{out}, the *power gain* is defined as

$$G_{dB} = 10\log_{10}\left(\frac{P_{out}}{P_{in}}\right) \tag{1.30}$$

The basic formula is similar, taking the logarithm of a power ratio and then multiplying by 10 (the "deci" part). What was the reference power in the previous example has now become the input power. This is not unreasonable, since the "reference" is at the input to the system we are considering.

Suppose a system has a power gain of 2. That is, the output power is twice the input power. The power gain in dB is

$$G_{dB} = 10\log_{10}2$$

$$\approx 3 \text{ dB} \tag{1.31}$$

Now suppose another system has a power gain of 1/2. In that case, the power gain in dB is

$$G_{dB} = 10\log_{10}\frac{1}{2}$$

$$\approx -3 \text{ dB} \tag{1.32}$$

Notice how these are the same values, but negated. This gives us a clue as to one of the useful properties of decibels: increasing the power is a positive dB figure, whereas decreasing is a negative dB figure. So what about the same power for input and output, when $P_{out} = P_{in}$? It is not hard to show that this gives a figure of 0 dB.

A common use of the decibel is to state the *voltage* gain of a circuit or system in decibels rather than the power gain. Suppose we have two power flows P_{out} and P_{in} as above and that they each drive a load resistance of R. We can determine the voltage at the input and output using $P = V^2/R$ as before, and noting that $\log x^a = a \log x$, the decibel ratio becomes

$$G_{dB} = 10\log_{10}\left(\frac{P_{out}}{P_{in}}\right)$$

$$= 20\log_{10}\left(\frac{V_{out}}{V_{in}}\right) \tag{1.33}$$

So, now we have a multiplier factor of 20× rather than 10×.

It is useful to keep some common decibel figures in mind. The most commonly encountered one is a doubling of power, and 3 dB corresponds approximately to a double ratio

$$3 \text{ dB} \approx 10\log_{10}2$$

The exact figure is 3.0103, but 3 is close enough for most practical use. Similarly

$$2 \text{ dB} \approx 10\log_{10}1.6$$
$$4 \text{ dB} \approx 10\log_{10}2.5$$

From these values, it is possible to derive many other dB figures fairly easily. For example, 6 dB is

$$6 \text{ dB} = 3 \text{ dB} + 3 \text{ dB}$$
$$\therefore 6 \text{ dB} \rightarrow 2 \times 2$$
$$= 4\times$$

and so the ratio is 4. Since adding logarithms corresponds to multiplication, it follows that subtracting corresponds to division. So, for example,

$$1 \text{ dB} = 4 \text{ dB} - 3 \text{ dB}$$
$$\therefore 1 \text{ dB} \rightarrow \frac{2.5}{2}$$
$$= 1.25\times$$

Finally, note that the dB when used as a difference represents a ratio, and not a normalized power. So, for example, using two power values referenced to 1 mW,

$$4 \text{ dBm} - 3 \text{ dBm} = 1 \text{ dB}$$

We have two power figures (in dBm) but the difference is a ratio and is expressed in dB. Remember, because of the logarithmic function, a seemingly small number – such as a power loss of 20 dB – in fact represents a 99% power loss.

1.7.3 Maximum Power Transfer

When a signal is received by an antenna, that signal is likely to be exceedingly small. It follows that we do not want to waste any of that signal in the transmission from the antenna to the receiver. Similarly, if a transmitter is connected to an antenna, ideally the maximum amount of power would be transferred, implying no loss along the connecting wires. How can this be achieved?

To motivate the development, consider a simple circuit as shown in Figure 1.22. The question may be framed for this case as: What value of load resistance R_L will give the maximum amount of power transferred to that load? The assumption is that the source has a certain resistance R_S, and in practice

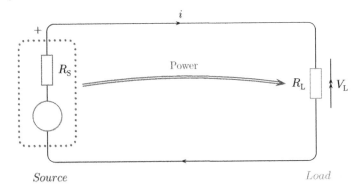

Figure 1.22 Transferring power from a source to a load. The source resistance R_S is generally quite small, and is inherent in the power source itself. We can adjust the load resistance R_L to maximize the power transferred.

this may be composed of the voltage source's own internal resistance or the equivalent resistance of the driving circuit.

For a simple circuit that has purely resistive impedances, we may write some basic equations governing the operation. The equivalent series resistance is

$$R_{eq} = R_S + R_L \tag{1.34}$$

Ohm's law applied to the circuit gives

$$V_S = i\, R_{eq}$$
$$= i\,(R_S + R_L)$$
$$\therefore i = \frac{V_S}{R_S + R_L}$$

and so the load power and current are

$$V_L = i\, R_L$$
$$\therefore i = \frac{V_L}{R_L} \tag{1.35}$$

The power dissipated in the load, which is our main interest, is

$$P_L = i\, V_L$$
$$= i^2 R_L$$
$$= \frac{V_S^2}{(R_S + R_L)^2}\, R_L \tag{1.36}$$

A simulation of this scenario, using only the basic equations for voltage, current, and power, helps to confirm the theory. Using the MATLAB code below, the power as the load resistance varies is calculated, with the result shown in Figure 1.23.

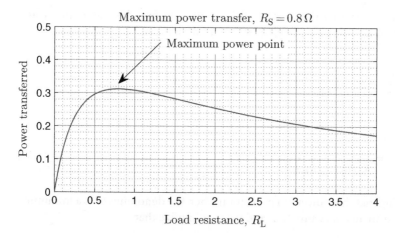

Figure 1.23 The power transferred to a load as the load resistance is varied. There is a point where the maximum amount of power is transferred, and this occurs when the load resistance exactly matches the source resistance.

```
% parameters of the simulation
Vsrc = 1;
Rsrc = 0.8;

% load resistance range
Rload = linspace(0, 4, 1000);

% equations
Req = Rsrc + Rload;
i = Vsrc./Req;
Pload = (i.*i).*Rload;

% plotting
plot(Rload, Pload);
xlabel('Load Resistance R_{load}')
ylabel('Power Transferred P_{load}')
```

From the figure, we can see that there is a point where the amount of power transferred is a maximum. Why does this occur? If the load resistance is very high, the current flowing through it will be low, and the voltage drop across it will be high. If the load resistance is low, the current flowing through it will be higher, but the voltage drop across it will be lower. Since the power dissipated in the load depends on both voltage and current, there is obviously an interplay between these factors.

How can we verify this analytically? We need to find the maximum P_L as a function of R_L. The governing equation was derived as

$$P_L = \frac{V_S^2}{(R_S + R_L)^2} R_L$$

$$= \frac{V_S^2}{R_S^2 + 2R_S R_L + R_L^2} R_L$$

$$= \frac{V_S^2}{R_S^2/R_L + 2R_S + R_L} \qquad (1.37)$$

Reasoning that this power is a maximum when the denominator is a minimum, we define an auxiliary function and try to minimize that

$$f(R_L) = R_S^2/R_L + 2R_S + R_L$$

$$\frac{df}{dR_L} = -\frac{R_S^2}{R_L^2} + 0 + 1 \qquad (1.38)$$

Setting this to zero, we have that $R_L = R_S$, and so the conclusion is that **the maximum amount of power is dissipated in the load if the load resistance equals the source resistance.** Equivalently, the maximum power is dissipated (transferred) when the source resistance equals the load resistance.

In a communication system, we might have an antenna (load) fed by a source and transmission line. Thus, the line resistance (actually, impedance, which is resistance at certain frequency) must match the source and load resistance.

Note that *maximum power transfer* does not equal *maximum efficiency*. Defining efficiency η as the power delivered to the load over the total power dissipated,

$$\eta = \frac{iR_L}{iR_S + iR_L}$$

$$= \frac{R_L}{R_S + R_L}$$

$$= \frac{1}{1 + \frac{R_S}{R_L}} \qquad (1.39)$$

If the source resistance were zero ($R_S = 0$), which is not really a practical scenario, the efficiency would be 100%. However, for some other resistance, if we arranged that $R_S = R_L$, then the efficiency would be 50%.

1.8 System Noise

Any real system is subject to the effects of extraneous noise. This may come from devices that deliberately radiate energy, such as radio or wireless transmissions, or nearby electronics such as computers and switch-mode power supplies, which radiate interference as an unintended but inevitable consequence of their operation. There is also noise present naturally – as a result of cosmic background radiation and from the thermal agitation of electrons in conductors. In this section, we briefly summarize some important concepts encountered when dealing with noise in a system.

A key result found in the early development of radio and electronics was that noise is present in any resistance that is at a temperature above absolute zero. Johnson (1928) is generally deemed to be the first to have experimentally assessed this phenomenon, which was further explained by Nyquist (1928). As a result, thermal noise is often termed Johnson Noise or Johnson–Nyquist Noise. The key result was that current was proportional to the square root of the temperature, and as a result the noise power $\mathcal{N}$ dissipated in a load is

$$\mathcal{N} = kTB \tag{1.40}$$

where T is the absolute temperature (in Kelvin), B is the bandwidth of the system being measured, and k is a constant due to Planck, but usually termed Boltzmann's constant, which has an approximate value of $k \approx 1.38 \times 10^{-23}$ J K^{-1}. Importantly, this result shows that noise power is dependent on temperature, but not on resistance. Furthermore, since the bandwidth employed in a particular application may not be known in advance, the noise power is often expressed as a power per unit bandwidth, or dBm/Hz. Following on from this, the noise voltage is then V/$\sqrt{\text{Hz}}$.

The amount of noise present in a system is not usually considered in isolation, but rather with respect to the size of the desired signal that carries information. Thus, the signal-to-noise ratio (SNR) is defined as the signal power divided by noise power and is usually expressed in decibels:

$$\frac{S}{N} = \frac{P_{\text{signal}}}{P_{\text{noise}}} \tag{1.41}$$

It is usually expressed as a dB figure:

$$\text{SNR}_{\text{dB}} = 10\log_{10}\left(\frac{P_{\text{signal}}}{P_{\text{noise}}}\right) \text{ dB} \tag{1.42}$$

Telecommunication systems are composed of numerous building blocks, such as amplifiers, filters, and modulators. An excessive amount of noise results in audible distortion for analog audio systems, and an increase in the bit error rate (BER) for digital systems. In extreme cases, digital systems may not function if the BER is over a maximum tolerable threshold. It is therefore

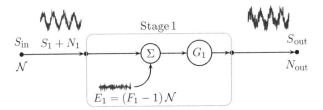

Figure 1.24 Modeling the noise transfer of a system. The noise at the input of the first block is $\mathcal{N} = N_1$, and this is used as a "noise reference" when subsequent blocks are added after the first. The quantity E is the excess noise added by the stage.

useful to know the effect of thermal noise on one particular block in isolation, and also the net result of cascading several blocks. This is done with the *noise factor* or *noise ratio*. The noise factor (or ratio) is defined as the SNR at the input terminals of a device, divided by the SNR at the output terminals. On the assumption that the block incorporates amplification, and the bandwidth of the block is not a limiting factor, then a noise ratio of unity would mean that no noise is added by the block, whereas a noise ratio of greater than one would imply that there is more noise at the output than the input (or, equivalently, the particular element reduces the SNR).

Often, the *noise figure* is expressed in dB, which is derived from the noise ratio as

$$F_{\text{dB}} = 10\log_{10}F \qquad (1.43)$$

Using the noise figure concept, an important step in analyzing block-level design is Friis's noise equation, first devised in Friis (1944) and covered in many textbooks in detail (for example, Haykin and Moher, 2009). To illustrate the basic idea, consider a block within a system as shown in Figure 1.24, which performs amplification of a signal by a factor of G_1. The input on the left may be an antenna, or some other receiver such as an optical sensor. Since any system block will add some noise to the overall system design, it is good to be able to quantify just how much noise is added.

Referring to Figure 1.24, we have an input signal S_{in} and thermal noise $\mathcal{N}$. These are assumed to be additive, with a resulting signal input $S_1 + N_1$ seen by the input of the block. It is assumed that the gain is greater than one and that the bandwidth is sufficient to pass the signal.

In the present context, we would like to know how much a given system degrades the SNR overall. To that end, we define a noise factor F, which pertains to how much noise is added when a signal passes through a system block. It is the SNR at the input, divided by the SNR at the output:

$$F = \frac{S_{in}/N_{in}}{S_{out}/N_{out}} \tag{1.44}$$

Referring to Figure 1.24, the signal output is simply the signal input multiplied by the gain of the block, so mathematically $S_{out} = G_1 S_1$. Assuming that noise is added to the signal, the input noise is also multiplied by the gain of the block. However, the block may also add its own noise, so we may write

$$N_2 = F_1 G_1 N_1 \tag{1.45}$$

where F_1 is a multiplicative factor greater than one. If $F_1 = 1$, it would imply a perfect block, which adds no additional noise. So, we could calculate the noise ratio as

$$\begin{aligned} R &= \frac{S_{in}/N_{in}}{S_{out}/N_{out}} \\ &= \left(\frac{S_1}{N_1}\right)\left(\frac{F_1 G_1 N_1}{G_1 S_1}\right) \\ &= F_1 \end{aligned} \tag{1.46}$$

So, the noise figure F is actually the noise ratio, defined as SNR at the input divided by SNR at the output.

It is useful in a practical sense to refer the output noise of cascaded blocks back to the noise appearing at the input. To follow the path of this noise, we write it as $\mathcal{N}$, where $\mathcal{N} = N_1$ is the noise at the input. Referring to Figure 1.24, we may rewrite noise at the output as

$$N_2 = G_1[\, \overbrace{(F_1 - 1)\mathcal{N}}^{\text{"excess noise"}} + \mathcal{N}\,] \tag{1.47}$$

This turns out to be useful in analyzing a cascade of two systems, as shown in Figure 1.25. The noise at the output of the second stage will be the input noise, multiplied by the gain factor, plus any additional noise from the system itself. This gives

$$\begin{aligned} N_3 &= F_1 G_1 G_2 \mathcal{N} + G_2(F_2 - 1)\mathcal{N} \\ &= G_2[F_1 G_1 \mathcal{N} + (F_2 - 1)\mathcal{N}] \end{aligned} \tag{1.48}$$

As a result, the overall noise figure (or noise ratio) is

$$\begin{aligned} F_{12} &= \left(\frac{S_1}{N_1}\right)\left(\frac{N_3}{S_3}\right) \\ &= \left(\frac{S_1}{\mathcal{N}}\right)\left\{\frac{G_2[F_1 G_1 \mathcal{N} + (F_2 - 1)\mathcal{N}]}{G_1 G_2 S_1}\right\} \\ &= F_1 + \frac{F_2 - 1}{G_1} \end{aligned} \tag{1.49}$$

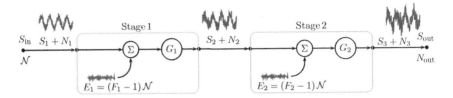

Figure 1.25 Analysis of two systems in cascade. The E values refer to the hypothetical noise added if referred back to the input of the first stage, whose noise is $\mathcal{N}$.

The significance of this is that **the first stage in a multistage system dominates the noise figure overall**. Subsequent stages contribute an amount lessened by the gain; in this case, the contribution of stage 2, which is $(F_2 - 1)$, is reduced by a factor equal to the gain of the previous stage G_1.

This could be extrapolated to any number of stages, for which the Friis equation for overall noise figure becomes

$$F = F_1 + \frac{F_2 - 1}{G_1} + \frac{F_3 - 1}{G_1 G_2} + \cdots + \frac{F_n - 1}{G_1 G_2 \ldots G_n} \tag{1.50}$$

Thus, it makes sense to maximize efforts to reduce the noise in the very first stage. Additionally, a high gain is helpful in the first stage, to reduce the effects of subsequent stages.

1.9 Chapter Summary

The following are the key elements covered in this chapter:

- The description of a waveform as a time-evolving quantity.
- The description of signal as comprising various frequency components, and how these components may be affected by filtering.
- Operations such as averaging, multiplication, and phase shifting, which may be applied to a waveform.
- One method of variable-frequency waveform generation: the DDS.
- The significance of power transfer, impedance matching, and noise in telecommunication system design.
- Thermal noise, and how noise may be characterized in a cascade of system blocks.

Problems

1.1 The decibel requires the calculation $10\log_{10}(P_{\text{out}}/P_{\text{in}})$. Using $P = V^2/R$ and assuming V_{out} is the voltage at the output, V_{in} the voltage at the

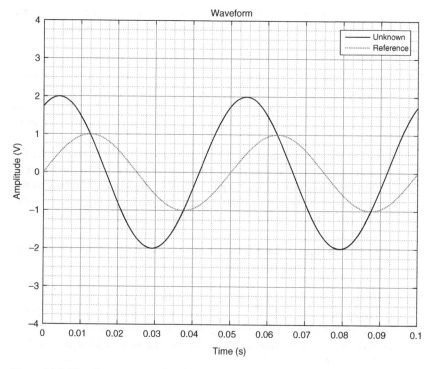

Figure 1.26 Waveform parameter problem.

input, and that the impedances of both are R Ω, show that an equivalent calculation is $20\log_{10}(V_{out}/V_{in})$.

1.2 The input to a Radio Frequency (RF) spectrum analyzer states that the input impedance is 50Ω, and that the maximum input power is $+10$dBm. What would be the maximum safe voltage in that case?

1.3 A copper communications line has a noise level of 1 mV RMS when a signal of 1 V RMS is observed. What is the SNR?

1.4 Determine the parameters (amplitude, phase, and frequency) of the waveform shown in Figure 1.26.

1.5 Given the mathematical description of a signal $x(t) = A \sin \omega t$, show that over one period $\tau = 2\pi/\omega$ the mean-square value is $\overline{x^2}(t) = A^2/2$. Hence show that the RMS value is $A/\sqrt{2}$. *Hint: Remember that the* arithmetic mean is really an average, so you integrate the square value over one period. You may need the trigonometric identity $\sin^2\theta = (1/2)(1 - \cos 2\theta)$.

1.6 Given a signal equation and the system transfer function, we can work out the output for both linear and nonlinear systems.

a) Given the system transfer function $y(t) = \alpha\, x(t)$, show that for an input $x(t) = A \sin \omega t$, the output is $y(t) = \alpha\, A \sin \omega t$. Is this system linear?

b) Given the system described by $y(t) = \alpha\, x(t) + \beta x^2(t)$, show that for an input $x(t) = A \sin \omega t$, the output can be simplified to the summation of a constant (or DC) term, a term at the same frequency as the input, and a term at twice the frequency of the input. Is this system linear? *Hint: You may need the trigonometric identity* $\sin^2\theta = (1/2)(1 - \cos 2\theta)$.

c) From the above results, can you infer what might happen if you had cubic-form transfer function, such as $y(t) = \gamma x^3(t)$?

1.7 Systems may be defined in terms of basic building blocks.

a) Given two series blocks as depicted on the left of Figure 1.3, what is the overall gain if each block's gain is given in decibels?

b) Would the same rule apply if the blocks were added in parallel? Why not?

1.8 The correspondence between dB and ratio is approximately

$$2 \text{ dB} \approx 1.6 \times$$
$$3 \text{ dB} \approx 2 \times$$
$$4 \text{ dB} \approx 2.5 \times$$

a) Explain why the dB figure goes up in equal increments of one, but the ratio figure goes up in differing increments (0.4 then 0.5).

b) Plot a graph of ratio r versus $10\log_{10} r$ for $r = 0.1$ to $r = 10$ in steps of 0.1, and explain the shape.

c) Plot a graph of ratio r versus $10\log_{10} r$ for $r = 10$ to $r = 100$ in steps of 1, and explain the shape. Compare the two graphs and explain their shapes as well as the values on the vertical axis.

1.9 Many concepts in telecommunications deal with very large or very small signals or cover a very wide range of values. In these cases, a *logarithmic* scale is useful rather than the usual linear scale. A good example is the decibel for measuring power. Suppose the frequency response of a certain system is defined by a function $g(f) = 1/(f + 1)$.

a) Explain what is deficient in the following approach, and suggest a better way.

```
f = 0.01:1:100;
g = 1./( f + 1);
plot ( f, g, 's-');
set ( gca, 'xscale', 'log');
grid ( 'on');
```

b) Noting that the exponents of 10 on the frequency axis go from -2 to $+2$, change the code to

```
r = -2:0.04:2;
f = 10.^r;
g = 1./( f + 1);
plot ( f, g, 's-');
set ( gca, 'xscale', 'log');
grid ( 'on');
```

Why does this give a proportional spacing of the data points, and hence a better plot?

c) Investigate the difference between the MATLAB functions linspace() and logspace() and briefly comment on why they are useful.

1.10 An amplifier has an SNR of 50 dB and Noise Figure of 3 dB. Determine the output SNR.

1.11 This question investigates the extension to two-stage systems as shown in Figure 1.25, in order to find an expression for the cascaded noise figure.

a) Draw a block diagram for this system, labeling all the "useful" signals and the unwanted noise signals.

b) Show mathematically that the noise factor for a three-stage system is

$$F = F_1 + \frac{F_2 - 1}{G_1} + \frac{F_3 - 1}{G_1 G_2}$$

2

Wired, Wireless, and Optical Systems

2.1 Chapter Objectives

On completion of this chapter, the reader should:

1) Be conversant with the basic principles of telecommunication transmission systems, which employ wired cabling, wireless or radio signals, and fiber-optic light transmission, and be able to explain the salient points of each approach.
2) Understand the importance of frequency and bandwidth in relation to a telecommunication system.
3) Be conversant with various digital line codes used for synchronization, and be able to explain their purpose.
4) Be able to explain the nature of a transmission line and how standing waves are produced.
5) Understand the general principles of radio propagation, and be able to explain the method by which antennas transmit or receive a signal.
6) Be able to explain the principles of optical communications, including light generation, propagation through optical fiber, reception, and synchronization.
7) Be able to apply knowledge of wireless and light propagation to transmission system loss calculations.

2.2 Introduction

As was discussed in Chapter 1, a waveform is a signal that takes on a certain amplitude, and that amplitude changes over time. The physical signal may be a voltage, light, or electromagnetic (EM) wave. This chapter deals with the physical aspects of transmission – either electrical voltages on a transmission line,

Communication Systems Principles Using MATLAB®, First Edition. John W. Leis.
© 2018 John Wiley & Sons, Inc. Published 2018 by John Wiley & Sons, Inc.
Companion website: www.wiley.com/go/Leis/communications-principles-using-matlab

radio waves propagating through air or space, and optical waves propagating within an optical fiber. There is some overlap in concepts between all of these – for example, the notion of loss or attenuation, which means that a signal level is reduced in size from that transmitted by the time it is received. This applies equally to wired systems using copper or other cabling, wireless systems employing radio signals, and optical fiber transmission. The way in which the loss occurs differs for each, but the loss of signal level is a common problem in all telecommunication systems, and must be well understood by the designer.

2.3 Useful Preliminaries

Telecommunication systems make extensive use of signals that are repetitive in nature. This means that there is a basic signal shape or pattern, and this shape is repeated over and over again, with known variations according to the information to be transmitted. It is these variations that convey information from one point to another. These variations occur over time, and their repetitive nature results in many other periodic signal components being produced as a by-product. Analyzing these signal components is crucial to understanding both how information may be conveyed and determining the limits on how fast information may be transmitted.

2.3.1 Frequency Components When a Signal Waveform Is Known

Suppose a waveform repeats over a time interval τ, called the waveform period. Mathematically, this means that

$$x(t) = x(t + \tau) \tag{2.1}$$

Such a waveform may be decomposed into its underlying *frequency components*. This concept means that the lowest-frequency component – termed a *fundamental* – exists, together with higher frequencies termed *overtones*, which are normally multiples of the fundamental frequency. Each integer multiple of the fundamental is termed a *harmonic*, and these are sometimes denoted as $1f, 2f, 3f, \ldots$ to make it clear which multiple is being referred to. Note that the fundamental is denoted as the first harmonic, as it is really just the first integer multiple of the fundamental. For a fundamental frequency of f_0 Hz, or ω_0 rad s^{-1} (radians per second), the defining relationships are

$$\omega_0 = 2\pi f_0 \tag{2.2}$$

$$\tau = \frac{1}{f_0} \tag{2.3}$$

$$\omega_0 = \frac{2\pi}{\tau} \tag{2.4}$$

The first of these shows how to convert Hz frequency f into radian frequency ω. The second states that the period τ is the reciprocal of the Hertz frequency f, and vice versa. The third relates radian frequency to the period. The kth harmonic may then be written as $\omega_k = k\omega_o$ using radian frequency, or $f_k = kf_o$ using Hertz frequency, for integer values $k = 0, 1, 2, \ldots$ The key question is then: Given a wave shape, how do we determine what underlying frequencies are present? Fourier's theorem states that *any* periodic function $x(t)$ may be decomposed into an infinite series of sine and cosine functions:

$$
\begin{aligned}
x(t) &= a_0 + a_1 \cos \omega_o t + a_2 \cos 2\omega_o t + a_3 \cos 3\omega_o t + \cdots \\
&\quad + b_1 \sin \omega_o t + b_2 \sin 2\omega_o t + b_3 \sin 3\omega_o t + \cdots \\
&= a_0 + \sum_{k=1}^{\infty} (a_k \cos k\omega_o t + b_k \sin k\omega_o t)
\end{aligned}
\tag{2.5}
$$

Notice that the frequencies are of the form $k\omega_o$, where k is an integer. The first component a_0 is in fact the special case of $a_0 \cos k\omega_o$ where $k = 0$. This is, in effect, just the average value. The coefficients a_k and b_k are determined by solving the following equations, evaluated over one period of the input waveform:

$$
a_0 = \frac{1}{\tau} \int_0^{\tau} x(t)\, dt
\tag{2.6}
$$

$$
a_k = \frac{2}{\tau} \int_0^{\tau} x(t) \cos k\omega_o t\, dt
\tag{2.7}
$$

$$
b_k = \frac{2}{\tau} \int_0^{\tau} x(t) \sin k\omega_o t\, dt
\tag{2.8}
$$

The integration limit is over one period, and so could be either 0 to τ or $-\tau/2$ to $+\tau/2$. Each covers exactly one period of the waveform, but with a different starting point (and consequently, a different end point).

To illustrate the application of the Fourier series, consider a square wave with period $\tau = 1$ s and peak amplitude ± 1 for simplicity (that is, $A = 1$) as depicted in Figure 2.1. The waveform is composed of a value of $+A$ for $t = 0$ to $t = \tau/2$, followed by a value of $-A$ for $t = \tau/2$ to $t = \tau$.

The coefficient a_0 is found from Equation (2.6) as

$$
\begin{aligned}
a_0 &= \frac{1}{\tau} \int_0^{\tau} x(t)\, dt \\
&= \frac{1}{\tau} \int_0^{\frac{\tau}{2}} A\, dt + \frac{1}{\tau} \int_{\frac{\tau}{2}}^{\tau} (-A)\, dt
\end{aligned}
\tag{2.9}
$$

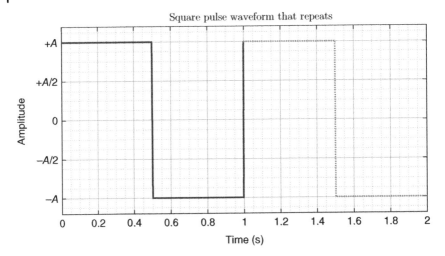

Figure 2.1 A square pulse waveform. The fundamental cycle from $t = 0$ to $t = 1$ is shown, and after that the same wave shape repeats forever.

Evaluating this, we find $a_0 = 0$. This seems reasonable, since the integration finds the net area and the division averages it over one period. Thus the integration is the average value, which may be seen by inspection to be zero. The coefficients a_k are found from Equation (2.7) as

$$a_k = \frac{2}{\tau} \int_0^\tau x(t) \cos k\omega_0 t \, dt$$

$$= \frac{2}{\tau} \int_0^{\frac{\tau}{2}} A \cos k\omega_0 t \, dt + \frac{2}{\tau} \int_{\frac{\tau}{2}}^\tau (-A) \cos k\omega_0 t \, dt$$

$$= \frac{2A}{\tau} \frac{1}{k\omega_0} \sin k\omega_0 t \Big|_{t=0}^{t=\frac{\tau}{2}} - \frac{2A}{\tau} \frac{1}{k\omega_0} \sin k\omega_0 t \Big|_{t=\frac{\tau}{2}}^{t=\tau} \tag{2.10}$$

Once again, this results in zero. This tells us that no cosine waves are required at all. Finally, the coefficients b_k are found from Equation (2.8):

$$b_k = \frac{2}{\tau} \int_0^\tau x(t) \sin k\omega_0 t \, dt$$

and may be shown to equal

$$b_k = \frac{2A}{k\pi}(1 - \cos k\pi) \tag{2.11}$$

When k is an even number $(2, 4, 6, \ldots)$, then $\cos k\pi = 1$, and hence the equation for b_k reduces to 0. When the integer k is an odd number $(1, 3, 5, \ldots)$, then $\cos k\pi = -1$, and hence this reduces to

$$b_k = \frac{4A}{k\pi} \qquad k = 1, 3, 5, \ldots \tag{2.12}$$

The completed Fourier series representation is obtained by substituting the specific coefficients for this waveform (Equation 2.11) into the Fourier series expansion (Equation 2.5), giving

$$x(t) = \frac{4A}{\pi} \left(\overbrace{1 \sin \omega_0 t}^{k=1} + \overbrace{\frac{1}{3} \sin 3\omega_0 t}^{k=3} + \overbrace{\frac{1}{5} \sin 5\omega_0 t}^{k=5} + \cdots \right) \tag{2.13}$$

This is illustrated in Figure 2.2. What this means is that the square wave may be represented by sine waves and only odd-numbered harmonics are present. The first sine wave has a period (and thus frequency) the same as the original, but has an amplitude of $4/\pi \approx 1.3$. The second sine wave has a frequency of three times the original, but is scaled in amplitude by

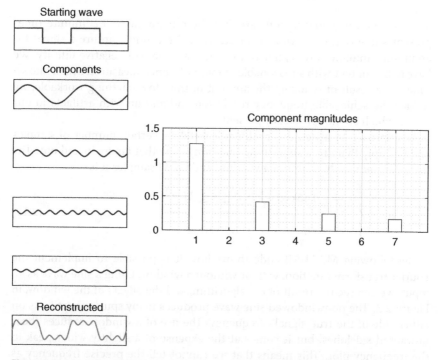

Figure 2.2 Approximating a square waveform with a Fourier series. The Fourier series approximation to the true waveform is shown; it has a limited number of components, but is not a perfect approximation.

$4/\pi \times 1/3 \approx 0.4$. A number of these sine waves must be added together before we have something looking at all like the original, but as shown in Figure 2.2, just three sine components produce a passable approximation to the waveform we started with.

2.3.2 Frequency Spectrum When a Signal Is Measured

The previous section showed how to calculate the Fourier components for a periodic waveform, whose shape is known in advance. But what happens if only a measured waveform is available, and the equation for it is unknown? In that case, it is necessary to measure or sample the waveform over time and calculate the Fourier transform.

Since the Fourier series equation used sine and cosine terms, we can combine them using complex numbers. Remembering that one of the basic properties of a complex number is that $e^{j\theta} = \cos\theta + j\sin\theta$, the Fourier transform then has the definition

$$X(\Omega) = \int_{-\infty}^{+\infty} x(t)e^{-j\Omega t}\,dt \tag{2.14}$$

This may be calculated in software, but the integral limits of "infinite time" present somewhat of a problem in practice. We cannot capture a signal for an infinite amount of time, nor can we start at a time of negative infinity. We have to be content with a reasonable amount of signal, measured over a known time. As a result of reducing the amount of time to something workable, we reduce the achievable frequency resolution and end up with artifacts in our result, which are not physically present.

An approach to lessen this problem is to take a certain number of samples and multiply them by a smoothing function – one that tapers in and out. The most common is the Hamming window, and for M samples it is

$$w_n = \begin{cases} 0.54 - 0.46\cos\frac{2n\pi}{M} & : 0 \leq n \leq M \\ \\ 0 & : \text{otherwise} \end{cases} \tag{2.15}$$

The following MATLAB code shows how it is possible to implement the Fourier transform function, with or without a window. Using a sine wave as an input, we can see the result of the algorithm, and the effects of the window, in Figure 2.3. The nonwindowed sine wave produces many spurious artifacts on either side of the true signal's frequency. The use of a window reduces these unwanted sidelobes, but it comes at the expense of a slightly wider peak in the frequency plot. This means that we cannot tell the precise frequency as accurately.

```
function [Xm, faxis, xtw] = CalcFourierSpectrum(xt, tmax,
fmax, UseWindow)

    %————————————————————————————————————————————————————————
    dt = tmax/(length(xt) - 1);
    t = 0:dt:tmax;
    %————————————————————————————————————————————————————————


    %————————————————————————————————————————————————————————
    xtw = xt;
    if( UseWindow )
        % window
        fw = 1/(2*tmax);

        % Hamming window
        fw = 1/(tmax);
        w = 0.54 - 0.46*cos(2*pi*t/tmax);
        xtw = xt.*w;
    end

    % continuous frequency range
    OmegaMax = 2*pi*fmax;
    dOmega = OmegaMax*.001;
    %————————————————————————————————————————————————————————


    %————————————————————————————————————————————————————————
    fvec = [];
    Xmvals = [];


    p = 1;
    for Omega = 0:dOmega:OmegaMax

        coswave = cos(Omega*t);
        sinwave = -sin(Omega*t);

        % perform the Fourier Transform via numerical
        % integration
        Xreal = sum(xtw.*coswave*dt);
        Ximag = sum(xtw.*sinwave*dt);
        mag = sqrt(Xreal*Xreal + Ximag*Ximag);

        % scale frequency to Hz, magnitude to maximum time
        fHz = Omega/(2*pi);
        mag = 2*mag/tmax;
```

```
        % save frequency and magnitude
        faxis(p) = fHz;
        Xm(p) = mag;
        p = p + 1;
    end
    %————————————————————————————————————————————
end
```

Figure 2.4 shows the situation where two sine components are present. Clearly it is able to resolve the presence of these two underlying components. They are not evident in the time waveform as measured, but the frequency plot of $|X(\omega)|$ reveals them quite clearly.

Now we have two useful analytical tools: the Fourier series, which tells us about periodic signals, and the Fourier transform, which is useful when we can measure a signal and need to determine the components present. Importantly for digital communications, we worked out that the Fourier series of a square pulse train has diminishing odd-numbered harmonics. In the case of the Fourier transform, measuring for a longer period of time gives greater frequency resolution, but a tapering or windowing function may be needed to give a clearer picture without spurious points in the frequency plot.

2.3.3 Measuring the Frequency Spectrum in Practice

The previous sections dealt with calculating the frequency spectrum, either from an equation or from the sampled data. If it is possible to sample the signal

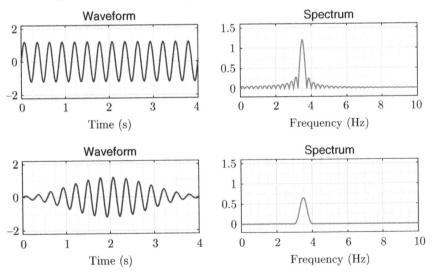

Figure 2.3 Using the Fourier transform to calculate the frequency magnitude of a signal. The use of a window to taper the signal provides a smoother picture, but less resolution.

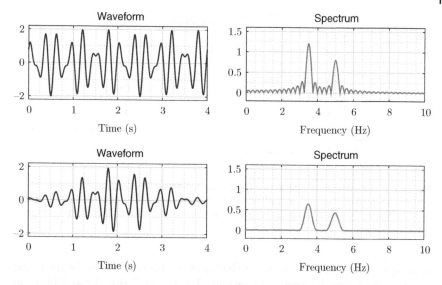

Figure 2.4 Using the Fourier transform to calculate the frequency magnitude where two underlying sinusoidal signals are present. It is able to resolve the presence of the two components.

digitally, then calculating the Fourier transform provides one way of determining the frequency spectrum of the signal. Another method that is employed at higher (radio) frequencies and used in a dedicated instrument called a *spectrum analyzer* is so-called swept-frequency analysis. The basic idea is to ascertain the power level in a narrow frequency band, then move the center of the band up a little, and repeat the measurement. The process is repeated until the desired frequency range has been scanned.

The main functional blocks required to do this are shown in Figure 2.5. In theory, it is necessary to have a bandpass filter, which is swept over the range of desired frequencies, measuring the average power as it goes. In practice, it is very difficult to produce a tunable bandpass filter. This is because of the relative range of frequencies: Consider, say, a 100 kHz signal band of interest, centered at a frequency of 1 GHz. The relative ratio of these is 10^4, which is a very large range, resulting in an impractical design. A workable solution is to replace tunable bandpass filters with a mixing arrangement. This achieves the same objective – namely, tunable bandpass filtering – but using a different means.

Referring to Figure 2.5, the ramp signal is a voltage that is proportional to the frequency we want to analyze. The oscillator then produces a sinusoidal signal whose frequency matches the desired frequency of analysis, according to the input voltage from the ramp. This is termed a Voltage-Controlled Oscillator (VCO) or, in the case of a digitally controlled oscillator, a Numerically Controlled Oscillator (NCO). The latter may use Direct Digital Synthesis (DDS)

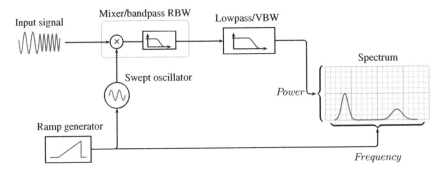

Figure 2.5 The principle of operation of a spectrum analyzer. The resolution bandwidth (RBW) filter is swept over the desired range and is implemented as a mixer (multiplier and lowpass filter). The video bandwidth (VBW) filter serves to smooth out the resulting display.

techniques (Section 1.6). The mixer/bandpass section multiplies the incoming signal with the specific frequency produced by the oscillator. As we will see in later sections, this moves the frequency components down to be centered at zero, but maintains their relative frequency spacing and power levels. The lowpass filter then removes higher-frequency components, which result from this mixing. The subsequent stage is another lowpass filter, which in effect averages out the result of the bandpass stage. The result is a display of power (vertically) relative to frequency (horizontally).

So there are two filtering stages – a bandpass stage (which is actually a multiplier and lowpass filter, referred to as a mixer) and a lowpass stage. Each performs a different function. Figure 2.6 shows the result when the swept passband is relatively wide. Four signals of interest are shown in panel "a," along with the ever-present noise at a low level. The product of the bandpass filter response and the input frequency is shown in panel "b." A snapshot only is shown, as the passband moves up in frequency from left to right. Panel "c" shows the cumulative result, as the passband is swept over the entire frequency range of interest. Finally, panel "d" shows the smoothed result of stage "c." It may be seen that the wider bandpass filter bandwidth does not permit discrimination of incoming peaks that are close together. What is needed is a narrower band filter.

The first filter (corresponding to stages "b" and "c") is termed the Resolution Bandwidth (RBW). It controls the frequency resolution and time taken to sweep across the desired frequency range. The second lowpass filter (corresponding to stage "d") is termed the Video Bandwidth (VBW). It controls the relative smoothness of the displayed frequency response.

Using a narrower bandpass filter as shown on the right of Figure 2.6, the two peaks at higher frequencies are successfully resolved. However, this comes at a price: Because the filter is much smaller, it takes much longer to sweep over

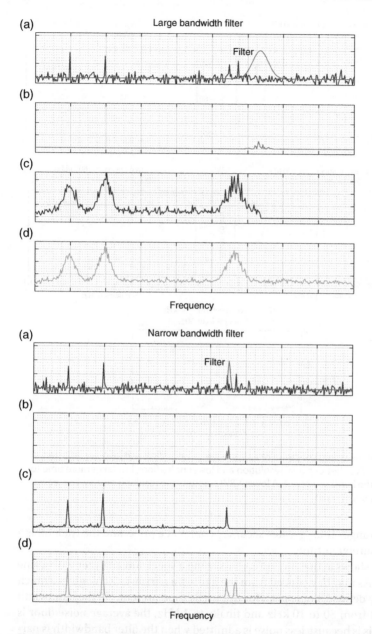

Figure 2.6 Spectrum analysis stages with a wide window as it progresses (top) and a narrow window (bottom). Progressively, we see the input signal and RBW bandpass filter (a), the bandpass filtered signal (b), the accumulated bandpass filtered signal as the sweep progresses (c), and the final result after VBW lowpass filtering (d). The two close peaks in the input are able to be resolved with the narrower filter on the right.

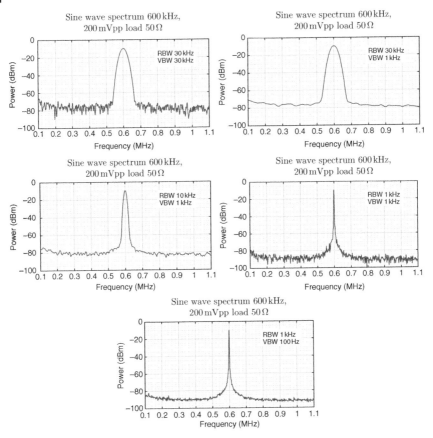

Figure 2.7 The measured spectrum of a sine wave, as both VBW and RBW are adjusted. A narrower RBW gives better signal resolution and lower noise floor, but takes more time to sweep across the band of interest. A lower VBW smooths the resulting display, but leaves the noise floor unchanged.

the desired range. Thus, greater accuracy comes at a price, and the tradeoff is a longer measurement time.

Figure 2.7 shows actual measurements using a spectrum analyzer with a sine wave input. As may be deduced from the figures, a narrower RBW gives a much more clearly defined single peak for the single tone present. As the bandwidth is decreased from 30 to 10 kHz and finally to 1 kHz, the average noise floor is reduced. This is because less noise is admitted when the filter bandwidth is narrower. The VBW filter does not provide any better definition of the frequency, nor does it reduce the noise floor. It does, however, smooth out the random nature of the noise, producing a "cleaner" display. These two settings are interrelated, and as mentioned, narrow RBW bandwidth requires a longer sweep

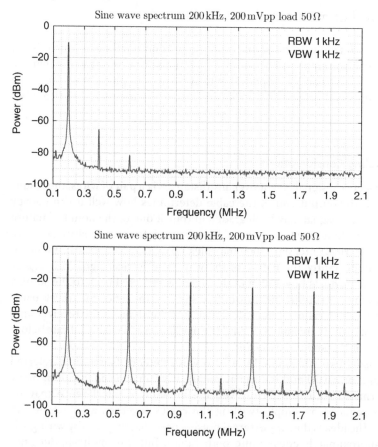

Figure 2.8 Top: measured spectrum of a "pure" sine wave. Note the spurious peaks at the first and second harmonics, due to imperfections in the waveform generation. Bottom: the spectrum of a square wave. Note the frequencies of the harmonics are integer multiples of the fundamental and that their amplitudes decay successively as 1/3, 1/5, ... if the decibel scale is converted to a ratio.

time. The measurements shown in Figure 2.7 are for a signal of 600 kHz, amplitude 200 mV peak to peak. It may be verified that this corresponds to −10 dBm with a 50Ω load.

Figure 2.8 shows input signals of 200 kHz over a wider frequency range. For the case of a sine wave input, it is observed that there is an additional spurious peak at twice the input frequency. This occurs in practice due to imperfections in the signal generator. Also shown in Figure 2.8 is the spectrum of a square wave signal, and as we noted earlier, the odd-harmonic magnitudes decrease in the successive ratio 1/3, 1/5, 1/9, ...

2.4 Wired Communications

This section introduces *baseband* communication methods, which is an approach typically employed in wired connections. That is not to say that the principles discussed are *only* applicable to systems with physical wiring, since many concepts and principles are also applicable to wireless and optical fiber systems.

2.4.1 Cabling Considerations

When an electrical signal is sent over any distance via a pair of wires, the resulting signal at the output end of the cable determines how well it can convey information. The signal may be degraded in transit due to the nonideal nature of the cable itself; it may be corrupted by external interference, since any wire acts as an electrical antenna, which may pick up extraneous signals; and probably most importantly, the nature of the way the electrical signal propagates along the cable affects how much of the signal is reflected back to the source, and in so doing, corrupting the transmitted signal as the reflection travels back. All of these considerations reduce the quality of the signal at the receiver and, in the case of digital transmission, limit the maximum achievable bit rate before a certain error threshold is exceeded.

Simply using two wires, one for a zero volt reference (sometimes referred to as ground or earth) and the signal voltage wire, may be adequate in many low-performance situations. However, the amount of interference may be somewhat reduced by simply twisting the cables together, as illustrated in Figure 2.9. The idea is that any external noise – perhaps from nearby wiring – is coupled approximately equally into both cables. But this, by itself, does not provide better performance. Using a differential signal or balanced mode of operation is required.

Figure 2.10 illustrates the concept of differential signaling. In the first case, a signal voltage in one wire is carried with respect to a reference or zero voltage in the second cable. The signal shown is a simple digital one, comprising two voltage levels. Now suppose a short burst of interference affects *one* of the cables, as shown. It is possible that this induced voltage spike may be sufficient to exceed the binary 1/0 threshold and thus corrupt the data transmission. However, consider the final case, where a differential voltage system is employed. One wire carries a positive-going voltage, but the other carries a negative-going voltage. The two wires are the reverse of each other, and the receiver uses the *difference* between the two voltages, rather than one line referenced to ground. As illustrated, a voltage spike superimposed on *both* cables will lead to no net change in the difference of the voltages. Hence, differential cabling is more immune to noise, especially when configured as a twisted pair. It also reduces interference

Straight cable

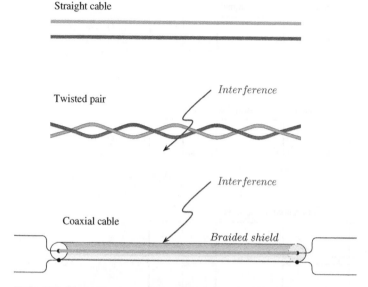

Figure 2.9 Illustrating some approaches to signal cabling. The use of a twisted pair helps to impart some noise immunity and is widely used in practice for Ethernet data cables. Coaxial cable is used for high-frequency applications such as antenna connections. It should not be confused with shielded or screened cable, which is composed of two or more wires with a separate outer shield conductor, which is not part of the circuit.

from one cable to an adjacent one (termed *crosstalk*), since any common-mode signal is superimposed on both cables.

Differential signaling brings with it some additional complications at the transmitter, since a circuit is required to drive the two wires symmetrically. The receiver requires a circuit to sense this difference, without the benefit of a ground reference.

Next, consider the coaxial cable (or "coax") in Figure 2.9. This type of cable uses a center core and a braided shield, which acts as the reference. In coax, the spacing between the center conductor and the shield is kept constant, to maintain the same electrical properties along the length of the cable. The shield, in effect, acts to keep out more of any external interference. However, this comes at a price: The manufacture of this type of cable is more involved, and hence it is more costly. It should be noted that single-wire, two-conductor or even four-conductor cable is available with a shield. This is not the same as coaxial cable, since coax is manufactured with precise geometry and spacing to keep the electrical properties constant along the length of the cable. Multicore shielded cables often find use in instrumentation applications, where very small voltages must be carried, but the signaling rate is usually not especially high.

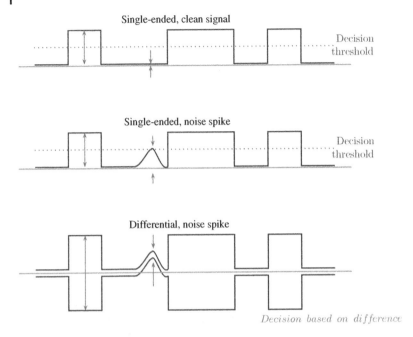

Figure 2.10 Differential or balanced signals are often used for transmission. Illustrated here is a sequence of digital pulses, affected by a short noise spike. Differential voltage driving is most effective where the noise is approximately equal in both wires, such as with a twisted pair.

Coaxial cables are always supplied with a rated impedance, and for optimal performance, this impedance must be matched to the circuitry both driving the cable and at the receiver. The characteristic impedance is usually given the symbol Z_0. We will see in subsequent sections why a mismatch between a driving circuit and the cable may lead to reduced signal voltages, with correspondingly increased susceptibility to external noise.

The most common classification is termed RG (originally referring to *Radio Gage*). Cable denoted RG58 has a 50 Ω impedance and is often used in laboratories. RG59 has a 75 Ω impedance and is used for video equipment. It has a similar (5 mm) diameter to RG58. RG6 cable also has a 75 Ω impedance but employs a solid core. As a result, this larger (7 mm diameter) cable is less flexible. It does, however, have superior characteristics for very-high-frequency signals.

2.4.2 Pulse Shaping

In transmitting digital data, it is necessary to transmit pulses to represent the binary 1's and 0's. This could be done simply by using positive and negative

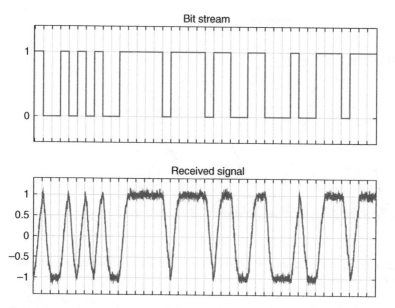

Figure 2.11 Transmitted pulse sequence and the corresponding received signal. Cable impairments and external interference combine to reduce the quality of signaling.

voltages. Ideally, these 1's and 0's would be transmitted as fast as possible, implying the shortest possible spacing between successive pulses. The physical cabling itself places a limitation on this maximum rate. The simple transmission of binary data in this way may be improved upon greatly by transmitting not just one bit at a time, but several bits. In baseband transmission, the simplest way to envisage this is to use multiple voltage levels to encode multiple bits at once. The first question to address, then, is what happens when we put a series of pulses on a transmission line? When launched into the cable, the pulses may have a square shape, but they become distorted during transmission. At the receiving end, the pulse shape may be rounded, with the start and end of individual pulses indistinct.

Consider Figure 2.11, which shows a bit stream to be transmitted. The 1's and 0's are transmitted serially and encoded as $+V$ and $-V$. The job of the receiver is to decode the voltage stream back into bits, and this involves (i) some type of thresholding of the voltage and (ii) some timing information to know *when* to apply that threshold. If either of these – timing or amplitude – is wrong, the receiver will make an incorrect decision about a 1 or 0 bit.

In Figure 2.11, two types of impairments to the voltage waveform are evident. One is that there is noise superimposed on the voltages. If this noise is large enough, then it may cause an incorrect decision to be made. Secondly, the rate of rise of the voltages is not instantaneous. If we wish to increase the bit rate,

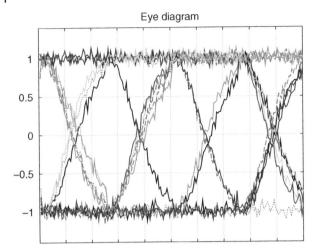

Figure 2.12 Using an eye diagram for ascertaining the timing and amplitude characteristics of a channel.

then we must transmit more bits per second. Examination of the figure will show that there are some places where this is marginal – the voltage has barely risen enough in one bit time before another bit has to be encoded.

For both wired and optical systems, a useful type of plot is to synchronize the capture of the waveform when receiving a random bit pattern, with the result as illustrated in Figure 2.12. Here the effects of the channel impairments become more obvious, and the "eye opening" gives a quick visual opinion on the channel quality.

To understand the mechanisms at work over a real channel, consider Figure 2.13, where we see some positive and negative pulses. The channel response is shown below the pulses, and in this case a simple second-order response has been used. Clearly, the short, sharp pulses have been turned into longer, slowly decaying pulses. However, a receiver could still decode the voltage stream, by sampling at the required pulse intervals τ. It would need some timing information to ascertain the phase of the pulses. After all, it could sample at the correct time intervals, but at the wrong place with respect to the start of the transmission. Also, a threshold would be required, since the channel may reduce the amplitude of the signal by some unknown proportion.

Now consider trying to increase the pulse rate. As shown, the earliest we might want to introduce a new pulse (for the next data bit) would seem to be at point "A," where the previous pulse has decayed away. We do not know how long this might take – however, this is a somewhat pessimistic assessment. If the receiver samples the peak of the waveform at point "B," then simply waiting until the pulse has died away seems to be necessary. Introducing the second

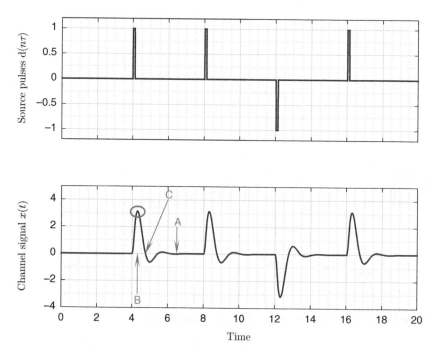

Figure 2.13 Ideal pulses (top) and their shapes when received (bottom). Smaller pulse spacing may mean that any given pulse waveform interferes with a later pulse.

pulse earlier than $t = 8$ may distort the magnitude of this second pulses' waveform and possibly lead to an incorrect 1/0 decision. But suppose now that we arranged the transmission such that the peak of the second pulse occurred at point "C." At point "C," the amplitude of the first pulse is exactly zero, and so it would not interfere with the amplitude of the second pulse *at that instant*. This means that we can send pulses at a much greater rate, since we do not have to worry about each pulse dying away before a new pulse is sent. Controlling the timing will be critical if this approach is to be used successfully.

Realizing that we could introduce a new pulse when the previous one is exactly zero due to the cable characteristics, it is reasonable to ask whether the pulse shape could be preshaped before transmission. Instead of a short, sharp pulse, what would happen if a particular waveform template "shape" was employed, which (conveniently) has zero-crossings corresponding to the rate at which we wish to send data? Such a pulse is shown in Figure 2.14 and is commonly referred to as a *sinc function*. It may be defined as

$$h(t) = \frac{\sin(\pi t/\tau)}{(\pi t/\tau)} \tag{2.16}$$

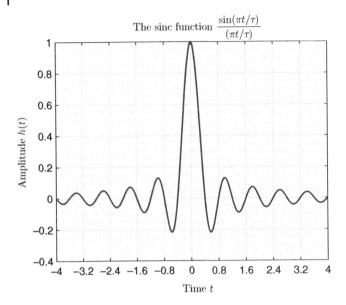

The sinc function $\dfrac{\sin(\pi t/\tau)}{(\pi t/\tau)}$

Figure 2.14 The sinc function with $\tau = 0.4$, centered about zero.

The equation shows that is a sine wave (a function of time), divided by a term proportional to time. Importantly for this application, the constant τ is the spacing of the zero-crossings, as shown in Figure 2.14.

This function appears as though it could satisfy the data transmission requirements: A smooth function that has zero-crossings at just the right places. For each bit to be transmitted, we translate a sinc function to $t = 0$ and make the peak either +1 or −1. The next bit could be transmitted at time $t = \tau$, and the fact that the pulse waveform now continues to oscillate about zero should not affect subsequent bit transmissions at $t = 2\tau, t = 3\tau, \ldots$, provided the timing is precise.

The sinc function, introduced in Woodward and Davies (1952, p. 41), finds its way into many digital transmission and pulse-shaping problems. Note that there are two definitions of the sinc function in the literature. One is $\text{sinc}(x) = (\sin x)/x$ and the other is $\text{sinc}(x) = (\sin \pi x)/\pi x$, so care must be taken in applying any equations directly.

The following MATLAB code illustrates how to generate the samples of a sinc function.

```
N = 1024*4;
Tmax = 10;

% note using 'negative' time here
```

```
dt = Tmax/((N-1)/2);
t = -Tmax: dt : Tmax;

% time of zero crossing
tau = 0.4;

% sinc
hsinc = sin(pi*t/tau + eps)./(pi*t/tau + eps);
plot(t, hsinc);
xlabel('time');
ylabel('amplitude');
```

There are some limitations with this basic idea, however, and we need to modify it a little. Firstly, the function seems to extend for positive and negative time. This is solved by simply incorporating a delay, in that we could delay the start of the pulse a little. Secondly, it would seem that the function extends forever in time. But since the amplitude decays reasonably rapidly, we might say that we could truncate the sinc function (at $t = \pm 4$ in Figure 2.14).

The frequency domain shows us that this pulse has constant energy up to a limiting frequency of 1.25 Hz, and it is no coincidence that this is $1/2\tau$, since the effective period of the underlying sine is 2τ. In the frequency domain, such a "brickwall" response is expressed as

$$H(\omega) = \begin{cases} K & : \quad -\frac{\omega_b}{2} \leq \omega \leq +\frac{\omega_b}{2} \\ 0 & : \quad \text{otherwise} \end{cases} \tag{2.17}$$

The limits are to $\pm \omega_b/2$, and using $\omega = 2\pi f$ and $f = 1/\tau$,

$$\frac{\omega_b}{2} = \pi f_b$$
$$= \frac{\pi}{\tau} \tag{2.18}$$

So we can rewrite the limits as

$$H(\omega) = \begin{cases} K & : \quad -\frac{\pi}{\tau} \leq \omega \leq +\frac{\pi}{\tau} \\ 0 & : \quad \text{otherwise} \end{cases} \tag{2.19}$$

Letting $K = 1$ for a constant, unity gain in the passband, the inverse Fourier transform is

$$h(t) = \frac{1}{2\pi} \int_{-\infty}^{\infty} H(\omega) \, e^{j\omega t} \, d\omega \tag{2.20}$$

With $H(\omega) = 1$, it becomes

$$
\begin{aligned}
h(t) &= \frac{1}{2\pi} \int_{-\pi/\tau}^{+\pi/\tau} 1 \ e^{j\omega t} \ d\omega \\[2mm]
&= \frac{1}{j2\pi t} \ e^{j\omega t} \Big|_{\omega=-\pi/\tau}^{\omega=\pi/\tau} \\[2mm]
&= \frac{1}{j2\pi t} \left(e^{j\frac{\pi t}{\tau}} - e^{-j\frac{\pi t}{\tau}} \right) \\[2mm]
&= \frac{1}{j2\pi t} \times 2j \sin\left(\frac{\pi t}{\tau}\right) \\[2mm]
&= \frac{1}{\pi t} \ \sin\left(\frac{\pi t}{\tau}\right) \\[2mm]
&= \frac{1}{\pi t} \ \frac{\sin(\pi t/\tau)}{(\pi t/\tau)}(\pi t/\tau) \\[2mm]
&= \frac{1}{\tau} \ \operatorname{sinc} \frac{\pi t}{\tau}
\end{aligned}
\tag{2.21}
$$

So the time-domain impulse response is

$$
h(t) = \frac{1}{\tau} \ \operatorname{sinc} \frac{\pi t}{\tau}
\tag{2.22}
$$

Finally, note that we want the value $h(t)$ to equal the sample at $t = nT$, and thus the gain should be unity. Comparing to $h(t)$ above, which has a $1/\tau$ factor, it is clear that we need a gain of τ for this to occur. As a consequence, the gain in the frequency domain is also τ.

We are now in a position to extend these ideas to a more practical pulse (one that does not extend to infinite time), yet maintains the periodic zero-crossing property, which is essential to avoid one pulse interfering with another – termed Intersymbol Interference (ISI).

Figure 2.15 shows the type of time response, which would be desirable for an ideal pulse-shaping filter (sinc-like, but decaying to zero), as well as the corresponding frequency response. If we taper the time response to zero, a smoother frequency response is the result. In Figure 2.15, the frequency response is tapered from A to B. This then will yield a sinc-like time function, but without extending to infinite limits.

So what form should the frequency response take? If we know that, then the time response should follow. What is required is to (i) define the frequency response and (ii) use Fourier techniques to determine what the corresponding time function should be. The first step is to mirror the frequency response as shown in Figure 2.16, so that we have a range of frequencies extending over a positive and negative domain, so as to permit evaluation of the double-sided integral.

To formulate the tapering problem, it is usual to introduce a rolloff factor, which we will call β, such that $0 < \beta < 1$. It is essentially a "tapering coefficient."

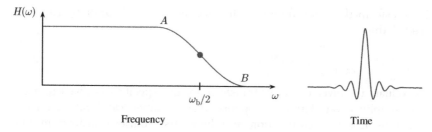

Figure 2.15 The frequency response of a raised cosine pulse (left) and the corresponding pulse shape (right).

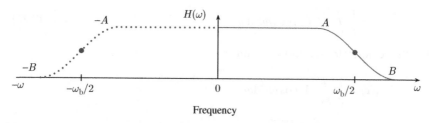

Figure 2.16 Calculating the frequency response of a raised cosine pulse in order to determine the required shape in the time domain.

According to the diagram, the rolloff is centered on $\omega_b/2 = \pi/\tau$. At point A, the frequency is $(1 - \beta)\omega_b/2 = (1 - \beta)\pi/\tau$. Similarly, at point B the frequency is $(1 + \beta)\omega_b/2 = (1 + \beta)\pi/\tau$. The difference between these two is $\Delta = B - A = 2\pi\beta/\tau$.

In the region $0 \rightarrow A$ there is a gain of unity, so $H_{0B}(\omega) = 1$. In the region $A \rightarrow B$ we arrange for a smooth taper from one to zero. Let it be of the form $1/2(1 + \cos x)$, where x goes from 0 to π, since that seems like a convenient "shape." In effect, this is just a "shoulder" rolloff, using a cosine function.

It is next necessary to work out the precise form of $H_{AB}(\omega)$. The frequency relative to A is $(\omega - A)$, and this is scaled by Δ to obtain a quantity that goes from 0 to 1. After that, we multiply by π to obtain a range suitable for a cosine function. Finally, the $+1$ and scaling of $1/2$ is because we want the function to go from 1 to 0, and a cosine would go from $+1$ to -1 in this range. Thus the cosine part is

$$\cos\left(\frac{\omega - A}{\Delta}\right)\pi = \cos(\omega - A)\frac{\pi}{\Delta} \tag{2.23}$$

and the resulting frequency response is

$$H_{AB}(\omega) = \frac{1}{2}\left[1 + \cos(\omega - A)\frac{\pi}{\Delta}\right] \tag{2.24}$$

To transform this from frequency to time, an inverse Fourier transform is needed; this is

$$h(t) = \frac{1}{2\pi} \int_{-\infty}^{\infty} H(\omega) \, e^{j\omega t} \, d\omega \tag{2.25}$$

To evaluate the inverse Fourier transform for this specific case, we can make some simplifications. First, the response will be symmetrical about the vertical axis. Thus, it is an even function, and the inverse Fourier transform may be reduced to

$$
\begin{aligned}
h(t) &= 2 \times \frac{1}{2\pi} \int_0^{\infty} H(\omega) \cos \omega t \, d\omega \\
&= \frac{1}{\pi} \int_0^{\infty} H(\omega) \cos \omega t \, d\omega
\end{aligned} \tag{2.26}
$$

For the region 0 to A, $H(\omega) = 1$, and so

$$
\begin{aligned}
h_{OA}(t) &= \frac{1}{\pi} \int_0^{A} 1 \; \cos \omega t \, d\omega \\
&= \frac{1}{\pi t} \; \sin \omega t \big|_{\omega=0}^{\omega=A} \\
&= \frac{1}{\pi t} \sin At
\end{aligned} \tag{2.27}
$$

For the region A to B, $H(\omega)$ is the cosine taper function, and so

$$h_{AB}(t) = \frac{1}{\pi} \int_A^{B} \frac{1}{2} \left[1 + \cos(\omega - A)\frac{\pi}{\Delta} \right] \cos \omega t \, d\omega \tag{2.28}$$

Combining these integrals, we get the final result for the time-domain response. This must be scaled by $1/\tau$ so that the time-domain gain is unity. The final result for the frequency response is

$$H(\omega) = \begin{cases} \tau & : \; 0 \leq \omega \leq (1-\beta)\frac{\pi}{\tau} \\ \frac{\tau}{2}\left[1 + \cos\left(\frac{\tau}{2\beta}\left(\omega - (1-\beta)\frac{\pi}{\tau} \right) \right) \right] & : \; (1-\beta)\frac{\pi}{\tau} \leq \omega \leq (1+\beta)\frac{\pi}{\tau} \\ 0 & : \; \omega \geq (1+\beta)\frac{\pi}{\tau} \end{cases} \tag{2.29}$$

and the required time response of the raised cosine filter is

$$h(t) = \left(\frac{\cos \pi \beta t/\tau}{1 - (2\beta t/\tau)^2} \right) \text{sinc}\left(\frac{\pi t}{\tau} \right) \tag{2.30}$$

where $\text{sinc}(x) = \sin x / x$. These time and frequency responses are shown in Figure 2.17, where we can see that a sharper cutoff in the frequency domain (decreasing β) results in a more oscillatory function in the time domain.

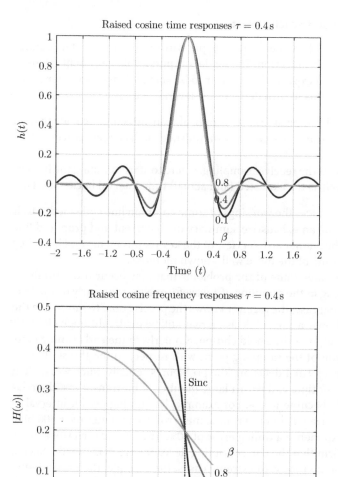

Figure 2.17 Illustrating the changing of parameters for the raised cosine pulse in the time domain (top) and corresponding frequency rolloff (bottom).

In fact, if we substitute $\beta = 0$, the impulse response reverts to a sinc function as previously discussed. Conversely, as β approaches unity, the frequency rolloff is much slower, and the time pulse decays more rapidly. This permits the designer to tailor the time-domain response to the bandwidth of a given channel.

2.4.3 Line Codes and Synchronization

Transmitting a digital data stream over copper wire or fiber-optic cable requires some method of synchronization. The sender outputs a voltage or light intensity for a certain time and expects the receiver to be able to deduce whether a 1 or 0 bit was intended. This in turn depends on several factors:

1) Whether the light intensity (in fiber optical transmission), or voltage level (electrical cable transmission), has diminished or has been corrupted by noise.
2) The preciseness of the receiver sampling: where to start sampling the incoming signal, as well as the expected (and actual) time interval between bits.

This section introduces the workings of some common line code methods. It is not intended to be an exhaustive summary of all current and proposed line coding methods, but rather an overview of the key requirements and how they are addressed in some of the more well-known methods.

Figure 2.18 illustrates some of the problems that may occur if the timing is incorrect – resulting in the sampling of a waveform either too early or too late. In the correct timing case, the receiver knows the precise time to sample. In the late timing case shown, a correct decision may still be made, although clearly it depends on the midpoint threshold chosen. Similarly, in the early timing case, the relative position of the sampling point with respect to the threshold may result in an incorrect 0 or 1 decision. The problem is actually much worse than indicated, since over time and with a large number of bit transitions, the relative skew will only become worse. For example, late timing per bit interval of only 0.1% may seem small, but after the same error accumulates over 500 bits, the last bit may experience a timing error of 50%. Thus, there is a critical need to somehow confirm that the correct timing decisions are being made. Such a system should also be robust to extreme cases, for example, the transmission of long runs of 1's or 0's. The receiver also needs to be able to determine which points on the waveform are the starting points, since if (for example) the timing was exactly one bit out at the outset of a block of binary data transmitted, then *every* bit may potentially be received incorrectly.

There are a number of other considerations too, depending upon the media that are is being employed. For example, wired connections such as Ethernet are often bundled together. The result is that EM interference from one signal pair could influence others (called *crosstalk*). Simply increasing the voltage levels (or power) so as to reduce the susceptibility to interference is not usually an option, since that approach would result in more interference to other cable pairs.

To address the timing and synchronization issues, a *line code* method is often employed in baseband data transmission. In baseband transmission, the data is directly encoded as a set of amplitude levels and not modulated on a sinusoidal carrier. Different line codes suit differing purposes and

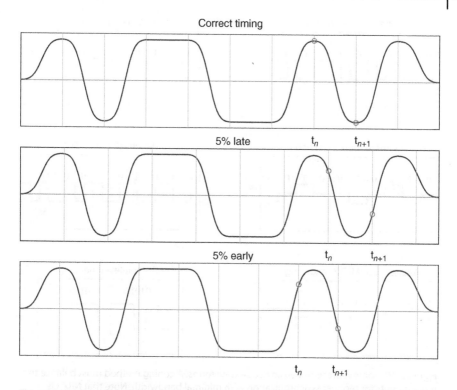

Correct timing

5% late t_n t_{n+1}

5% early t_n t_{n+1}

t_n t_{n+1}

Figure 2.18 The effect of sampling a waveform early or late. Incorrect timing at the receiver results in sampling the waveform's amplitude at the wrong time with respect to the transmitter, and hence the resulting sample value may be incorrect.

operational considerations. A selection of line coding methods are illustrated in Figure 2.19. The *Biphase Mark Encoding* method uses an encoding rule such that there is always a reversal of polarity at the start of each bit, with a 0 encoded as no mid-bit transition and a 1 encoded as a mid-bit transition. This has advantages if the physical wiring is reversed, since the resulting polarity inversion would not affect the bit decisions.

Two related methods are *NRZ-I* and *Alternate Mark Inversion* (AMI). NRZ-I inverts the waveform upon transitioning to a 0 bit, although NRZ-I may also be defined to invert on 1's rather than 0's. AMI requires three distinct levels, and encoding for AMI specifies that every 1 bit flips the voltage level with respect to the level at the last 1 bit. This has the advantage that violation of the coding rule can be more easily detected – two successive positive pulses should not occur, nor should two successive negative pulses.

One widely used encoding method for Ethernet over Unshielded Twisted Pair (UTP) at rates of 10 Mbps is *Manchester Encoding*. This belongs to a general class termed *biphase* coding methods and is in effect a type of phase

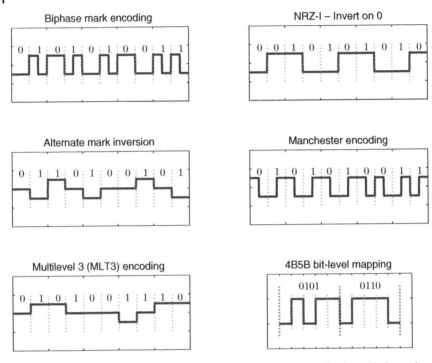

Figure 2.19 Some representative line code waveforms. A coding method must balance the requirements for receiver synchronization with minimal bandwidth. Note that NRZ-I is shown for invert on zero convention (as used in USB).

modulation. For wired Ethernet at 10 Mbps, the conventions for bit encoding in each bit cell (or timing interval) are that a 0 is encoded as a transition from a high to a low voltage level (H → L), with a 1 encoded as L → H. This convention could of course be reversed, and some sources give only one or the other definition. The possible confusion arising is discussed in Forster (2000). In operation, transmit and receive cable pairs are used for differential voltage transmission. This results in greater noise immunity, since any external noise coupled into a pair of wires is likely to affect both of them simultaneously – the result being that the *difference* is largely unaffected.

Ethernet over inexpensive UTP cabling has a number of shortcomings. The twists in the copper cable, as well as the number of twists in a given length, are important for noise rejection. Cables for UTP Ethernet are rated according to their category or Cat number. Cat-3, for example, has a notional bandwidth of 16 MHz, whereas Cat-5 extends to 100 MHz. The designations for the most common Ethernet deployment over copper UTP are 10BASE-T and 100BASE-TX for 10 and 100 Mbps rates, respectively.

The frequency spectrum of a given modulation scheme is also important since the cabling must accommodate the necessary bandwidth. Additionally,

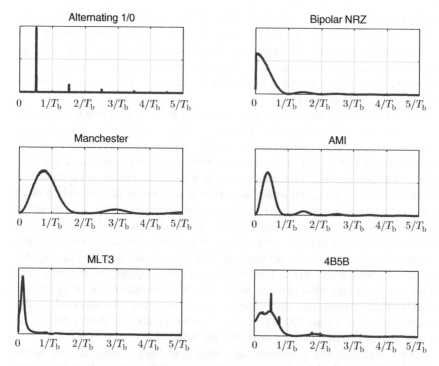

Figure 2.20 Spectra of some common line codes, derived from encoding a very long string of random binary data. The alternating 1/0 spectrum is shown for reference: It has a primary component at half the bit rate, with discrete harmonics at successively lower power levels.

the transmission line acts as an antenna to radiate out power at high frequencies. If there was only a positive voltage for binary 1 and a negative voltage for binary 0, then the highest frequency seen on the line would result from an alternating bit stream (101010...). A long string of 1's (or 0's) could cause the receiver to slip by one or more bit intervals, resulting in a stream of incorrect bits. In 10 Mbps Manchester encoding, we see that there is a voltage transition in the middle of each bit signaling interval. Furthermore, the direction of the transition unambiguously defines the original bit as 0 or 1. This addresses the issues of correct bit decoding and also correct synchronization. The encoding rule means, though, that radiated power at a frequency of 10 MHz may be significant.

Figure 2.20 shows the frequency spectra resulting from a long random sequence of binary data for these encoding methods. The worst-case frequency scenario results from the alternating 1/0 sequence, where there is a strong component at *half* the input rate, with subsequent harmonics at a lower amplitude. To increase this bit rate to 100 Mbps (as used in Ethernet), simply using the same encoding is not possible, since that would necessitate a higher bandwidth cable, and would also imply radiation in the vicinity of 100 MHz. Instead, two successive coding methods are applied. First, a multilevel scheme termed MLT-3 is employed. As illustrated, this uses three voltage levels, with the principle that a 0 is encoded as no voltage change, with a 1 forcing a transition to the next voltage level ($-A$, 0 or $+A$ in succession, where A is some amplitude level).

The advantage of this encoding is that the average frequency content is reduced, since 0 bits do not require any voltage transitions. A worse-case scenario would be a continuous input of 1's, which would result in a voltage waveform of $0, +1, 0, -1, 0, \ldots$. This is effectively 1/4 of the input bit rate. For a 100 Mbps input rate, this bit pattern would result in a fundamental frequency of 25 MHz. However, that comes at a price: the possible loss of synchronization at the receiver. This is addressed by using a subsequent encoding method termed 4B5B. This encodes each block of 4 input bits into 5 output bits, such that the output bits are carefully selected to optimize the transitions. The worst-case scenario of 25 MHz is then translated up slightly, to $25 \times (5/4) = 31.25$ MHz. Gigabit Ethernet must push these limits even further. It uses four pairs of wires, with a five-level amplitude encoding termed PAM5 (Pulse Amplitude Modulation, $4 + 1$ levels). The four allowable levels encode 2 bits at a time (00, 01, 10, 11). Synchronization is then achieved by using an extrapolated version of 4B5B, called 8B10B.

Some real sampled waveforms are shown in Figure 2.21. The 10 Mbps waveform shows the Manchester encoding using differential voltage transmission: When one line increases, the other decreases in mirror-image fashion. The 100 Mbps Ethernet traces show that MLT encoding is employed, with three distinct voltage levels. The reduction in sharpness of the waveforms illustrates the type of impairment that occurs with respect to the theoretical waveforms.

2.4.4 Scrambling and Synchronization

In some applications of digital transmission, the use of scrambling of the binary data is desirable. Since the end user may transmit any data pattern, or indeed no data at all for a given interval, the possibility arises that either a continuous 1 or 0 stream has to be sent, or else that a particular pattern is repetitively sent. This may be undesirable from several viewpoints – constant values make synchronization at the receiver more difficult, DC voltage levels may produce wander on transmission lines, and radio-frequency interference (RFI) may be induced over certain frequency ranges in nearby cabling.

Bit scramblers help to reduce the severity of some of these problems. These are used broadly for similar reasons to line codes as described in Section 2.4.3. However, whereas line codes introduce more voltage levels than binary digits (a one-to-many mapping), the use of a scrambler on a binary sequence is a

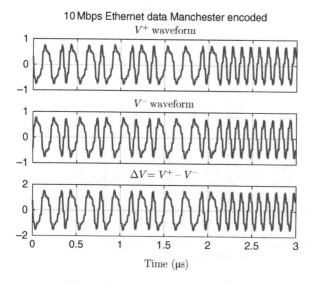

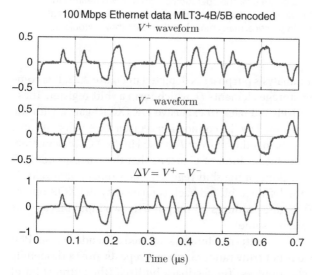

Figure 2.21 Two captured portions of Ethernet waveforms at 10 Mbps (Manchester) and 100 Mbps (4B5B/MLT). Note the differing scales for each time axis.

one-to-one mapping (the number of output bits normally equals the number of input bits). Of course, a matching descrambler is also required. It should be pointed out that scrambling in this sense does not produce any encryption of the data itself, although similar circuit structures appear in both data scramblers and various types of digital encryption systems.

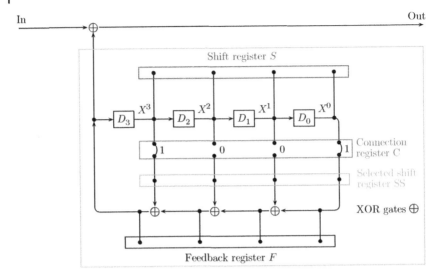

Figure 2.22 A scrambler using only 4 bits. The operation of each block is defined in Figure 2.23. The exclusive OR (XOR) operator (shown as ⊕) produces a 1 output if either of the inputs (but not both) is 1. In practice, many more bits than that shown would be employed.

A simple scrambler for analysis purposes is depicted in Figure 2.22. It is composed of a set of one-bit storage elements D, which form a shift register. These bit values are then selectively fed back to exclusive OR (XOR) gates, with the specific bit selection determined by the connection register C. A value of 1 in a given bit position of C means that the bit from D is used in the XOR calculation, whereas a value of 0 in the bit position of C means that the corresponding D bit is not used. At each clocking of the shift register, a new binary data value is formed by $D_3 D_2 D_1 D_0$. The leftmost bit fed back is formed by the XOR operations indicated. Each XOR gate produces a 1 output if its two input bits are not identical (and 0 if they are the same).

This forms a pseudorandom pattern of integers, termed a Pseudo-Noise (PN) sequence. This sequence is not truly random, since it repeats after a (hopefully large) number of shift clock pulses. The feedback bit itself (the leftmost bit of F) forms a Pseudo-Random Binary Sequence (PRBS).

The invert, shift, AND, OR, and XOR bit operations required are illustrated in Figure 2.23. These are shown with 4-bit example values, contained in registers R_n. The rightmost bit of R is the least significant bit (LSB), while the leftmost bit of R is the most significant bit (MSB). The shift operator moves (shifts) each bit one position to the left (or right) at each step, governed by a synchronizing clock. The inversion operator converts 0 to 1 and 1 to 0. The logical AND operator produces an output of 1 only if both inputs are 1 and 0 otherwise. The

$$\begin{array}{ll} R_0 & 1000 \\ R_1 = \overrightarrow{R_0} & 0100 \\ R_2 = \overrightarrow{R_1} & 0010 \\ R_3 = \overrightarrow{R_2} & 0001 \end{array}$$

Shift right

$$\begin{array}{ll} R_0 & 0001 \\ R_1 = \overleftarrow{R_0} & 0010 \\ R_2 = \overleftarrow{R_1} & 0100 \\ R_3 = \overleftarrow{R_2} & 1000 \end{array}$$

Shift left

$$\begin{array}{ll} R_0 & 1001 \\ \hline \overline{R_0} & 0110 \end{array}$$

Invert (not)

$$\begin{array}{ll} R_0 & 1001 \\ R_1 & 0101 \\ \hline R_0 \cdot R_1 & 0001 \end{array}$$

Logical AND

$$\begin{array}{ll} R_0 & 1001 \\ R_1 & 0101 \\ \hline R_0 | R_1 & 1101 \end{array}$$

Logical OR

$$\begin{array}{ll} R_0 & 1001 \\ R_1 & 0101 \\ \hline R_0 \oplus R_1 & 1100 \end{array}$$

Logical XOR

Figure 2.23 Binary operations required to implement the scrambler. Note the mathematical operators used for various cases.

logical OR operator produces an output of 1 if either (or both) inputs are 1 and 0 otherwise. The XOR operator produces a 1 output if either of the inputs (but not both) is 1.

Using these definitions and the diagram of a 4-bit scrambler, we can formulate the equations governing the shift and feedback stages. Referring again to Figure 2.22, we can define the mask register M that is used for selecting a particular bit position, the selected shift feedback SS, and the feedback itself starting from the rightmost or LSB as

$$\begin{aligned} M &= 0001 \\ \mathrm{SS} &= S \cdot C \\ F &= \mathrm{SS} \cdot M \end{aligned}$$

For a given shift stage, we must calculate the leftmost feedback bit. To do this, we step over each feedback bit from right to left by shifting the mask M and XORing the necessary bits to form the output bit:

$$\begin{aligned} M &= \overleftarrow{M} \\ \mathrm{SS} &= S \cdot C \\ F &= F \mid [(\overleftarrow{F} \oplus \mathrm{SS}) \cdot M] \\ S &= \vec{S} \end{aligned}$$

The sequence of outputs is tabulated in Figure 2.24. The input data is then XORed with the feedback PRBS output bit to form the final output to the next transmission stage.

The scrambler may be implemented in software, although invariably this is more cumbersome than implementing it directly in hardware. To help understand the overall operation, the following shows how a simple B-bit feedback shift register may be created. It is designed to directly implement the bit equations as given. The `uint8` data type forms an 8-bit quantity, which is sufficient for this simple 4-bit example (the upper 4 bits are not used).

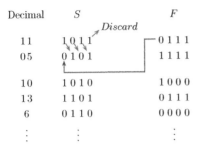

Figure 2.24 Step-by-step operation of the feedback register.

Bit shifting is performed using bitshift(reg,1) to shift left by one bit position and bitshift(reg,-1) to shift right. The bitand(), bitor(), and bitxor() functions perform the AND, OR, and XOR operations, respectively.

```
B = 4;                  % bits in shift register
sregin = uint8(11);     % shift register seed.  11 dec = 1011
                        % bin
creg = uint8(09);       % connection register.  9 dec = 1001 bin
MSBmask = uint8(08);    % binary 1000
LSBmask = uint8(01);    % binary 0001

sreg = uint8(sregin);
fprintf(1, 'start: sreg=%s (%d)\n', dec2bin(sreg, B), sreg);

ssave = [];
NS = 20;

for t = 1:NS
    ssave = [ssave sreg];

    % select rightmost bit of shift register
    M = LSBmask;
    ssreg = bitand( bitand(sreg, creg), M);
    freg = ssreg;

    for b = 2:B
        M = bitshift(M, 1);
        ssreg = bitand(sreg, creg);

        tmpreg = bitxor( bitshift(freg,1), ssreg );
        tmpreg = bitand(tmpreg, M);

        freg = bitor(freg, tmpreg);
    end
```

```
    % main  shift  register  shifted  right  and  OR  in  MSB
    sreg = bitor(bitshift(sreg, -1), bitand(freg, MSBmask));

    fprintf(1,'after:sreg=%s (%d)\n', dec2bin(sreg,B),sreg);
end
fprintf(1, 'Sequence: (%d terms) ', NS);
fprintf(1, 'seed %s  connection %s\n', dec2bin(sregin, B),
                                       dec2bin(creg, B));
fprintf(1, '%d ', ssave);
fprintf(1, '\n');
```

The descrambler can be formed by exactly the same means. Consider the feedback shift register to begin with. A second identical copy, with the same feedback taps and started at the same time, will produce the same sequence of values in the S register as well as for the fed-back bit of F. Furthermore, the XOR operation repeated a second time – this time with the scrambled bit stream and receiver's F register MSB – will recover the original bit values in order.

The shift register S must be initialized to a starting point or seed. For the seed of 1011 binary and connection vector 1001 binary, the sequence produced is

$$11 \quad 5 \quad 10 \quad 13 \quad 6 \quad 3 \quad 9 \quad 4 \quad 2 \quad 1 \quad 8 \quad 12 \quad 14 \quad 15 \quad 7 \quad 11 \quad 5 \quad 10 \quad 13 \quad 6 \quad \cdots$$

Notice the important fact that the sequence repeats itself after a time. To investigate this, using the same connection and changing the starting seed to 0110, we find that the sequence generated is produced is

$$6 \quad 3 \quad 9 \quad 4 \quad 2 \quad 1 \quad 8 \quad 12 \quad 14 \quad 15 \quad 7 \quad 11 \quad 5 \quad 10 \quad 13 \quad 6 \quad 3 \quad 9 \quad 4 \quad 2 \quad \cdots$$

This is the same sequence with a shifted starting point, a fact that is very important in terms of synchronizing the transmitter and receiver.

The problem of synchronization is important, and given the above it would seem that the transmitter and the receiver would need to be given the same starting point and start at the same bit position. Consider the arrangement shown in Figure 2.25. If the initial seed is loaded incorrectly, the binary output shown in Figure 2.26 results. The situation appears hopeless, since bits are continually descrambled incorrectly. If the seed is correct, though, and a burst of errors occurs on the transmission path, then as shown in the figure, the decoded bit stream will be able to recover as the error(s) propagates through the system.

An interesting variation, termed the *self-synchronizing scrambler*, is shown in Figure 2.27. The only real change is that the feedback path is altered, such that the input to the shift register comes from the output bit stream, not the shift register itself. Using the descrambler of Figure 2.28 – which is essentially the same as the scrambler – some interesting behavior results. As Figure 2.29 shows, even starting with the wrong seed, the output shortly locks to the correct sequence. In effect, the incorrect data is shifted out of the system after

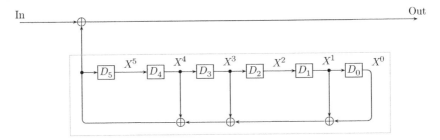

Figure 2.25 A slightly longer scrambler based on a feedback shift register. Interestingly, the descrambler is exactly the same.

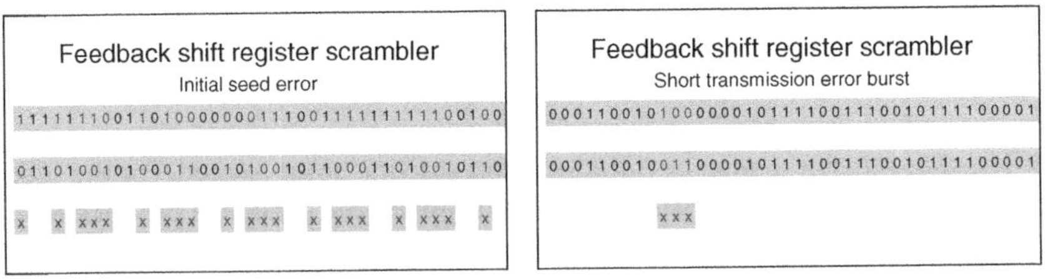

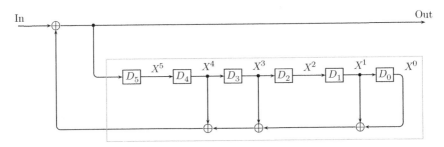

Figure 2.26 Scrambler sequencing, with an initial seed error (left) and a run of errors (right).

In   Out

Figure 2.27 A self-synchronizing scrambler. The essential change is to move the input of the shift register so that it comes from the output bit stream.

propagating through the shift register. One shortcoming, though, is that errors in transmission propagate further than the non-self-synchronized design.

It is clear that each type has strengths and weaknesses. In both cases, the feedback taps must be chosen carefully; otherwise the sequence will repeat after a short interval. The correct choice of feedback taps for a given number of bits in the shift register produces what is termed a *maximal-length sequence*.

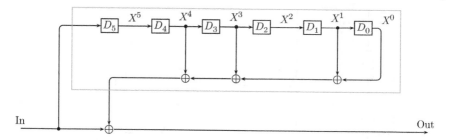

Figure 2.28 A self-synchronizing descrambler, which follows a similar arrangement to the self-synchronizing scrambler.

Self-synchronized scrambler
Initial seed error
0011001100000010011001011100110101110101
1100001100000010011001011100110101110101
xxxx

Self-synchronized scrambler
Short transmission error burst
00001101010110000101111001101010101000100
000011010000111100101111001101010101000100
x x xx

Figure 2.29 Self-synchronizing scrambler errors, showing a seed error (left) and transmission burst error (right).

2.4.5 Pulse Reflection

Digital data, comprising 1's and 0's, is transmitted using a set of amplitude levels. This section investigates what happens when an pulse is sent into a long cable. A voltage pulse sent out may get reflected back, thus corrupting the forward-traveling waveform. This problem can be eliminated, or at least greatly reduced, by following some simple guidelines. This is very important in the design of reliable, high-speed data communication systems. This treatment only considers terminating impedances that do not alter the phase of the input, apart from a simple inversion. This means that the impedances are pure resistances only and are not dependent upon the input frequency.

The experimental setup that will be used to help explain line reflection is depicted in Figure 2.30. It consists of a voltage source that is capable of delivering a positive voltage pulse lasting for a known time. Note that this model includes a series resistance Z_s, whose value will turn out to be very important.

To set some practical parameters, a common coaxial (coax) cable of type RG-58U with length $L = 30$ m is employed. The length is chosen to be reasonably long so that the propagation time (traveling time) is measurable. At radio frequencies (RFs), cables have a characteristic impedance that is assumed to be independent of the actual frequency. It is usually denoted by Z_o. The RG-58U

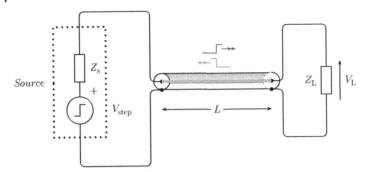

Figure 2.30 Experimental setup for reflection tests on a transmission line.

cable has a characteristic impedance $Z_0 = 50\ \Omega$, and the physical construction of the cable gives a velocity of propagation of approximately 2×10^8 m s^{-1} (around 2/3 of the speed of light). The source impedance is set to be $Z_s = 50\ \Omega$. The goal is to change the load impedance Z_L, in order to see the effect on the pulse propagation.

Consider two extreme cases: (i) with the end of the cable short-circuited and (ii) with the end open-circuited. This corresponds to $Z_L = 0$ and $Z_L = \infty$, respectively. Figure 2.31 shows what is expected for each of these cases. Both the theoretical forward-traveling pulse and the reflected voltage pulse are shown, since that helps in understanding the physical system. Of course, in a real cable, it is not possible to measure these pulses separately – only the net sum of the two.

The input pulse has a magnitude of 2 V, for a duration of 500 ns (from $t = 100$ ns to $t = 600$ ns). This time was chosen to be longer than the expected pulse propagation time from one end of the cable and back. Initially, as the pulse rises, it sees a voltage divider consisting of the source impedance and the cable impedance. Since these are equal, only half the input voltage step is measured, or one volt in response to an input pulse of 2V magnitude.

Consider the short-circuited load case initially, for which $Z_L = 0$, as shown in the left column of Figure 2.31. Since velocity v is distance over time, $v = d/t$, and so the time taken to travel the length of the cable must be $t = d/v = 30/(2 \times 10^8) = 150$ ns. This is the one-way trip time – the time it takes for the leading pulse edge to travel from the source to the load termination. Since the voltage is measured at the source, the edge of the reflected pulse appears 300 ns afterward. In the diagram, the forward rise starts at $t = 100$ ns, with the reflection occurring at $t = 100 + 300 = 400$ ns. Note that the reflected pulse is a *negative* pulse going *back toward the source*.

The result measured at the source is a positive rise, followed by a fall in the voltage. Since, in this case, the forward voltage equals the reflected voltage, they cancel for a time, and the net result is zero. But this only exists for a certain

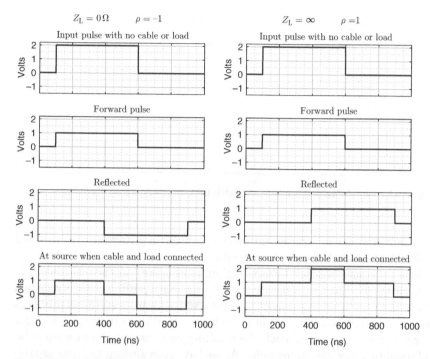

Figure 2.31 Pulse reflection with short-circuit (left) and open-circuit (right) termination. The cable length is 30 m. The reflection coefficient ρ determines the relative amount of reflection, as a proportion of the incoming wave at the end of the cable.

interval, since when the input pulse is taken away, and the reflected pulse is still traveling back, a negative voltage is measured at the start of the cable.

For the open-circuit case illustrated in the right-hand column of Figure 2.31, the timing considerations are similar; however the reflected voltage is *positive*. As a result of the positive reflection, the measured voltage at the cable input rises to twice the input voltage in the time during which the forward and reflected pulses overlap.

The amount of reflection is governed by both the terminating impedance and the transmission line impedance. To see why this is so, consider Figure 2.32, which shows a long transmission line with impedance Z_o and a terminating load impedance Z_L. At the instant the switch is thrown, a current pulse will travel through the wire, resulting in a measurable voltage at any point. It is important to remember that the time intervals are very short (of the order of nanoseconds or less), with relatively longer transmission lines (of the order of meters or longer). A standard DC or steady-state AC analysis of the circuit – which is evidently just a voltage-divider – would give an output voltage of $V_{out} = Z_L/(Z_L + Z_o)V_{in}$. But what happens before the steady-state condition is

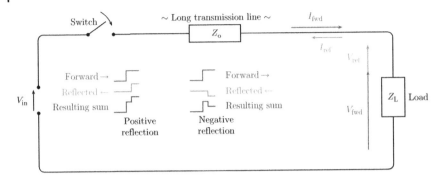

Figure 2.32 Waveforms at the instant a switch is thrown, driving a long transmission line.

reached? The current pulse must travel some distance, giving rise to a measured voltage pulse, which travels along the line.

When the pulse reaches the termination, some energy may be reflected backward toward the source. If the situation is such that a positive pulse is reflected back, then the resulting sum of forward and backward waveforms produces a two-step rise, as shown. If, on the other hand, the situation is such that a negative pulse is reflected back, the resulting sum consists of a positive pulse followed by a negative-going pulse. As we have seen already, the time separation between the forward-going pulse and the return pulse after reflection is governed by the propagation time.

What determines whether the reflection is positive or negative? At the instant of switch-on, the forward voltage and current are related by

$$V_{fwd} = I_{fwd} Z_o \qquad (2.31)$$

The reflected voltage and current are related by

$$(V_{fwd} + V_{ref}) = (I_{fwd} - I_{ref}) Z_L \qquad (2.32)$$

Note the minus for the current, as it is traveling in the reverse direction as shown in the diagram. From this equation,

$$
\begin{aligned}
Z_L &= \frac{V_{fwd} + V_{ref}}{I_{fwd} - I_{ref}} \\
&= \frac{V_{fwd}(1 + V_{ref}/V_{fwd})}{I_{fwd}(1 - I_{ref}/I_{fwd})} \\
&= \left(\frac{V_{fwd}}{I_{fwd}}\right)\left(\frac{1 + V_{ref}/V_{fwd}}{1 - I_{ref}/I_{fwd}}\right)
\end{aligned}
\qquad (2.33)
$$

The *reflection coefficient* with symbol ρ is defined as the ratio of reflected to incident forward-traveling voltage (or current):

$$\rho = \frac{V_{ref}}{V_{fwd}}$$
$$= \frac{I_{ref}}{I_{fwd}} \tag{2.34}$$

Substituting for ρ from Equation (2.34) and Z_o from Equation (2.31), the load impedance as given in Equation (2.33) is then expressed as

$$Z_L = Z_o \left(\frac{1+\rho}{1-\rho} \right)$$

Algebraically rearranging gives

$$\rho = \frac{Z_L - Z_o}{Z_L + Z_o} \tag{2.35}$$

This result is very important, since it relates the amount of reflection to the transmission line and terminating impedances. The amount of reflection can be calculated exactly.

To use this equation in practice and see how it explains the results shown in Figure 2.31, we can return to the short-circuit case where $Z_L = 0 \ \Omega$, and so

$$V_{ref} = \left(\frac{0 - 50}{0 + 50} \right) V_{fwd}$$
$$= -1 \times V_{fwd} \tag{2.36}$$

This also explains why the net result is zero volts when the reflected pulse reaches the source – the two voltages cancel exactly. For the open-circuit case, $Z_L = \infty$, and so

$$V_{ref} = \left(\frac{\infty - 50}{\infty + 50} \right) V_{fwd}$$
$$= \left(\frac{1 - 50/\infty}{1 + 50/\infty} \right) V_{fwd}$$
$$= +1 \times V_{fwd} \tag{2.37}$$

As a result, the voltage peaks when the reflected pulse returns.

Now consider two other intermediate cases, as illustrated graphically in Figure 2.33. In the $Z_L = 25 \ \Omega$ case, the load impedance is half the characteristic impedance, and the reflected voltage is

$$V_{ref} = \left(\frac{25 - 50}{25 + 50} \right) V_{fwd}$$
$$= \frac{-1}{3} \times V_{fwd} \tag{2.38}$$

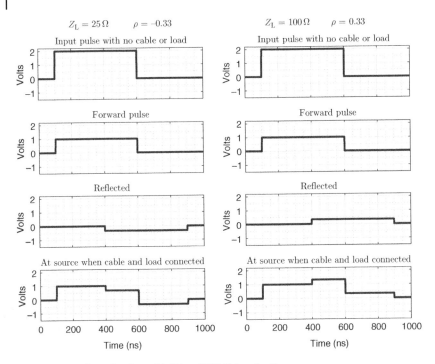

Figure 2.33 Pulse reflection with 25 and 100 Ω terminations.

Thus the reflection coefficient is −1/3, and so one-third of the amplitude is reflected back, but inverted in polarity. In the $Z_L = 100\ \Omega$ case, the load impedance is twice the characteristic impedance. Following a similar calculation,

$$
\begin{aligned}
V_{ref} &= \left(\frac{100 - 50}{100 + 50} \right) V_{fwd} \\
&= \frac{+1}{3} \times V_{fwd}
\end{aligned}
\tag{2.39}
$$

The reflection coefficient is +1/3 in this case. Interestingly, the proportionality is the same, but the polarity is opposite.

For each of these cases, there is a reflection. This leads to an interesting conclusion: If it is possible to cancel the reflection, the forward pulse will be able to continue without alteration. But how is it possible to achieve zero reflection? If the load impedance equals the cable impedance, then $Z_L = Z_o$, and our calculations give

$$
\begin{aligned}
V_{ref} &= \left(\frac{50 - 50}{50 + 50} \right) V_{fwd} \\
&= 0 \times V_{fwd}
\end{aligned}
\tag{2.40}
$$

This achieves the desired situation, with zero reflection and hence no distortion.

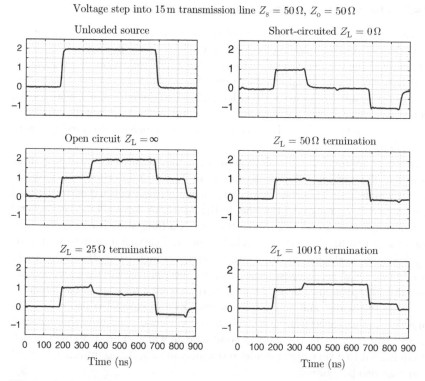

Figure 2.34 Experiments for reflection in a transmission line: pulse input with various termination impedances.

So how well does this model reality? Figure 2.34 shows the measured source voltage driving a real 15 m cable. The driving source pulse is shown. Taking the short-circuit load case, the voltage peak exists from $t = 200$ to $t = 350$ ns for a total of 150 ns. Since $v = d/t$, the one-way trip time is $t = d/v = 15/(2 \times 10^8) = 75$ ns.

When the input voltage step is taken away, the observed voltage at the source dips and becomes negative. Each of the other cases ($Z_L = 0, Z_L = \infty, Z_L = Z_o/2, Z_L = 2Z_o, Z_L = Z_o$) may also be explained using the same theory as our simulated situation. The difference, of course, is that we cannot physically separate the forward and reflected pulses, since they coexist simultaneously on the transmission line. In addition, there are several imperfections that may be observed in the real experiment: The rises are not infinitely fast, and of course the impedances are not *perfectly* matched – just as close as is practicable.

It is of course reasonable to assume that this reflection might occur again, when the reflected pulse returns to the source. This is why both the source impedance and load impedance must be matched to the cable impedance.

2.4.6 Characteristic Impedance of a Transmission Line

A transmission line – whether twisted pair or coaxial, or other – has a certain impedance to current passing through it. As shown in the previous section, this impedance may be used to calculate the amount of reflection of a waveform, which in turn is a very important quantity if we wish to transmit the maximum amount of power from the source to load. How can the impedance of a cable be determined? Moreover, it would seem to depend on the length of the cable. But in fact, the unexpected conclusion is that the impedance is largely independent of the length.

In order to motivate the derivation of a model for a transmission line, consider the conceptual representation of Figure 2.35. Here, we have an input, which "sees" a certain impedance when initially connected to the load impedance Z_0 by means of the line. In the figure, we have allowed for several characteristics of a real line. Since the line is a long stretch of parallel conductors, it will have a certain parallel-connected capacitance represented by C_p. Since any real wire has inductance, a series inductor L_s is also used. There would be a small series resistance, which represents losses in the current path, and this is represented by R_s. Finally, there may be a very small conductance between the wires, and this is represented by the parallel resistance R_p. In practice, the series resistance is likely to be very small, and thus we neglect it (equivalent to $R_s = 0$). The parallel conductance will also be very small, which is equivalent to a very high parallel resistance. Ideally, this is infinite ($R_p = \infty$) and may also be ignored.

That leaves the inductance and capacitance. Although quite small, they cannot be neglected at higher frequencies. As we will show, the ratio of these is important, and since it is a ratio, we cannot mathematically neglect either quantity. The inductance effectively limits the rate of change of current di/dt in the line, whereas the capacitor acts to limit the rate of change of voltage dv/dt.

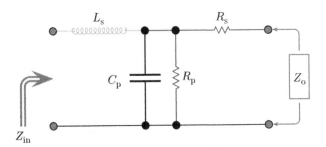

Figure 2.35 An electrical model of a short section of wire. It consists of series inductance and resistance, as well as parallel capacitance and resistance.

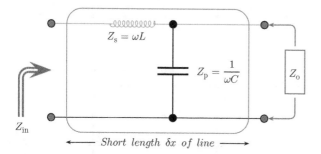

Figure 2.36 Simplified case of cable impedance, neglecting the series resistance (effectively zero) and parallel resistance (effectively infinite).

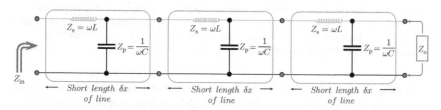

Figure 2.37 Lumping several small segments in series using the inductance/capacitance model for each segment separately.

As usual, the impedance of the inductor is $\omega L = 2\pi f L$, and that of the capacitor $1/\omega C = 1/2\pi f C$, and clearly both are dependent upon frequency. So, if we agree to neglect the series and parallel resistances (as being zero and infinite, respectively), we have the situation shown in Figure 2.36 for a very short segment of cable. A real cable can be modeled as a number of these connected together, as shown in Figure 2.37. Of course, the question of just how many segments are present remains to be answered.

Suppose there was no parallel capacitance, only the series inductance (Figure 2.38, top). If we (hypothetically) do this, then as the line gets longer and longer, the impedance will grow more and more. But suppose there was no series inductance, only the parallel capacitance (Figure 2.38, lower). If we (hypothetically) do this, then as the line gets longer and longer, more and more of the current is shunted back, and the impedance becomes lower and lower. So these two aspects – the series inductance and parallel capacitance – tend to cancel each other in a real situation.

Since the cable could be any length at all, we need to replace the actual values of inductance and capacitance with values per unit length. This would be L H m^{-1} and C F/m, respectively. Then, for a short cable length δx, the series and parallel impedances are $Z_s = \omega L\ \delta x$ and $Z_p = 1/\omega C\ \delta x$, respectively.

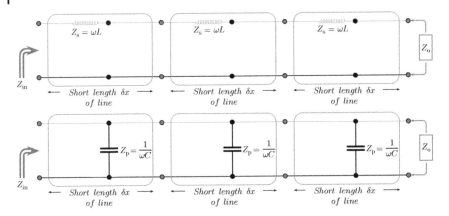

Figure 2.38 A hypothetical lossless line with the parallel capacitance neglected (top) and the series inductance neglected (lower). If we imagine that there is no capacitance in parallel and only a coil, then the series inductance adds cumulatively. If we imagine that there is no inductance in series and only parallel capacitance, then the capacitance adds cumulatively.

Looking from the left into the transmission line terminated in Z_0, we would see the impedance Z_s in series with the parallel combination of Z_p and the termination of Z_0. Since we wish to "see" this as the characteristic impedance Z_0, we can equate Z_0 with the series/parallel combination

$$Z_0 = Z_s + \left(\frac{Z_p Z_0}{Z_p + Z_0} \right)$$

$$Z_p Z_0 + Z_0^2 = Z_p Z_s + Z_0 Z_s + Z_p Z_0$$

$$Z_0^2 = \omega L \, \cancel{\delta x}^{\,0} \cancel{Z_0} + \frac{\omega L \, \cancel{\delta x}}{\omega C \, \cancel{\delta x}} \tag{2.41}$$

As we subdivide the cable into shorter lengths, $\delta x \to 0$, and the first term vanishes. Additionally, the ω cancels in the second term, as does the δx. We are then left with

$$Z_0 = \sqrt{\frac{L}{C}} \tag{2.42}$$

The significance of this is as follows. Firstly, it is *independent of the length*, since the length terms cancelled. Secondly, it is *independent of frequency*, since the frequency terms also cancelled. So as a result, the characteristic impedance is only dependent upon the ratio of inductance to capacitance. Since both of these are constant for a given length of cable, we arrive at the remarkable conclusion that a cable has a characteristic impedance, which is independent of its length.

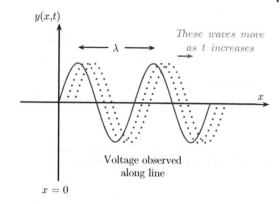

Figure 2.39 A wave traveling along a wire, effectively being delayed over time.

2.4.7 Wave Equation for a Transmission Line

The waveform traveling along a transmission line is a function of time t and also the particular point where it is measured – the distance x from the source. Since the source is described by $A \sin \omega t$, it is not unreasonable to assert that, neglecting any losses, the net wave amplitude at any point $y(x, t)$ is just the source signal with a delay factored in. This delay is just a net negative change in the phase of the waveform. The delay in phase will be

$$\beta = \frac{2\pi}{\lambda} \tag{2.43}$$

with units of radians per meter (rad m^{-1}). This may be understood by referring to Figure 2.39. The observed waveform *along* the transmission line is shown, with successive waveforms dotted as time progresses and the wave travels along. Thus, we imagine the source $A \sin \omega t$ being delayed by a phase angle equal to

$$\text{Phase delay due to distance} = \beta \ \frac{\text{radians}}{\text{meter}} \times x \text{ meters} \tag{2.44}$$

This adds to the phase change due to time

$$\text{Phase advance over time} = \omega \ \frac{\text{radians}}{\text{second}} \times t \text{ seconds} \tag{2.45}$$

Using both time t and distance x, the wave equation may be modified to become

$$y(x, t) = A \sin(\omega t - \beta x) \tag{2.46}$$

This does mean that, at $t = 0$ or when ωt is some multiple of 2π, the equation will have the form $-A \sin \beta x$, which is a negative sine wave; this may be confusing. To see why this is so in a physical sense, Figure 2.40 illustrates the input waveform, together with the observed traveling wave as it moves along the line to the right.

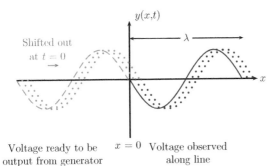

Figure 2.40 A wave described by $A\sin(\omega t - \beta x)$ shown along the length of the line x at time instant $t = 0$.

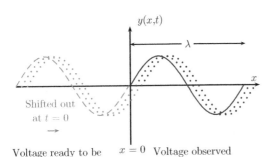

Figure 2.41 A wave described by $A\sin(\beta x - \omega t)$ shown along the length of the line x at time instant $t = 0$.

Alternatively, if we preferred to model it as a positive-going sine from $t = 0$ along the line, we could write the equation as

$$y(x, t) = A\sin(\beta x - \omega t) \tag{2.47}$$

This is illustrated in Figure 2.41. Here, we see that for a positive sine, the generator initially dips negative, corresponding to the $-\omega t$ term in the equation.

So which form is correct? Either can be used, as long as it is used consistently. As with all mathematical models, it depends on the assumptions, and in this case, the assumption is about when the generator starts, which defines when $t = 0$ occurs. What is more important is that the reflected waveform at the load end of the cable will return and add or subtract according to the reflection coefficient to give the net observed amplitude. The reflected waveform is always relative to the forward-traveling waveform, but the phase or starting point of the forward waveform is arbitrary. In a similar way, defining these waves as cosine rather than sine would be equally valid, provided we used that same form in calculating the relative phase of the reflection.

2.4.8 Standing Waves

Section 2.4.5 dealt with a pulse launched into a transmission line, and it was demonstrated that the reflection from the load at the far end travels back and

Figure 2.42 A simple wave traveling left to right and its reflection that travels back in the opposite direction. The net waveform that is observed at any point along the transmission line is the sum of the two.

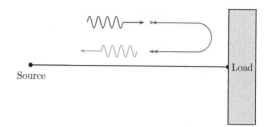

affects the net voltage waveform observed close to the source (or indeed, anywhere along the transmission line). A logical extension of this is to consider a sinusoidal waveform, which is continuous (rather than a discrete pulse).

At a given point along a transmission line, we may measure a voltage with frequency $\omega = 2\pi f$, which changes over time as a function of ωt. But suppose we now look *along* the transmission line. This could be imagined as taking a large number of simultaneous measurements with a probe at various intervals along the line. Each probe measures a specific voltage at a specific position (or length away from the source). This voltage will change over time – or will it?

As the wavefront travels and reaches the load, it will be reflected (Figure 2.42). The forward-traveling wave (conventionally shown moving from left to right) eventually meets the load. At the load, the power may be absorbed completely, but this only happens for a matched load, as shown in Section 2.4.5. Usually, some of the wave is reflected, with the same or reduced amplitude, possibly with a phase change. The phase change may in fact mean that the wave is inverted in polarity. This reflected wave travels back toward the source (right to left in the diagram). In doing so, it adds to the forward-traveling wave. Assuming that the waveform at the input is continuous (that is, not a pulse or burst), the wave traveling back will also be continuous. Thus we see a voltage pattern *along* the length of the line, and this pattern depends on how the forward and backward waves add. Precisely how they add is determined by the amplitude and phase of the reflected wave with respect to the forward wave.

In short, we can say that the reflection characteristics – change in amplitude (or gain/attenuation) and change in phase (time shift relative to frequency) – determine the pattern observed along the length of the line.

Consider Figure 2.43, with the left side showing successive sine waves along the length of the 1 m length of cable. The source is on the left, and the load is on the right. Thus the forward-traveling waveform (top left panel) reaches the load and is reflected as shown in the plot immediately below. However, over time, if the input is continuous, a measuring instrument will record the net sum of forward-plus-reflected waveforms. Translating some of these net waveforms to the right-hand side, we see what resembles another sine wave, defined over distance rather than time. The maximum and minimum amplitudes of this waveform are shown on the lower right. The interesting result is that the magnitude of the sine wave observed – which is still oscillating with

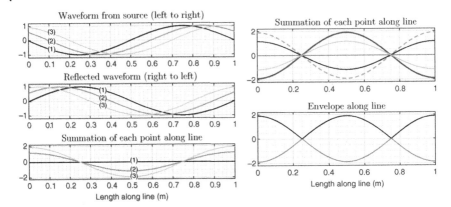

Figure 2.43 Formation of a standing wave, when the reflection has gain of unity and phase shift of zero. On the left, we see the forward wave (top), the reflected wave (middle), and net sum of these (bottom). On the right, we see a snapshot of what happens over time with a few waves traveling (top) and the upper and lower envelopes that result over a period of time (bottom right).

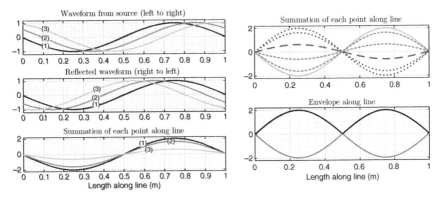

Figure 2.44 Formation of a standing wave, when the reflection has gain of unity and phase shift of 180°.

frequency ω – changes, depending upon the *position* where we take the measurement. At some points, the amplitude of the oscillation is large, but at some points it is actually zero. That is to say, there is no net voltage waveform at precisely those locations (this is at positions $x = 0.25$ and $x = 0.75$ in the figure).

What is important is the *envelope* of these maxima and minima. For the case shown in Figure 2.43, the reflected waveform was equal in amplitude and had no phase change with respect to the incoming wave. From the earlier pulse experiments, we deduce that the load impedance Z_L is infinite, and the reflection coefficient is $\rho = +1$, resulting in lossless reflection. The maximum magnitude is observed at the load and is twice the incoming voltage. For the waveform

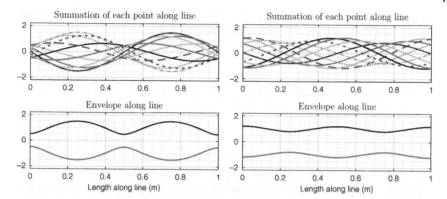

Figure 2.45 A traveling wave occurs for cases when the reflection is incomplete. Here we illustrate with a reflection coefficient magnitude of 0.5 and phase of 180° (left) and a reflection coefficient magnitude of 0.2 and phase of 0° (right). From the envelope of all waves thus generated, the standing wave ratio may be determined.

illustrated, the frequency is $f = 1$, and the velocity is $v = 1$. As a result, the wavelength $\lambda = 1$. We observe that the minimum magnitude (which in this case is actually zero) occurs at a distance $d = \lambda/4$ from the load end (right-hand side). Another maximum occurs at a distance $d = \lambda/2$ from the load. The positions where the maximum and minimum occur are related to the wavelength, and clearly there are multiple maxima and minima at fixed separations when traveling from the load back toward the source.

This case is an open-circuit termination, and the opposite case is when the termination is a short circuit. Figure 2.44 illustrates how successive forward waves and reflected waves combine over time for the short-circuit case. There is one clear difference evident: the position of the maxima and minima. The relative spacing of these is the same at $\lambda/2$ between each maxima or between each minima and $\lambda/4$ between a given maximum and the next minimum.

Other interesting cases occur when the reflection is not 100%. Figure 2.45 shows two cases, with $\rho = 0.5$ and $\rho = 0.2$, with phases of 180° and 0°, respectively. The difference here is that the minima are not zero, but some higher value. In these cases, the waveform appears to move along the length of the line; hence the term *traveling wave* is used. In the previous cases, the waveform appears still when viewed along the line, hence the term *standing wave*. The ratio of the maximum to minimum is termed the Voltage Standing Wave Ratio (VSWR):

$$\text{VSWR} = \frac{V_{\max}}{V_{\min}} \tag{2.48}$$

If the minimum is zero, then the VSWR becomes infinite. If we extrapolate the previous figures, as the reflection coefficient reduces to zero, the ratio of maximum to minimum approaches one (since they become equal). This may also be tied back to the reflection coefficient ρ using

$$\text{VSWR} = \frac{1 + |\rho|}{1 - |\rho|} \tag{2.49}$$

Note the use of the absolute value in the above formula. It is instructive to reconsider some of the earlier cases using this formula. If $\rho = 0$, then the VSWR becomes unity. That is, no reflections occur. If 100% reflection occurs, then $\rho = 1$ and the VSWR becomes infinite. This gives us a useful and practical way to determine whether or not the load is matched – we can measure the amplitude of the waveform at multiples of $\lambda/4$ and calculate the VSWR.

Taking this further, we can substitute for ρ from Equation (2.35), which relates impedances to the reflection coefficient, into Equation (2.49), which relates reflection coefficient to standing wave ratio to mathematically relate the VSWR to the characteristic and load impedances. One slight problem is that since the above equation uses the absolute value of the reflection coefficient $|\rho|$, it is necessary to deal with the positive and negative cases separately.

First, for positive ρ,

$$\begin{aligned}
\text{VSWR} &= \frac{1 + |\rho|}{1 - |\rho|} \\
&= \frac{1 + \rho}{1 - \rho} \quad \text{for} \quad \rho > 0 \\
&= \frac{1 + (Z_L - Z_o)/(Z_L + Z_o)}{1 - (Z_L - Z_o)/(Z_L + Z_o)} \\
&= \frac{Z_L}{Z_o} \quad \text{for} \quad \rho > 0
\end{aligned} \tag{2.50}$$

Note that this makes sense, in that if $Z_L > Z_o$, then ρ is positive, and the VSWR is greater than one (since VSWR is defined as a ratio of maximum to minimum, it must be a quantity greater than or equal to one). For negative ρ,

$$\begin{aligned}
\text{VSWR} &= \frac{1 + |\rho|}{1 - |\rho|} \\
&= \frac{1 - \rho}{1 + \rho} \quad \text{for} \quad \rho < 0 \\
&= \frac{1 - (Z_L - Z_o)/(Z_L + Z_o)}{1 + (Z_L - Z_o)/(Z_L + Z_o)} \\
&= \frac{Z_o}{Z_L} \quad \text{for} \quad \rho < 0
\end{aligned} \tag{2.51}$$

Note that this also makes sense, in that if $Z_L < Z_o$, then ρ is negative, but the VSWR is still greater than one. These equations also demonstrate that for matched loads where $Z_L = Z_o$, the VSWR is unity, as expected. In the case of no reflection, $\rho = 0$, and

$$\text{VSWR} = \frac{1+0}{1-0} = 1 \tag{2.52}$$

This implies $V_{\max} = V_{\min}$, and

$$\frac{Z_L - Z_o}{Z_L + Z_o} = 0 \tag{2.53}$$

from which it may be deduced that $Z_L = Z_o$, and the load impedance equals the characteristic impedance. For the case where $\rho = -1$,

$$\begin{aligned}
\text{VSWR} &= \frac{1 + |\rho|}{1 - |\rho|} \\
&= \frac{1+1}{1-1} \quad \text{for} \quad \rho = -1 \\
&\rightarrow \infty
\end{aligned} \tag{2.54}$$

In terms of impedances,

$$\begin{aligned}
\frac{Z_L - Z_o}{Z_L + Z_o} &= -1 \\
Z_L - Z_o &= -Z_L - Z_o \\
\therefore \quad Z_L &= 0
\end{aligned} \tag{2.55}$$

That is, the transmission line is terminated in a short-circuit load. For $\rho = +1$,

$$\begin{aligned}
\text{VSWR} &= \frac{1 + |\rho|}{1 - |\rho|} \\
&= \frac{1+1}{1-1} \quad \text{for} \quad \rho = +1 \\
&\rightarrow \infty
\end{aligned} \tag{2.56}$$

To determine the load impedance, we have to proceed a little differently, since the possible algebraic solution of $Z_o = 0$ is not realistic (and also does not solve for Z_L):

$$\begin{aligned}
\frac{Z_L - Z_o}{Z_L + Z_o} &= 1 \\
\frac{1 - Z_o/Z_L}{1 + Z_o/Z_L} &= 1 \\
Z_L &\rightarrow \infty
\end{aligned} \tag{2.57}$$

That is, the "load" is actually an open circuit.

Returning to the formation of a standing wave by means of reflection, the wave equation requires factoring in a distance x along the transmission line. This distance affects the actual wave at any given point along the line due to the propagation time of the waveform.

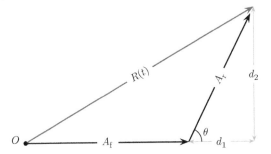

Figure 2.46 Calculating the magnitude of reflection at a given point.

Letting the forward peak amplitude be A_f and the reflected peak amplitude be A_r, then a diagram at some instant in time of the amplitudes may be drawn as in Figure 2.46. The angle θ is composed of the reflection phase φ with the time delay required for a wave point to travel forward, be reflected, and back. Thus the phase delay is $2\beta d$, where d is the distance back *from* the load. Applying some geometry to Figure 2.46, we have that the distances $d_1 = A_r \cos\theta$ and $d_2 = A_r \sin\theta$, where $\theta = 2\beta d + \varphi$ as reasoned above. The net magnitude sum of these, R, is

$$
\begin{aligned}
R^2 &= [A_f + A_r \cos(2\beta d + \varphi)]^2 + [A_r \sin(2\beta d + \varphi)]^2 \\
&= A_f^2 + 2A_f A_r \cos(2\beta d + \varphi) + A_r^2\cos^2(2\beta d + \varphi) \\
&\quad + A_r^2\sin^2(2\beta d + \varphi) \\
&= A_f^2 + A_r^2 + 2A_f A_r \cos(2\beta d + \varphi) \\
&= A_f^2\left[1 + \left(\frac{A_r}{A_f}\right)^2 + 2\left(\frac{A_r}{A_f}\right)\cos(2\beta d + \varphi)\right]
\end{aligned}
$$

$$
\therefore \quad R = A_f\sqrt{1 + \left(\frac{A_r}{A_f}\right)^2 + 2\left(\frac{A_r}{A_f}\right)\cos(2\beta d + \varphi)} \tag{2.58}
$$

Using $\rho = A_r/A_f$ this becomes

$$
R = A_f\sqrt{\rho^2 + 2\rho\cos(2\beta d + \varphi) + 1} \tag{2.59}
$$

This gives us a way to plot the VSWR envelope, given the reflection coefficient ρ, reflection phase φ, and the distance d from the load. Some cases are shown in Figures 2.47 and 2.48. Note that these are read in reverse, in that the horizontal axis is the distance *from the load*. The scale is in terms of quarter wavelengths, in accordance with our finding that the maxima/minima occur at multiples of this distance.

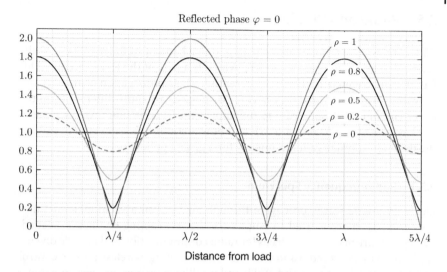

Figure 2.47 VSWR calculated envelope magnitudes, in-phase reflection case. Note that the scale is reversed by convention, showing the distance back *from* the load.

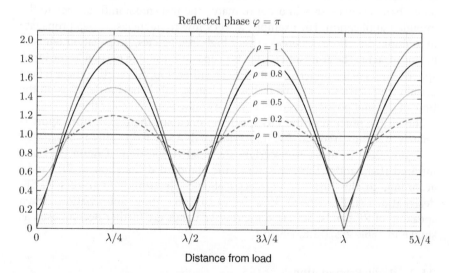

Figure 2.48 VSWR calculated envelope magnitudes, out-of-phase reflection case. Once again, the distance scale shows the distance back *from* the load.

2.5 Radio and Wireless

Radio and wireless signals are transmitted in a band of frequencies generically referred to as Radio Frequency (RF), which is a part of the EM spectrum. Different radio frequencies behave differently – their propagation characteristics (how they travel and how they become dispersed) are quite different from the lower end of the RF spectrum to the upper end. This section reviews some of the key concepts pertaining to RF and how a communication system utilizes the most appropriate radio frequency for a given application.

2.5.1 Radio-frequency Spectrum

Electromagnetic waves propagate through air or even a vacuum; sometimes the term *free space* is used to emphasize that no cabling or fiber is needed. Such waves are of course the basis for radio communication and include diverse applications such as radio and television broadcasting, wireless remote control and telemetry, Bluetooth and WiFi, and satellite transmission, just to name a few. The EM wave is characterized by its frequency or (equivalently) the wavelength of the propagating wave. The range of frequencies and wavelengths, and their relationship, is illustrated in Figure 2.49 in relation to the visible light spectrum. Commonly grouped frequency bands have broadly similar propagation characteristics, and as such a knowledge of the expected behavior for the different bands is necessary in order to match the transmission frequency to the intended application. For example, at very low frequencies, waves propagate in multiple directions, whereas at very high frequencies, propagation tends to be line of sight. The atmosphere affects propagation, and in particular the ionospheric layers of the upper atmosphere can either stop, diffract (bend), or reflect certain frequencies. For some frequency bands, this behavior depends on the time of day and possibly even solar activity (sunspots and solar flares).

The visible portion of the EM spectrum occupies the region from approximately 620 nm (red) to 380 nm (violet) (NASA, n.d.). This is illustrated in Figure 2.49 in order to ascertain the relative location of radio (RF) and infrared (IR) bands. Table 2.1 shows the standard delineation of RF bands, according to the Institute of Electrical and Electronics Engineers (IEEE) designations (IEEE, 1997a). It is useful to know these abbreviations when speaking of RF bands. Standard microwave bands are further decomposed as shown in Table 2.2 (IEEE, 1997b).

2.5.2 Radio Propagation

The scale in Figure 2.49 is not linear, but logarithmic. That is to say, the frequency increments do not move upward by addition, but rather by multiplying each one by a factor (in this case, the factor is 10×). The radio/wireless

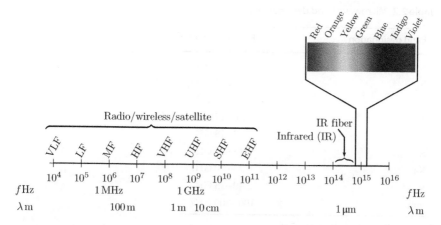

Figure 2.49 The portion of the electromagnetic spectrum important for telecommunications. Radio, wireless, and satellite systems use the frequency ranges shown. At extremely high frequencies, infrared (IR) is used in fiber optics. Still higher in frequency is the visible light spectrum.

Table 2.1 Radio-frequency (RF) band designations.

Abbreviation	Frequency range	Description
ULF	< 3 Hz	Ultra low frequency
ELF	3 Hz–3 kHz	Extremely low frequency
VLF	3 kHz–30 kHz	Very low frequency
LF	30 kHz–300 kHz	Low frequency
MF	300 kHz–3 MHz	Medium frequency
HF	3 MHz–30 MHz	High frequency
VHF	30 MHz–300 MHz	Very high frequency
UHF	300 MHz–3 GHz	Ultra high frequency
SHF	3 GHz–30 GHz	Super high frequency
EHF	30 GHz–300 GHz	Extremely high frequency
—	300 GHz–3 THz	Submillimeter

Source: Adapted from IEEE (1997a).
The center block indicates the range of the most commonly used terrestrial (earth-bound) frequencies.

Table 2.2 Microwave band designations.

Band designation	Frequency range
L	1–2 GHz
S	2–4 GHz
C	4–8 GHz
X	8–12 GHz
Ku	12–18 GHz
K	18–27 GHz
Ka	27–40 GHz
V	40–75 GHz
mm	100–300 GHz

Source: Adapted from IEEE (1997b).
Some variations exist throughout the world, with some
nonstandard terminology also commonly encountered.

signals shown may be generated over a very wide range, from kHz up to GHz – corresponding to a proportional range of $10^9 : 10^3 = 10^6$ or more. Although some aspects of telecommunications (such as modulation) are broadly similar across this range, other aspects (such as antenna design) vary significantly from low to high frequencies.

A number of factors dictate the practical use of the various frequency bands. Propagation at the lower end of this range, up to MHz regions, occurs due to surface waves on the Earth, whereas the higher frequencies at UHF and beyond tend toward line-of-sight transmission. Furthermore, effects such as attenuation by water vapor in the atmosphere, diffraction (bending) around obstacles and in the upper layers of the atmosphere, and reflection from natural features such as seawater and man-made obstructions, may become significant for certain frequency bands.

In the early days of radio, propagation (rather than modulation) was the primary consideration (Barclay, 1995). There was a limited ability to generate certain frequencies, and reception was even more difficult. Thus the term *detection* of an on/off-type signal came into use (and this term is still used today). The very earliest radio transmitters such as spark transmitters generated EM energy across a wide range of frequencies. This produced a very small power per unit of bandwidth and was thus very inefficient. This in turn severely limited the range over which wireless signals could be reliably received. Later, with the need to share the common radio spectrum by multiple users, the generation of RF needed to be much more controlled, and various organizations such as the International Telecommunication Union (ITU) evolved to administer the use of the spectrum. National regulators also play a large part in defining acceptable use of RF within countries, both in terms of maximum bandwidth and permitted radiated power levels.

Directionality is also important: Some applications require omnidirectional (all-direction) coverage, whereas other applications such as point-to-point links require highly directional transmission to minimize signal wastage. In addition, the bandwidth of the radio channel relative to the center frequency means that at higher frequencies, a greater bandwidth is available as a proportion of the available spectrum. Finally, the physical size of antennas is usually an important consideration, since optimal antenna sizing for resonant antennas is proportional to the wavelength. Higher frequencies with a shorter wavelength (and thus smaller antennas) are preferred for mobile communications. Typical frequency ranges and common uses include the following:

VLF and LF The Very Low Frequency and Low Frequency ranges where propagation through seawater is possible; these bands may be utilized for underground, undersea, and intercontinental radio transmission and ship-to-shore communications. The bandwidth in this region is limited, atmospheric noise is significant, and very large antenna structures are required (Barclay, 2003).

MF So-called Medium Frequency ranges are characterized by ground waves propagating up to about 30 MHz (ITU, n.d., p. 368) and enable regional coverage in daylight hours, often with longer distances possible at night (Barclay, 2003).

HF The High Frequency band is characterized by very long distance ranges, both continental and intercontinental, which is possible due to reflection and diffraction in the Earth's atmospheric layer known as the ionosphere. Antenna sizes, though still large, start to become more manageable at these frequencies.

VHF A very widely used band is Very High Frequency (VHF). It is used for land and marine communications, emergency communications, and radio navigation, although antenna sizes of the order of meters make it more suitable to fixed installations. Diffraction of waves and reflection of waves may be problematic in this frequency region.

UHF The Ultrahigh Frequency area has become much more widely utilized due to the ability of transmitter and receiver systems to cope at frequencies that were previously not easily attainable. It is commonly used for television, mobile radio such as short-range emergency services, and cellular (mobile) phones. The limited coverage creates problems in rural areas, but this may actually be an advantage in the case of mobile communications, which require frequency reuse between smaller transmission areas or cells. Diffraction of waves and reflection off obstacles including buildings becomes more pronounced at these frequencies.

SHF   Satellite communications employ Super High Frequency, but attenuation (loss) due to water vapor above about 10 GHz is a significant problem (Barclay, 2003).

The main conclusions are that the particular frequency band used depends firstly on the propagation characteristics required (short or long range), secondly on the size of antenna (smaller antennas for shorter wavelengths), and finally on the available modulation technology. A much more extensive discussion of the various bands and their propagation characteristics may be found in Barclay (2003).

2.5.3 Line-of-sight Considerations

The focus of this section is line-of-sight transmission of radio waves. Understanding this idea helps understand why transmitters intended to cover large areas (typically radio or television transmitters or mobile communication base stations) must be mounted at a height well above the ground. This is not simply to get above buildings and mountains; the curvature of the Earth's surface is also a consideration.

Consider the visible horizon or the line from a tower that grazes the Earth as depicted in Figure 2.50. This shows a somewhat simplified cross-sectional view, where the assumption is that the Earth is a sphere of constant radius with no surface features such as hills or valleys. The diagram is not to scale, since R is the radius of the Earth and h_{tx} is the height of the transmitter, and the latter would be insignificant if drawn to scale. In order to determine the maximum transmission distance d_{tx}, the equations of the triangle with a right angle at G show that

$$
\begin{aligned}
R^2 + d_{tx}^2 &= (R + h_{tx})^2 \\
&= R^2 + 2Rh_{tx} + h_{tx}^2 \\
\cancel{R^2} + d_{tx}^2 &= \cancel{R^2} + 2Rh_{tx} + h_{tx}^2 \\
d_{tx} &= \sqrt{2Rh_{tx} + h_{tx}^2}
\end{aligned}
\tag{2.60}
$$

Since the radius of the Earth is substantially greater than the transmitter height ($R \gg h_{tx}$), the second term h_{tx}^2 can be ignored to yield

$$
d_{tx} \approx \sqrt{2Rh_{tx}}
\tag{2.61}
$$

Figure 2.50 Radio horizon calculations for a spherical Earth with no surface features. The maximum transmission distance d_{tx} is determined by the height of the transmitter h_{tx} and the mean radius of the Earth R (diagram not to scale).

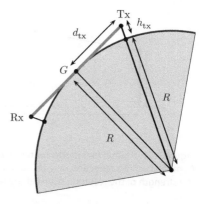

It is more convenient to measure the tower height h_{tx} in meters, but with the resulting distance d_{tx} to be in kilometers (km). Using these scalings and approximating the radius of the Earth as $R \approx 6370$ km,

$$d_{tx} \approx \sqrt{2 \times 6370 \times \frac{h_{tx}}{1000}}$$
$$d_{tx} \approx 3.6\sqrt{h_{tx}} \tag{2.62}$$

Since there may be some ground-plane propagation as well as diffraction effects, and since this is a very coarse initial approximation, the actual distance is slightly more than this, resulting in a common approximation

$$d_{tx} \approx 4\sqrt{h_{tx}} \tag{2.63}$$

If, for example, the antenna height is 9 m, the line-of-sight distance is approximately 12 km. Note that doubling the tower height does not double the distance, due to the $\sqrt{h}$ term. If the situation requires two towers, the added distance d_{rx} for height h_{rx} may be calculated in a similar manner. The total radio horizon is then

$$d = d_{tx} + d_{rx}$$
$$d \approx 4\sqrt{h_{tx}} + 4\sqrt{h_{rx}} \tag{2.64}$$

Finally, if the transmitter and receiver heights are equal, then the total distance is $d = d_{tx} + d_{rx} \approx 8\sqrt{h_{tx}}$. It is worth reiterating that this is an approximation only, but that it does serve to provide a guideline as to how far we may reasonably expect line-of-sight systems to operate over. Clearly, both the location and height of transmission antenna are important considerations in practice.

2.5.4 Radio Reflection

Suppose now we have a radio wave that is reflected from a surface, such as the ground or seawater, or from a structure, as illustrated in Figure 2.51. If the path

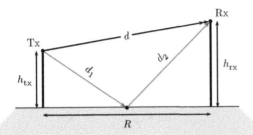

Figure 2.51 A simplified model for radio reflection calculations. The direct path *d* differs from the path via reflection $d_1 + d_2$, so that if the signal is not attenuated upon reflection, the strength of the resulting signal at the receiver may be altered.

difference is a multiple of half a wavelength, then there may be reinforcement (if the effective total path difference is a multiple of one wavelength) or cancellation (if the effective total path difference is an odd multiple of half a wavelength). Note that the phase of the reflection at point *R* also has to be taken into account in calculating the effective path difference.

The difference in total distance traveled is

$$\Delta d = (d_1 + d_2) - d \tag{2.65}$$

In the diagram, moving h_{tx} across to the receiver side to form a right triangle gives

$$d^2 = R^2 + (h_{tx} - h_{rx})^2$$
$$= R^2 \left[1 + \frac{(h_{tx} - h_{rx})^2}{R^2} \right]$$
$$\therefore \quad d = R \sqrt{\left[1 + \frac{(h_{tx} - h_{rx})^2}{R^2} \right]} \tag{2.66}$$

A necessary mathematical approximation at this point is that for small values of x, $\sqrt{1+x} \approx 1 + x/2$. The use of this is justified here since R is considered to be much greater than $(h_{tx} - h_{rx})$, and so $x = [(h_{tx} - h_{rx})/R]^2$ is small. This results in

$$d \approx R \left[1 + \frac{(h_{tx} - h_{rx})^2}{2R^2} \right] \tag{2.67}$$

Using the ray diagram geometrically again, reflecting h_{rx} into the Earth to form another right triangle with d_1 and d_2 forming a straight line,

$$(d_1 + d_2)^2 = R^2 + (h_{tx} + h_{rx})^2$$

$$= R^2 \left[1 + \frac{(h_{tx} + h_{rx})^2}{R^2} \right]$$

$$= R^2 \left[1 + \frac{(h_{tx} + h_{rx})^2}{R^2} \right]$$

$$\therefore \quad (d_1 + d_2) = R \sqrt{\left[1 + \frac{(h_{tx} + h_{rx})^2}{R^2} \right]} \qquad (2.68)$$

Recognizing a similar form, and again applying the small x approximation, gives

$$d_1 + d_2 \approx R \left[1 + \frac{(h_{tx} + h_{rx})^2}{2R^2} \right] \qquad (2.69)$$

When the approximations for d and $(d_1 + d_2)$ are substituted into the path difference equation (2.65), the result is

$$\Delta d = (d_1 + d_2) - d$$

$$\approx R \left[1 + \frac{(h_{tx} + h_{rx})^2}{2R^2} \right] - R \left[1 + \frac{(h_{tx} - h_{rx})^2}{2R^2} \right]$$

$$= R \left[\frac{(h_{tx} + h_{rx})^2 - (h_{tx} - h_{rx})^2}{2R^2} \right]$$

$$= \frac{2 h_{tx} h_{rx}}{R} \qquad (2.70)$$

The phase difference is the path difference multiplied by the phase constant $2\pi/\lambda$:

$$\varphi = \frac{2\pi}{\lambda} \Delta d \qquad (2.71)$$

If Δd is comparable in magnitude with λ, the phase becomes an appreciable fraction of 2π. This result shows that even small changes in the height of either the transmitting or receiving antenna (resulting in a change in Δd) with respect to the wavelength λ may result in phases changes that either reinforce or cancel the received wave.

2.5.5 Radio Wave Diffraction

As with reflection, radio waves may be subject to *diffraction* in certain circumstances. A useful aid in understanding what occurs due to diffraction of a wave is Huygens' principle.

Consider a water wave on a pond attempting to pass through an opening or aperture, as depicted in Figure 2.52. The wave travels toward the barrier and will be stopped where the barrier physically exists – but what will happen to the portion that passes through the opening? These waves travel through directly, but

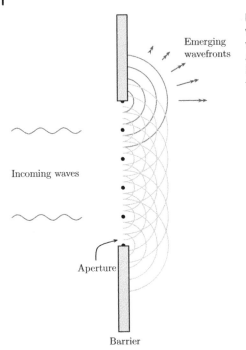

Incoming waves

Emerging wavefronts

Aperture

Barrier

Figure 2.52 A wave that meets a barrier with an opening. Each point where the wave passes through may be thought of as a new source of wavefronts, which interfere with each other. This produces the phenomenon of diffraction.

also form small "wavelets," which continue traveling. Eventually these smaller wavelets interfere with each other. According to the Huygens–Fresnel principle, each point on the wavefront acts as a new source of waves, as illustrated. Figure 2.53 depicts what happens when these multiple wavefront sources interact. Depending upon the relative phases with which the wavefronts arrive at any given point, the amplitude observed may be increased or decreased. The relative phase differences occur due to the differences in the distance traveled.

Combining all the various wavelets results in the situation depicted in Figure 2.54. If there is no diffraction at all, then a receiver placed immediately after the barrier but slightly away from the direct line from the aperture should receive no signal at all. However, this is not what is observed in practice. As the images illustrate, as the wavefront expands, and assuming Huygens' principle applies, then areas in the "shadow" region will receive some signal.

The concept of diffraction is useful in planning the path between a transmitter and receiver. If there is an obstacle that is not necessarily in the line of sight, but close to it, then diffraction effects will occur. Figure 2.55 shows what may occur due to knife-edge diffraction. The obstacle midway along the path produces some diffraction, which in turn means that there are some addition and cancellation of the waves traveling toward the receiver. It is useful to be able to estimate the values of h when this may become an issue. As might be expected, the answer is related to the wavelength λ.

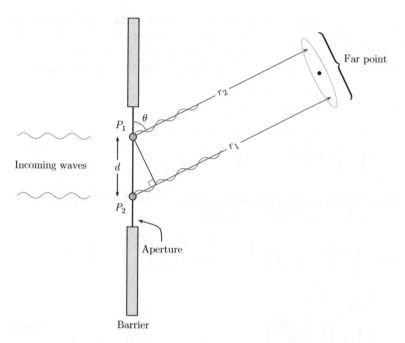

Figure 2.53 Considering just two points, the physical path difference results in a wave that reaches the observer, which appears to be from one point. However, the waves interfere according to their phase relationship. The wavelength λ relative to the aperture d is clearly important.

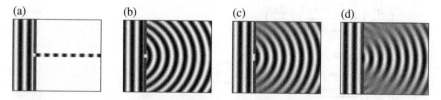

Figure 2.54 Diffraction at an aperture. Image (a) is shown assuming no diffraction; (b), (c), and (d) illustrate the situation as the aperture gradually increases.

Figure 2.55 Illustrating knife-edge diffraction in line-of-sight transmission and the resulting Fresnel zone.

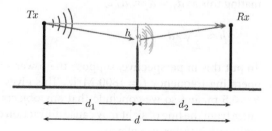

As with reflection, the path difference with respect to the wavelength is the critical parameter. Referring to Figure 2.55, the path difference may be calculated as the sum of the paths taken due to diffraction, less the direct path d. Using the right triangle formed with h at one side, this difference becomes

$$\Delta d = \left(\sqrt{d_1^2 + h^2} + \sqrt{d_2^2 + h^2} \right) - (d_1 + d_2)$$

$$= \left[d_1 \sqrt{1 + \left(\frac{h}{d_1} \right)^2} + d_2 \sqrt{1 + \left(\frac{h}{d_2} \right)^2} \right] - (d_1 + d_2) \tag{2.72}$$

Once again, making use of the approximation $\sqrt{1 + x} \approx 1 + x/2$ by letting $x = (h/d_1)^2$ and then $x = (h/d_2)^2$ in turn yields

$$\Delta d = \left\{ d_1 \left[1 + \frac{1}{2} \left(\frac{h}{d_1} \right)^2 \right] + d_2 \left[1 + \frac{1}{2} \left(\frac{h}{d_2} \right)^2 \right] \right\} - (d_1 + d_2)$$

$$= \frac{h^2}{2d_1} + \frac{h^2}{2d_2}$$

$$= \left(\frac{h^2}{2} \right) \left(\frac{d_1 + d_2}{d_1 d_2} \right) \tag{2.73}$$

This is the path difference, which equates to a phase difference of $(2\pi/\lambda)\,\Delta d$. If the phase difference is $180°$ or π radians, which would result in cancellation, then

$$\left(\frac{h^2}{2} \right) \left(\frac{d_1 + d_2}{d_1 d_2} \right) \left(\frac{2\pi}{\lambda} \right) = \pi \tag{2.74}$$

Solving for h gives

$$h = \sqrt{\frac{d_1 d_2 \lambda}{d_1 + d_2}} \tag{2.75}$$

In the case where the diffracting point is in the middle, then $d_1 = d_2$. Approximating this as $d_1 = d_2 \approx d/2$,

$$h \approx \frac{1}{2} \sqrt{\lambda d} \tag{2.76}$$

To put this in perspective, suppose the tower separation is $d = 1$ km and the operating frequency $f = 1800$ MHz. This gives a wavelength $\lambda \approx 16$ cm and $h \approx 6.5$ m, and so we conclude that any objects within this approximate distance from the line of sight may cause diffraction of the radio wave, thus causing multipath interference effects.

2.5.6 Radio Waves with a Moving Sender or Receiver

One important consideration in mobile communications is the frequency shift that is produced when either the sender or receiver is moving. The basic concept is similar to the shift in frequency, which may be observed when an emergency vehicle with a siren is moving toward or away from us. The apparent frequency (or pitch of the sound) is either increased (if the vehicle is moving toward us) or decreased (if the vehicle is moving away from us). Although the wave in this example is acoustic pressure in nature, similar concepts apply at other frequencies. This is termed the Doppler effect and is discussed in many physics texts (for example, Giancoli, 1984).

In radio systems, this results in a small shift of the carrier frequency at the receiver. Since synchronization with the carrier is critical to most types of receiver, this may potentially cause a problem. Clearly, the frequency shift is related to the speed of the sender (or receiver) when considered relative to the radio frequency. Since radio waves travel with a very high but constant velocity, the apparent change in wavelength or frequency must be derived.

For radio waves, the velocity v in free space is conventionally denoted by c, the speed of light. Air may be taken as approximating free space, and the slight reduction in velocity may be safely neglected. As usual, the wavelength λ and the frequency f are related by $v = f\lambda$, and the reciprocal of frequency is the period τ of the waveform; thus $f = 1/\tau$. Let the Doppler-shifted wavelength be $\tilde{\lambda}$ with corresponding frequency $\tilde{f}$. In free space, the distance traveled in a time interval τ would be $v\tau$.

The first Doppler case is depicted in the uppermost diagram of Figure 2.56, where the receiver is stationary and the source is moving *toward* the receiver with velocity v_s. The distance between wave crests would be $v\tau = \lambda$ less the distance traveled in the same time, which would be $v_s\tau$. So the new wavelength is apparently

$$\tilde{\lambda} = v\tau - v_s\tau$$

$$= (v - v_s)\left(\frac{1}{f}\right) \tag{2.77}$$

Since $f = v/\lambda$,

$$\tilde{f} = v/\tilde{\lambda}$$

$$= \frac{v}{(v - v_s)}f$$

$$= \left(\frac{1}{1 - v_s/v}\right)f \tag{2.78}$$

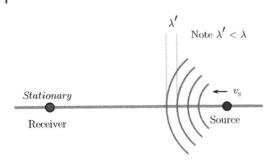

Figure 2.56 Doppler cases 1,2 (top) and 3,4 (lower).

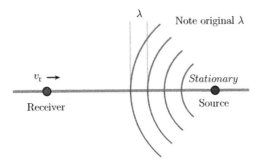

If the receiver remained stationary, but the source moved *away* from the receiver with velocity v_s, then substituting $-v_s$ for v_s leaves

$$\tilde{f} = \left(\frac{1}{1 + v_s/v}\right) f \tag{2.79}$$

The other two cases pertain to a stationary source with a moving receiver, as depicted in the lower diagram of Figure 2.56. With a stationary source and a receiver moving *toward* the source with velocity v_r, the wavelength λ in space is the same. However, the moving receiver will see more crests in a given time, meaning that the velocity of the wave is (apparently) faster. So the new wave velocity is

$$\tilde{v} = v + v_r \tag{2.80}$$

It follows that the apparent frequency would be

$$\tilde{f} = \frac{\tilde{v}}{\lambda}$$
$$= \frac{v + v_r}{\lambda} \tag{2.81}$$

Figure 2.57 Doppler left/right conventions.

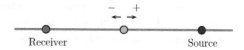

Since $\lambda = v/f$,

$$\tilde{f} = \frac{v + v_r}{\lambda}$$

$$= \left(\frac{v + v_r}{v}\right) f$$

$$= \left(1 + \frac{v_r}{v}\right) f \tag{2.82}$$

The final case is where the source is still stationary, but the receiver is moving *away* from the source with velocity v_r. Again, logic dictates that we can just substitute $-v_r$ for v_r to leave

$$\tilde{f} = \left(1 - \frac{v_s}{v}\right) f \tag{2.83}$$

This gives four equations to juggle in order to cover all eventualities. However, if we adopt the conventions as shown in Figure 2.57 of positive velocity to the right (and so negative to the left), these four cases may be combined into one formula:

$$\tilde{f} = \left(\frac{v + v_r}{v + v_s}\right) f$$

$$= \left(\frac{1 + v_r/v}{1 + v_s/v}\right) f \tag{2.84}$$

Considering that the relative speed of any vehicle is likely to be quite small with respect to the radio wave velocity, we might be tempted to assume that the terms v_s/v and v_r/v are negligible. However, with the use of UHF frequencies for mobile communications, a vehicle that is traveling may result in a frequency offset which is not insignificant. With a receiver traveling at $100\,\mathrm{kmh^{-1}}$, $v_r \approx 25\mathrm{ms^{-1}}$ and $v_r/c \approx 10^{-7}$. This results in a frequency shift of the order of 100 Hz, and the receiver phase-locking must be able to track this variation.

2.5.7 Sending and Capturing a Radio Signal

This section introduces some of the basic principles behind capturing and transmitting a radio signal. Perhaps unsurprisingly, many of the approaches are related, with a common body of theory to support them. Some of the key types of antennas, and notions of why antennas act as they do, are introduced. Like many fields, this aspect has its own particular terminology, and a useful reference is the IEEE Standard Definitions of Terms for Antennas (IEEE, 2013).

An important starting point is to remember that for a given transmitter power P_t, if the signal were to radiate out equally in all directions, then the net flux per unit area would be spread over the area of the sphere $4\pi r^2$, where r is the radius of the expanding sphere. Thus the flux would be proportional to $P_t/(4\pi r^2)$, giving rise to the fact that the power decays as the square of the distance. Many applications, however, require much more directional transmission. Furthermore, real antennas tend to favor one direction over another, depending on the antenna design.

Before introducing some of the theoretical concepts, it is helpful to examine what is both the most fundamental and widely used antenna structure – the *dipole*. A *half-wave dipole* is depicted in Figure 2.58, which shows the dipole arrangement as if looking down on the dipole arms. If the dipole is mounted such that the arms are horizontal, then the emanating radio signal is said to be horizontally polarized. This means that the electric field oscillates in the horizontal sense and there is less reflection from water and land. Vertical polarization is also possible. It is typically employed for radio and mobile communications and takes advantage of ground-wave propagation as well as surface reflection. Although this may seem preferable from the point of view of having a wider coverage area, reflections of the radio signal may be problematic. The direction of an antenna (vertical or horizontal) indicates the type of polarization in use, and the receiver's antenna must match the transmitter's polarization direction; otherwise the received signal will be greatly attenuated (in theory, there would be no received signal induced at all). Circular polarization is also employed in some circumstances, typically for satellite links. In this case, the electric field's polarization rotates as it travels.

Essentially, a dipole antenna is nothing more than two arms that carry the current from the transmitting electronics via a cable feed in order to be radiated out. In reverse, it may be used to receive a radio signal. Clearly, it is desirable to radiate out as much power as possible. A less obvious aspect is the directionality of that radiation, since a given application may aim to cover a very wide geographical area (for example, broadcast TV), or alternatively we may wish the power to be concentrated in a very narrow beam (for example, point-to-point telecommunications).

In Figure 2.58, the total length of the two dipoles L is shown as half a wavelength. This is not coincidental – essentially, the reason for this is to create a standing wave along the length of the dipole. Each end of the open-ended dipole is effectively an open circuit, and thus no current flows. However, at the feed point (where the incoming cabling attaches to the antenna), we expect the current to be maximized. For the particular frequency being used, a standing wave is produced along the length of the dipole, by virtue of the mechanisms discussed in Section 2.4.8. The resulting current distribution along the length of the dipole is also depicted in the figure.

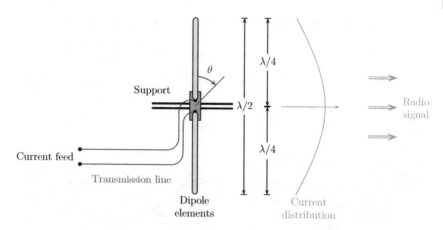

Figure 2.58 A basic half-wave dipole. The support beam is electrically insulated from each arm of the dipole. Note that the angle toward a receiver θ is measured from the dipole arms, and thus the direction of maximum intensity or sensitivity is perpendicular to the dipole arms ($\theta = 90°$). The total length of the dipole is L, and in this case it is $L = \lambda/2$. As a result, each arm is a quarter wavelength.

With the angle θ conventionally measured from the dipole arm as shown, the direction of maximum radiation is at an angle of $90°$. But how does the radiation pattern vary around the dipole? It is also reasonable to ask how sensitive a receiving dipole is when rotated at some angle with respect to the transmitter. To analyze this, the phase constant β (defined as the angle swept out per wavelength) is again employed. It is

$$\beta = \frac{2\pi}{\lambda} \text{ rad m}^{-1} \tag{2.85}$$

It may then be shown that the field intensity $E(\theta)$ at angle θ is then (Kraus, 1992; Guru and Hiziroğlu, 1998)

$$E(\theta) = K \frac{\cos(\beta(L/2)\cos\theta) - \cos\beta(L/2)}{\sin\theta} \tag{2.86}$$

where L is the total dipole length and K is a proportionality constant, which may be ignored for the purposes of studying the power pattern (since K does not affect the shape). For a so-called "half-wave dipole," $L = \lambda/2$. Other lengths for L result in a suboptimal radiation pattern.

Graphing the power $P(\theta) = E^2(\theta)$ gives the pattern as depicted in Figure 2.59. In part (a) the location of the dipole itself is shown to emphasize the directions of maximum and minimum radiation. The intensity has been normalized to unity. Moving at any angle either way from $\theta = 90°$ results in a smaller radiation power and reduced sensitivity. An important parameter that characterizes any antenna is the half-power beamwidth, which is the angle at which half of

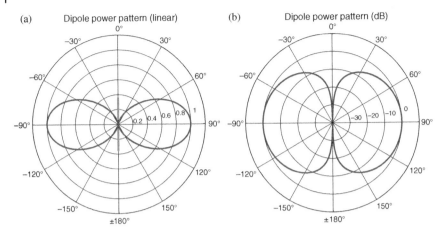

Figure 2.59 The normalized dipole pattern. From this, we may determine the relative field strength of a transmission at a given angle or, alternatively, when used as a receiving antenna, the sensitivity when aligned with respect to a transmitter. (a) Linear scale. (b) Decibel (logarithmic) scale.

the maximum power is found. Commonly, such power patterns are shown in decibels, as plotted in part (b). It is important to realize that these depict the same antenna, just in a different way.

Finally, it may be noticed that exactly half of the antenna power is directed toward the back of the antenna. This may be desirable if the aim is to reach a broad geographic region, for example. But if the aim is to focus the energy toward one particular area, then this represents a significant wastage. Acting as a receiving antenna, it would mean that a significant sensitivity occurs in the direction opposite the desired aim of the antenna, possibly resulting in increased interference (from reflections, for example).

To further investigate just why such an arrangement produces the radiation pattern as shown, consider Figure 2.60a, which shows a theoretically small slice of conductor and some far point in space P. The electric field $E(\theta)$ at the faraway point P is assumed to be proportional to the current I_0 flowing in the element. The current is sinusoidal with the usual equation $\cos \omega t$. There will be a phase delay at the point P, corresponding to the radial distance r. This phase delay is equal to the phase constant β multiplied by the distance r. Thus the sinusoidal field then becomes proportional to $I_0 \cos(\omega t - \beta r)$. This includes the phase delay βr and the current I_0 in the conductor element.

The radiating field travels outward, and the intensity of the field is assumed to decay in proportion to r. The field is composed of vectors in space as shown in Figure 2.60b. The component received by a receiving antenna positioned at P is thus proportional to $\sin \theta$, as the perpendicular component continues to

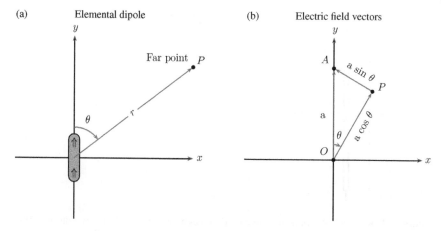

Figure 2.60 An elemental or Hertzian dipole (a) consists of a hypothetical current-carrying element. It is used as the basis for modeling more complex antenna types. The electric field vectors are decomposed into orthogonal (perpendicular) components (b).

travel outward. Combining these concepts, we have the equation that models the electric field intensity as

$$E_\theta = \frac{K}{r} I_0 \cos(\omega t - \beta r)\ \sin\theta \tag{2.87}$$

Notice that the $\sin\theta$ component is consistent with the radiation field as shown earlier, in that when $\theta = 0$ the field is zero and when $\theta = 90°$ the field is maximized.

The next step in understanding a half-wave dipole is to consider it as a multitude of hypothetical or *elemental dipoles*. As depicted in Figure 2.61, we may consider the net field at any given point by summing all the individual field contributions from a very large number of very small elemental dipoles. With reference to the figure, at point P, which is assumed to be in the *far field*, the angles θ and θ_0 are approximately the same. From the point of view of *distance* from the original and dipole, in terms of the field strength decay term, $1/r$ is about the same as $1/r_0$. However, from the point of view of phase differences, the small difference between r and r_0 cannot be neglected, since it is comparable to the wavelength λ and thus the dimensions of the antenna. That is, the difference between the quantities βr_0 and βr *is* significant.

It is possible to assume the current distribution of the form outlined earlier, for which a mathematical model is $I_0 \cos \pi y/L$, where y is the distance from the origin of the dipole to the location of one individual elemental dipole. The integration of this field is shown in many texts (for example, kraus, 1992, Guru and Hiziroglu, 1998) and results in the field equation (2.86).

Alternatively, we may create a numerical estimation of the field strength as follows. We first use the meshgrid function to create a grid of (x, y) pairs, and

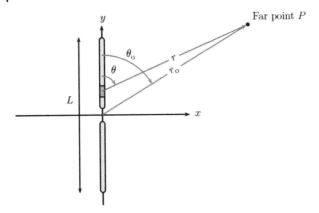

Figure 2.61 A half-wave dipole may be considered as a multitude of elemental dipoles. The resulting field is the summation of all the individual small dipole contributions.

the resulting intensity will be computed for each point within the matrix $P(x, y)$. The index of the dipole's location is determined by variable id (for dipole index).

```
x = -2:0.1:2;
y = -2:0.1:2;
Nx = length(x);
Ny = length(y);

L = 0.1;    % use to simulate a point source
L = 1.0;    % use to simulate a longer antenna

[X, Y] = meshgrid(x, y);
id = find( (X == 0) & ((Y <= L/2) & (Y >= -L/2)) );

omega = 2*pi;
dt = 0.01;
tmax = 4;

lam = 1;
beta = 2*pi/lam;
```

After the initial setup, we iterate over all points, in space (x, y) and time t, resulting in an intensity $I(x, y, t)$. The code below computes an animation of the electric field. The distance $R(x, y)$ is computed as $\sqrt{x^2 + y^2}$ for each point on the plane. The field intensity $I(x, y)$ is computed using Equation (2.87). The relative scaling before plotting is merely to produce an acceptable visualization.

Representative field plots are shown in Figure 2.62 for the elemental dipole and in Figure 2.63 for the half-wave dipole. Note that the antenna conductor current is shown as part of the plot for reference purposes, but that it is not part of the radiated field as such (it is shown to indicate the location and relative orientation).

```
for t = 0:dt:tmax
    Isum = zeros(Ny, Nx);
    Idip = zeros(Ny, Nx);

    for i = 1:length(id)
        xo = X(id(i));
        yo = Y(id(i));

        % cosine current profile
        Io = cos(pi*yo/L);

        % use for Hertzian point dipole with small L
        %Io = 4;

        % save dipole current at this point
        Idip(id(i)) = Io;

        dX = X - xo;
        dY = Y - yo;

        % distance plot from current point
        R = sqrt(dX.^2 + dY.^2);
        theta = atan2(dX, dY);
        I = Io*cos(omega*t - beta*R).*sin(theta);

        % division by zero if R=0, but this is only on the
        % dipole itself
        i = find( abs(R) < eps );
        R(i) = 1;
        I = I./R;
        Isum = Isum + I;
    end

    figure(1);
    set(gcf, 'position', [20 90 450 300]);
    meshc(X, Y, Idip);

    figure(2);
    set(gcf, 'position', [500 90 450 300]);
    IsumDisp = Isum*1 + Idip*40;
    mesh(X, Y, IsumDisp);
    set(gca, 'zlim', [-20 40]);
```

```
        figure(3);
        set(gcf, 'position', [1000 90 400 300]);
        IsumDisp = abs(Isum)*10 + Idip*200 + 80;
        image(IsumDisp);
        colormap(parula(255));
        axis('off');
        drawnow
        pause(0.05);
end
```

As noted above, the basic dipole operates symmetrically in opposing directions. But for many practical applications, this is not desirable, and a method of favoring one direction over the other is preferred. The earliest and best-known design of this type is the Yagi antenna (Yagi, 1928; Pozar, 1997). As shown in Figure 2.64, this consists of a driven dipole element, with other elements that are not directly connected, but rather coupled via the electric field. There is usually one *reflector*, which is used to reflect the incident energy, although the reflector may in fact be a more elaborate corner reflector structure. The *directors* serve to direct the energy in the wanted direction. A minimum of one director is used, often a larger number for enhanced directivity. The separation of the reflector and directors from adjacent elements is critical and is usually a quarter of a wavelength.

So far, the discussion has concentrated on theoretical predictions of antenna patterns. Figure 2.65 shows experimental measurements of the half-wave

(a)   (b)

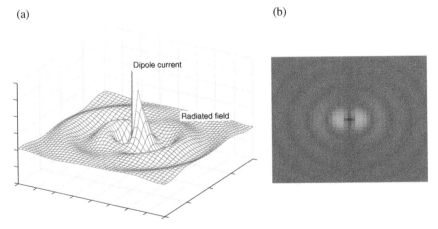

Figure 2.62 Elemental or Hertzian dipole – snapshot at an instant in time. A surface plot (a) shows the intensity as the height, while the image visualization (b) shows a false-color representation.

(a)   (b)

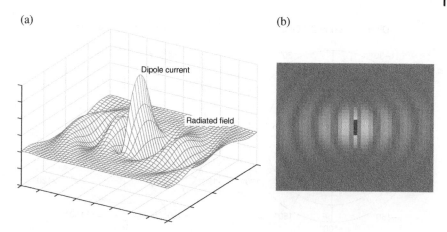

Figure 2.63 A half-wave dipole using the same method of calculating the field. Only the conductor current profile has changed compared with the elemental dipole. A surface plot (a) shows the intensity as the height, while the image visualization (b) shows a false-color representation.

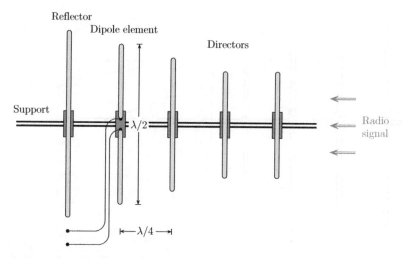

Figure 2.64 A half-wave dipole, when combined with one or more directors and a reflector, forms a Yagi antenna.

dipole and Yagi antennas, both at the wireless LAN frequency of 2.4 GHz. Clearly, the half-wave dipole is quite symmetrical and shows the pattern predicted by theory. The Yagi is somewhat more directional, also as expected.

One aspect of the antennas discussed so far is that they are essentially tuned to one particular wavelength. So, although nearby frequencies may be received, their coupling into the antenna is not as strong. For relatively close

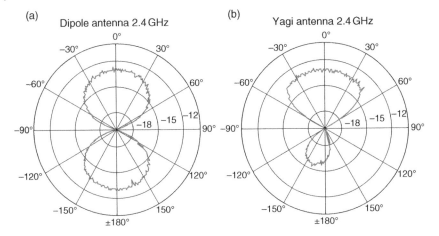

Figure 2.65 Experimental measurements of antennas at 2.4 GHz. (a) Dipole antenna. (b) Yagi antenna.

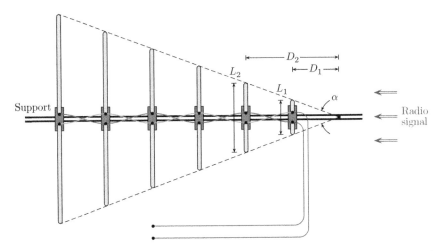

Figure 2.66 A log-periodic antenna formed by multiple half-wave dipoles. Note the reversal of the interconnections between successive dipoles, which effects a phase reversal.

frequencies, this may not present a problem in practice. However, where a large number of channels at larger frequency spacings are to be received, a simple dipole or Yagi structure can be improved upon. This is done using the so-called log-periodic structure, originally proposed in DuHamel and Isbell (1957), with the common log-periodic dipole array first described in Isbell (1960) and analyzed in Carrel (1961). As illustrated in Figure 2.66, there are several key design features embodied in the log-periodic dipole array. First, the dipole lengths are defined, along with their spacing, such that

$$\tan\frac{\alpha}{2} = \frac{L_1/2}{D_1} \tag{2.88}$$

$$\tan\frac{\alpha}{2} = \frac{L_2/2}{D_2} \tag{2.89}$$

and so,

$$\frac{D_1}{D_2} = \frac{L_1}{L_2} \tag{2.90}$$

This pattern is repeated for all elements, which span the range from highest frequency (shortest dipole) to lowest frequency (longest dipole). The second important aspect is that the driven dipoles are interconnected in a phase-reversed fashion. The idea of this arrangement is that only one dipole is resonant for a given band and the remaining dipoles will, in pairs, tend to cancel each other out due to phase reversal.

The antenna types discussed so far have a broad direction of radiation. However point-to-point antennas, especially at microwave frequencies, often require a very directional transmit and receive arrangement. In this case, a parabolic reflector is useful. Strictly speaking, the parabolic reflector serves to focus the RF energy and is not an antenna structure in the way a dipole converts RF into a current. The parabolic reflector simply serves to focus the RF energy. The incoming rays R in Figure 2.67 are focused at a point F. The opening of the dish D and curvature h define the shape and must be chosen in an appropriate ratio.

A parabola is formally defined as the set of points equidistant from the focus F with coordinates $(f, 0)$ and the line $x = -f$. This means that we may write the equation

$$x + f = \sqrt{(x - f)^2 + y^2} \tag{2.91}$$

which may be simplified to

$$y^2 = 4fx \tag{2.92}$$

Given this equation and referring to Figure 2.67, the point at which $y = D/2$ requires $x = h$, and so

$$\left(\frac{D}{2}\right)^2 = 4fh$$

$$f = \frac{D^2}{16h} \tag{2.93}$$

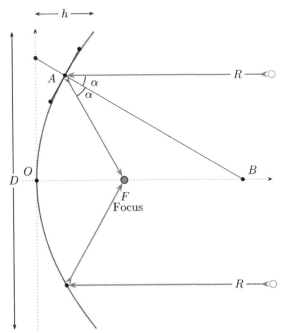

Figure 2.67 Illustrating the focus of parallel waves encountering a parabolic reflector. The tangent at A results in equal angles α, so the focus is always at F irrespective of which horizontal waves we consider.

This determines the focal length from the dish opening D and height h. Clearly, this is vital in sizing the dish itself and also in placing the receiving element at the focal point.

Why does a parabola focus at one particular point? By drawing a tangent at point A, which is perpendicular to the line AB, the angles α on each side of the line mean that, for simple reflection, the same focal point is always reached, irrespective of the vertical position of the incident line R. This, of course, is critical to the application of focusing the radio waves.

All of the antenna types discussed thus far have a fixed pattern of transmission or reception. For a fixed antenna geometry, changing the direction requires physically moving the antenna, so as to rotate the beam direction to that desired. But there is another way in which we can steer the beam without physically moving the antenna at all.

Consider Figure 2.68, which illustrates two small but theoretically isotropic (all-direction) radiating elements, separated by a distance d. The phase difference along the lines R_1, R_2 is $d \cos \theta$. The aim is to adjust the phases of the signals fed to s_r (right of M) and s_l (left of M) such that a maximum (or minimum) is achieved for some directional angle θ. There are several parameters that can be changed in this arrangement: the number of radiating elements, their relative spacing, the relative drive current phase for each, and the amplitude of each.

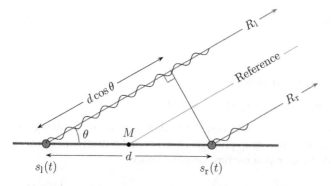

Figure 2.68 An antenna array composed of two ideal sources. The receiver is at some angle θ to the line connecting the sources.

To introduce the idea of electrically steerable arrays, consideration will be limited to two radiating elements, with equal current amplitudes and symmetric phases.

For a symmetrical array, it is easier to develop the required theory by considering the midpoint (where there is no radiator) to have a reference phase of zero. Let the feed signals be of the basic form $\cos \omega t$, which is a sinusoid of frequency $\omega = 2\pi f$. The assumption of symmetry in this case means that the phases of the left- and right-hand elements are equal but opposite. To model the symmetry, let the left-hand element be of the form $\cos(\omega t - \varphi)$, and so the right-hand becomes $\cos(\omega t + \varphi)$. The resulting addition is then

$$
\begin{aligned}
f(t, \varphi) &= \cos(\omega t - \varphi) + \cos(\omega t + \varphi) \\
&= (\cos \omega t \cos \varphi + \sin \omega t \sin \varphi) + (\cos \omega t \cos \varphi - \sin \omega t \sin \varphi) \\
&= (\cos \varphi + \cos \varphi) \cos \omega t + (\sin \varphi - \sin \varphi) \sin \omega t \\
&= 2 \cos \varphi \cos \omega t
\end{aligned}
\tag{2.94}
$$

This result shows that the addition of pairs of symmetrical-phase sinusoids is another sinusoid that has an amplitude proportional to $\cos \varphi$. This means that control of the resulting field may be effected by changing the relative phase φ of each element's feed current. This may be done by the driving electronics for each element.

Returning to the antenna array, using the midpoint M as a reference, $s_r(t)$ will be phase-advanced with respect to the midpoint M. The distance traveled with respect to the midpoint will be less by an amount $\beta d/2 \cos \theta$, resulting in a phase advance. The signal at $s_l(t)$ will be retarded in phase with respect to the midpoint M by an amount equal to $-\beta d/2 \cos \theta$.

Furthermore, we can arrange to feed each element with a current that has phase symmetry. Letting the feed current be ψ, the total phases may be expressed as the summations

$$\varphi_r = \frac{\beta d}{2} \cos \theta + \psi \tag{2.95}$$

$$\varphi_l = -\frac{\beta d}{2} \cos \theta - \psi \tag{2.96}$$

The resulting field is then approximated as

$$E(\theta) = K[I_o \cos(\omega t + \varphi_l) + I_o \cos(\omega t + \varphi_r)] \tag{2.97}$$

Since only the shape or pattern with respect to θ is of interest, omitting the constant K as well as the current I_o gives the normalized pattern of radiation (field at a given angle θ) for equal current amplitudes as

$$f(t, \theta) = \cos\left(\omega t - \frac{\beta d}{2} \cos \theta - \psi\right) + \cos\left(\omega t + \frac{\beta d}{2} \cos \theta + \psi\right) \tag{2.98}$$

This may be recognized as being of a similar form to the sum of cosines in Equation (2.94) and so simplifies to

$$f(t, \theta) = 2 \cos\left(\frac{\beta d}{2} \cos \theta + \psi\right) \cos \omega t \tag{2.99}$$

This sinusoid has an amplitude pattern that is found by removing the $\cos \omega t$ term to become

$$f(\theta) = \left|\cos\left(\frac{\beta d}{2} \cos \theta + \psi\right)\right| \tag{2.100}$$

where we have omitted the multiplier of 2 so as to normalize the pattern and taken the magnitude only since a negative amplitude peak is identical to a positive amplitude peak for the purposes of determining the peak over time.

Now all that remains is to select the radiator spacing d and the relative phase of the input currents ψ. Note that the phase ψ could easily be zero, a positive (advance) or negative (delay) – this is a matter for the design and governs the resulting field pattern $f(\theta)$ according to Equation (2.100).

To utilize a specific example, suppose the separation is half a wavelength ($d = \lambda/2$) and the feed currents are in-phase ($\psi = 0$). Then, the array pattern may be simplified to

$$\begin{aligned}
f(\theta) &= \left|\cos\left(\frac{\beta d}{2} \cos \theta + \psi\right)\right| \\
&= \left|\cos\left(\frac{2\pi}{\lambda} \frac{\lambda}{2} \frac{1}{2} \cos \theta + 0\right)\right| \\
&= \left|\cos\left(\frac{\pi}{2} \cos \theta\right)\right|
\end{aligned} \tag{2.101}$$

This power pattern $P(\theta) = f^2(\theta)$ is shown in Figure 2.69a. The direction of the peak radiation is perpendicular to the axis of the radiators. This may be understood since the phase delay due to physical separation is $\beta d = (2\pi/\lambda)(\lambda/2) = \pi$.

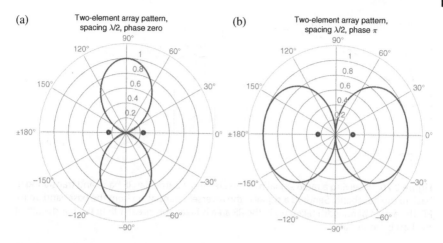

Figure 2.69 Changing the relative phases produces the two patterns illustrated. Note that the $\theta = 0$ axis is along the horizontal, corresponding to the axis of the radiator elements, whose relative positions are indicated. (a) $\psi = 0, d = \lambda/2$. (b) $\psi = \pi, d = \lambda/2$.

This is exactly half a wavelength if traveling along the $\theta = 0$ axis, and thus cancellation occurs.

Next, suppose the separation remains $d = \lambda/2$ and the feed currents are arranged so as to try to cancel the half-wavelength difference found. The phase delay resulting from this value of d is still half a wavelength along the axis, and so this means that we must arrange for a further half-wavelength phase difference. This may be arranged by setting $\psi = \pi/2$, so as to have a $+90°$ phase advance as well as a $-90°$ phase lag, which is equivalent to 180° in total, or half a wavelength. As a result, the array pattern may be simplified to

$$f(\theta) = \left| \cos\left(\frac{\pi}{2} \cos\theta + \frac{\pi}{2} \right) \right| \tag{2.102}$$

This pattern is shown in Figure 2.69b. The direction of the peak radiation is along the axis of the radiators, as we set out to achieve.

It is of course possible to extrapolate this concept for more than two radiating (or receiving) elements. Combining both differing feed current amplitudes and phases for a number of elements gives a great deal of flexibility in terms of the resulting signals, which add together in the far field, and thus control over the radiation pattern produced.

2.5.8 Processing a Wireless Signal

When a radio signal is to be sent, it requires conversion of the *baseband* signal into a *passband* channel. For practical reasons, this is usually done by first converting the baseband into an *Intermediate Frequency* (IF). Consider that an

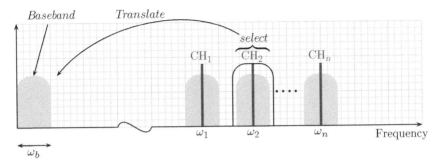

Figure 2.70 Receiving a radio signal means selecting a particular RF band and translating it back to the baseband. Sending a signal is the reverse – translating from the baseband up to RF. The actual bandwidth taken up in the RF area is invariably greater than the bandwidth of the baseband signal.

audio signal may extend from a few hundred Hz to several kHz, but a radio signal on which this audio is superimposed may be of the order of 100 MHz or higher. It follows that we need to convert from the lower frequency to the higher frequency, which involves a change in frequency from around 10^3 cycles per second to 10^8, which is quite a large ratio. Of course, the audio signal propagates in air as pressure waves, whereas the radio signal propagates as an EM wave. Receiving a radio signal is the reversal of what we have just done: converting from the RF to the original or baseband frequency range. Again, for practical reasons, this is usually done by first converting the received passband signal into an IF. So what is required is *upconversion* when the signal is to be sent and *downconversion* of the signal when it is received.

Upconversion and downconversion involve basically the same mechanisms, and so we start by considering the case where the RF signal is available and needs to be recovered. The basic requirement for the radio reception problem is outlined in Figure 2.70. Here we have the transmitted passband signal of bandwidth ω_b, which is centered on one of several possible carrier frequencies $\omega_1, \omega_2, \ldots$, also termed channels. Transmission is essentially the reverse: translation from the baseband up to RF.

The bandwidth around the RF channel is typically more than the baseband bandwidth – how much more depends on the type of modulation used. The simplest type of modulation, *Amplitude Modulation* (AM), uses twice the bandwidth of the baseband signal. Inspecting the diagram, it may be observed that a critical factor is that one channel must not interfere with another. There may also be a small unused channel space in between (a guard interval), since real filters to separate the channels are not perfect.

Historically, the Tuned Radio Frequency (TRF) receiver was the first method used, no doubt because it is the most straightforward approach. As shown in

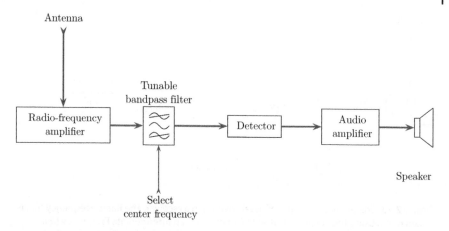

Figure 2.71 The Tuned Radio Frequency (TRF) receiver is essentially just a bandpass filter followed by a detector. The very weak received RF signal is first amplified, and the particular band of interest is selected using a filter. The information signal must be selected from that, and originally this was just a "detector" before more sophisticated modulation methods were devised.

block diagram form in Figure 2.71, this method is really just a logical implementation of the concept: to have a tunable filter that selects out the particular band or channel of interest. Despite its simplicity, it has one significant drawback, which makes it difficult to use in practice. The problem is that designing a bandpass filter that operates at a very high frequency is difficult. The center frequency ω_n in Figure 2.70 is much higher than the bandwidth ω_b. The ratio ω_n/ω_b is called the Q factor of the filter, and high-Q filters are difficult to realize in practice.

To address this problem, initial designs used multiple filters. This meant that several sets of filters had to be adjusted simultaneously – a very difficult task requiring a skilled operator. For this reason, TRF designs are not commonly employed, especially at high RF.

The invention of the *heterodyne* or mixing approach changed the thinking around radio design completely. In heterodyne receivers, a local oscillator (LO) signal is combined with the incoming radio signal. The combination is essentially a multiplication and lowpass filter operation. This serves to shift the frequency down to an IF. There are two advantages to this approach. The first is that it is far easier to create an oscillator at a particular frequency than it is to create a tunable filter with a very high Q factor. Secondly, the idea is to use a lower but constant IF, thus reducing the constraints on subsequent stages (amplification, filtering, demodulation) since they all operate at a fixed IF.

A block diagram of this approach is shown in Figure 2.72. This is called a *heterodyne* or mixing receiver and is generally ascribed to Armstrong (1921), though like many inventions, there were other, often parallel, contributors.

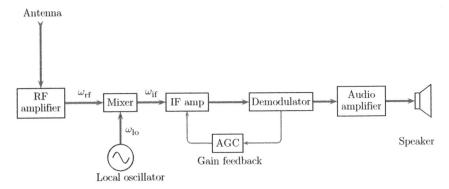

Figure 2.72 The general principle of heterodyning in a receiver. The Radio Frequency (RF) is mixed down using the Local Oscillator (LO) to produce an Intermediate Frequency (IF), which is then demodulated according to the modulation method used at the transmitter. The final stage shown is the Audio Frequency (AF) output. The Automatic Gain Control (AGC) feedback loop is used to adjust the output amplitude to maintain a constant output irrespective of the presence of strong or weak radio signals.

The LO at frequency ω_{lo} is tuned and applied to a mixer, with the output being a signal with a frequency equal to the difference between the LO and the incoming RF. Subsequent stages all operate at the IF, which is fixed. A feedback path may be included to compensate for varying channel conditions, which affect the strength of the incoming signal; this is the Automatic Gain Control (AGC). Multiple IF stages may be incorporated, in which case the design is termed *superheterodyne* or simply *superhet*.

We can draw a similar frequency allocation for the heterodyne receiver, as in Figure 2.73. In the diagram, the LO tracks below the RF, and the entire signal content over the band of interest is translated down to the IF, which is the difference in frequencies. Thus, in this case, $\omega_{if} = \omega_{rf} - \omega_{lo}$. This type of translation is termed low-side injection; it is possible to have the LO at a higher frequency than the RF, termed high-side injection.

The frequency reduction to the lower IF is achieved by a process termed *mixing*. The functions required of a mixer are depicted in Figure 2.74. The multiplication operation is key to the downconversion, but as we will see, higher frequencies are also produced; hence a lowpass filter is required.

To understand how mixing works on signals, consider two signal waveforms that are close but not identical in frequency, as shown in Figure 2.75. Multiplication of the two waveforms shown produces the signal shown in the lowest panel. It consists of two underlying signals. One is a much higher frequency, and closer inspection will reveal that this *frequency* is in fact the *sum of the two original frequencies*. The envelope around this higher frequency is evidently of a much lower frequency, and closer inspection will reveal that this *frequency* is the *difference between the two input frequencies*.

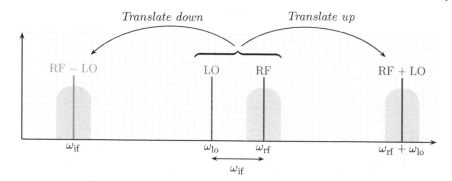

Figure 2.73 Downconversion from RF to an intermediate frequency, with low-side injection (LO less than RF). Sum and difference frequencies are generated as a result. Note that the LO must be tuned to be below the desired RF by an amount equal to the IF.

Figure 2.74 A signal mixer for downconversion consists of an oscillator and a signal multiplier, followed by a lowpass filter. The difference frequency will always be lower, and hence it is removed by an appropriately designed lowpass filter.

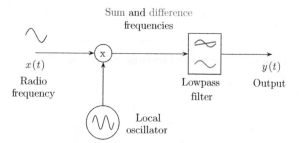

Thus we may say that multiplication of two sinusoidal signals produces both a sum frequency and a difference frequency. Referring back to Figure 2.73, we can see that when the RF signal is mixed with the LO signal, it produces signals as indicated at RF + LO and RF − LO. The lowpass filter removes the higher (sum) frequency components to leave only the difference (lower) frequency.

To show why this works, consider that the RF and LO signals are defined by $A_{rf} \cos \omega_{rf} t$ and $A_{lo} \cos \omega_{lo} t$, respectively. To find a formula for the product of these two waveforms, recall that the expansion of the product of two cosine functions is

$$\cos x \cos y = \frac{1}{2}[\cos(x + y) + \cos(x - y)] \tag{2.103}$$

Replacing x by $\omega_{rf} t$ and y by $\omega_{lo} t$, we have that

$$A_{rf} \cos \omega_{rf} t \times A_{lo} \cos \omega_{lo} t = \frac{1}{2} A_{rf} A_{lo} [\cos(\omega_{rf} + \omega_{lo})t + \cos(\omega_{rf} - \omega_{lo})t]$$

$$= \frac{1}{2} A_{rf} A_{lo} \cos(\omega_{rf} \pm \omega_{lo})t \tag{2.104}$$

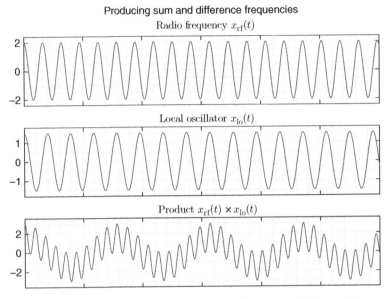

Figure 2.75 Converting a signal by multiplication. The sum and difference frequencies are produced.

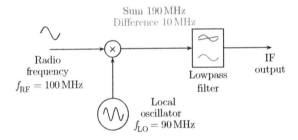

Figure 2.76 An ideal downconversion mixer example. The multiplication gives sum and difference frequencies, and the lowpass filter passes only the lower (difference) component.

This is clearly the sum and difference frequencies; keeping the difference only yields the IF:

$$\cos \omega_{if} t = \cos(\omega_{rf} - \omega_{lo})t \tag{2.105}$$

From this derivation, it is clear that

$$\omega_{if} = \omega_{rf} - \omega_{lo} \tag{2.106}$$

Applying this theory with some concrete figures, suppose we have a mixer with the hypothetical parameters shown in Figure 2.76. The sum frequency of 190 MHz is rejected by the lowpass filter, leaving only the IF signal at 10 MHz. This is exactly as it should be.

Figure 2.77 Mixer example with an image frequency present.

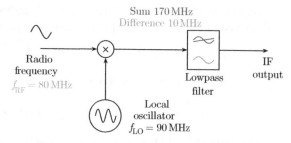

Figure 2.78 Illustrating how image signals may be generated in the frequency domain. The spacing between the local oscillator (LO) and desired radio frequency (RF) determines the region where an image frequency will interfere, if one is present.

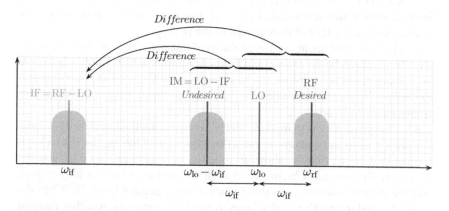

But suppose that there happens to be a second unwanted RF signal presenting itself at a frequency of 80 MHz at the input. This is not too far distant from the desired 100 MHz input, and Figure 2.77 illustrates this scenario.

The signal at 80 MHz would mix with the LO to produce sum and difference frequencies of 170 and 10 MHz, respectively. The 170 MHz would be rejected by the filter, leaving the IF of 10 MHz passing through. However, we stated that the 80 MHz signal was *not* the desired frequency (it was the 100 MHz signal that we wanted). So as a result, undesired frequency components from the 80 MHz frequency band will leak through to the IF and be processed. This unwanted frequency is termed an *image frequency*, and it must be removed (or at least minimized).

Considering the illustrations for this case, we can deduce that this situation occurs when the sum of, or difference between, the LO and an RF is equal to the IF. In this case, $100 - 90 = 90 - 80 = 10$. The difference between the wanted RF and unwanted image frequency is $2 \times$ IF, as shown graphically in Figure 2.78.

To consider this mathematically, we can multiply the incoming RF by the LO, and then convert the product of cosines into the sum of cosines, so that

$$x_{if}(t) = A_{rf} \cos \omega_{rf} t \cos \omega_{lo} t + A_{im} \cos \omega_{im} t \cos \omega_{lo} t$$

$$= \frac{A_{rf}}{2} \cos(\omega_{rf} + \omega_{lo})t + \frac{A_{rf}}{2} \cos(\omega_{rf} - \omega_{lo})t$$

$$+ \frac{A_{im}}{2} \cos(\omega_{lo} - \omega_{im})t + \frac{A_{im}}{2} \cos(\omega_{lo} + \omega_{im})t \qquad (2.107)$$

The first term (RF + LO) is quite high and not in the IF passband; likewise the last term (LO + IM) is well above the IF. The second term (RF − LO) is the wanted term, which is translated to the IF. However the third term (LO − IM) is also equal to the IF, and thus passed to the IF stage.

So what can be done about the problem of images? There are several strategies. First, a broader RF filter could be incorporated, which follows the tuning up and down of the LO. Since the bandwidth available to reject the image is of the order of twice the IF bandwidth, this filter has fewer design constraints and is thus easier to build.

Another possibility is to increase the IF, thus making the image further away. However, this to some extent negates the advantage of having the IF at a lower frequency. Considering the above arguments, another approach is to have two IF stages. The first has a higher IF so as to ensure the image is a long way off, and the second downconverts that higher IF into a second lower IF. While this is widely used in practice, it also leads to greater complexity. Another method, termed phasing image reject, is based on the Hartley modulator (Razavi, 1998). Figure 2.79 shows how this is implemented. Here, we need two oscillators in quadrature, or 90° out of phase with each other. Mathematically, these are cosine and sine for in-phase and quadrature-phase, respectively.

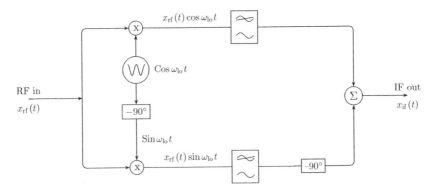

Figure 2.79 The Hartley image rejection approach. It relies not on filtering to reject the image, but on generating waveforms with a precise phase relationship (not difficult) as well as phase-shifting another waveform (usually more difficult to achieve).

To analyze how this works, suppose the incoming RF signal consists of *both* a wanted RF signal and an unwanted image signal. The upper branch shown in Figure 2.79 is then formed by

$$x_{rf}(t)\cos\omega_{lo}t = A_{rf}\cos\omega_{rf}t\cos\omega_{lo}t + A_{if}\cos\omega_{im}t\cos\omega_{lo}t$$

$$= \frac{A_{rf}}{2}[\cos(\omega_{rf}-\omega_{lo})t + \cos(\omega_{rf}+\omega_{lo})t]$$

$$+ \frac{A_{im}}{2}[\cos(\omega_{lo}+\omega_{im})t + \cos(\omega_{lo}-\omega_{im})t] \qquad (2.108)$$

after lowpass filtering

$$= \frac{A_{rf}}{2}\cos(\omega_{rf}-\omega_{lo})t + \frac{A_{im}}{2}\cos(\omega_{lo}-\omega_{im})t \qquad (2.109)$$

The lower branch shown in Figure 2.79 is

$$x_{rf}(t)\sin\omega_{lo}t = A_{rf}\cos\omega_{rf}t\sin\omega_{lo}t + A_{if}\cos\omega_{if}t\sin\omega_{lo}t$$

$$= \frac{A_{rf}}{2}[\sin(\omega_{rf}+\omega_{lo})t - \sin(\omega_{rf}-\omega_{lo})t]$$

$$+ \frac{A_{im}}{2}[\sin(\omega_{lo}+\omega_{im})t - \sin(\omega_{lo}-\omega_{im})t] \qquad (2.110)$$

after lowpass filtering

$$= -\frac{A_{rf}}{2}\sin(\omega_{rf}-\omega_{lo})t + \frac{A_{im}}{2}\sin(\omega_{lo}-\omega_{im})t \qquad (2.111)$$

and finally after a $-90°$ phase shift

$$= \frac{A_{rf}}{2}\cos(\omega_{rf}-\omega_{lo})t - \frac{A_{im}}{2}\cos(\omega_{lo}-\omega_{im})t \qquad (2.112)$$

Adding the signals in each branch (Equations (2.109) and (2.111)), the right-hand side term pertaining to the image $A_{im}/2\cos(\omega_{lo}-\omega_{im})t$ cancels, while the other terms involving the RF and IF are identical, and so we are left with

$$x_{if}(t) = A_{rf}\cos(\omega_{rf}-\omega_{lo})t \qquad (2.113)$$

Thus the image signal has been completely rejected, leaving only the desired signal. The disadvantage is that two phase-locked oscillators are required, together with a precise $-90°$ phase shift of one of the signal products. If these conditions are not met, the image signal will not be cancelled completely.

The third approach used is the so-called *direct-conversion* receiver (also called *homodyne*). Consider the earlier arguments for reducing the image signal. Increasing the LO toward the RF would mean decreasing the IF. If we decrease the IF to the point where it is nonexistent (that is, zero), there would be no image, and the IF then being zero means the signal of interest is now centered on the baseband. This does eliminate the advantage of using an IF as

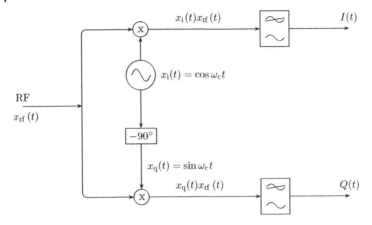

Figure 2.80 Direct downconversion with quadrature signals: *I* is the cosine component and *Q* is the sine component.

a constant lower frequency for subsequent stages to work with. It has come to be employed as the so-called zero-IF or *direct-conversion* receiver.

Direct conversion requires a great deal of linearity and processing at much higher frequencies. However, it is ideally suited to digital data demodulation. Figure 2.80 shows the basic idea. Rather than multiplying by a single LO, there are in effect two LOs. The frequency of the oscillator ω_c is exactly at the carrier frequency, and the waveforms are exactly 90° out of phase. This is said to be in *phase quadrature* (or just *quadrature*) and may be easily achieved by phase-shifting or digitally delaying the oscillator's cosine signal to produce a sine signal.

The result is that there are now two signal outputs: the in-phase component or $I(t)$ (which is related to the cosine signal) and the quadrature component $Q(t)$, which is related to the sine signal. The downconverted IQ signals are then demodulated directly, which is explained further in Section 3.8. For digital transmission, the presence of two signals, I and Q, permits a great many data bits to be encoded at once, thus substantially increasing the data rates possible in digital data transmission. This is dealt with in much more detail in Chapter 3.

2.5.9 Intermodulation

While multiplication of signals turns out to be very useful for performing modulation and demodulation, sometimes it can occur when we do not want it to. This may occur in practice because real-world devices are not perfectly linear (the idea of linear and nonlinear signal operations was introduced in Section 1.4.3). This means that for a given input signal, an amplifier ideally would produce an output signal that is the input multiplied by a constant factor.

However, electronic devices in practice invariably have some degree of nonlinearity.

Suppose that the output $y(t)$ of a system with input $x(t)$ is given by

$$y(t) = G_1 x(t) + G_2 x^2(t) + G_3 x^3(t)$$
$$= y_1(t) + y_2(t) + y_3(t) \tag{2.114}$$

If the system is purely linear, then only the first term would exist, and so $G_2 = G_3 = 0$. The output is then $y(t) = G_1 x(t)$, where G_1 is the gain multiplier. But suppose that, due to real-world imperfections, G_2 is a small but nonzero value and G_3 is an even smaller but still nonzero value. In this case, there is a contribution to the signal output whereby the signal amplitude is squared or cubed. What happens in this case? Let the input be two sinusoidal tones:

$$x(t) = A_1 \sin \omega_1 t + A_2 \sin \omega_2 t \tag{2.115}$$

The output is

$$\begin{aligned} y(t) = {} & G_1(A_1 \sin \omega_1 t + A_2 \sin \omega_2 t) \\ & + G_2(A_1 \sin \omega_1 t + A_2 \sin \omega_2 t)^2 \\ & + G_3(A_1 \sin \omega_1 t + A_2 \sin \omega_2 t)^3 \end{aligned} \tag{2.116}$$

The linear output (first term) is

$$y_1(t) = G_1(A_1 \sin \omega_1 t + A_2 \sin \omega_2 t) \tag{2.117}$$

This has the same frequencies as the input, exactly as we would hope. The second-order nonlinear contribution to the output is

$$y_2(t) = G_2(A_1 \sin \omega_1 t + A_2 \sin \omega_2 t)^2 \tag{2.118}$$

Using $(a + b)^2 = a^2 + 2ab + b^2$, this contribution to the output becomes

$$y_2(t) = G_2(A_1^2 \sin^2 \omega_1 t + 2A_1 A_2 \sin \omega_1 t \sin \omega_2 t + A_2^2 \sin^2 \omega_2 t) \tag{2.119}$$

Now using $\sin x \sin y = 1/2[\cos(x - y) - \cos(x + y)]$ and $\sin^2 x = 1/2(1 - \cos 2x)$, we can expand the right-hand side to show that

$$\begin{aligned} y_2(t) = {} & \frac{G_2 A_1^2}{2}(1 - \cos 2\omega_1 t) \\ & + \frac{2G_2 A_1 A_2}{2}[\cos(\omega_1 - \omega_2)t - \cos(\omega_1 + \omega_2)t] \\ & + \frac{G_2 A_2^2}{2}(1 - \cos 2\omega_2 t) \end{aligned} \tag{2.120}$$

Thus we have components at frequencies $2\omega_1, 2\omega_2$, and $|\omega_1 \pm \omega_2|$. As with harmonics of original frequencies, other components at new frequencies $|\omega_1 \pm \omega_2|$ are present. This constitutes distortion of the output waveform. If G_2 is small, then these contributions will be small.

The third-order nonlinear contribution to the output is

$$y_3(t) = G_3(A_1 \sin \omega_1 t + A_2 \sin \omega_2 t)^3 \tag{2.121}$$

Using the expansion $(a + b)^3 = a^3 + 3a^2 b + 3ab^2 + b^3$,

$$y_3(t) = G_3(A_1^3 \sin^3 \omega_1 t + 3A_1^2 A_2 \sin^2 \omega_1 t \sin \omega_2 t$$
$$+ 3A_1 A_2^2 \sin \omega_1 t \sin^2 \omega_2 t + A_2^3 \sin^3 \omega_2 t) \tag{2.122}$$

To determine what frequency components are present, it is evidently necessary to employ trigonometric expansions for $\sin^2 x \sin y$ and $\sin^3 x$. After grouping common terms into individual sine functions, it becomes evident that a number of frequency components are present:

$$\sin^3 \omega_1 t = \frac{3}{4} \sin \omega_1 t - \frac{1}{4} \sin 3\omega_1 t \tag{2.123}$$

$$\sin^2 \omega_1 t \sin \omega_2 t = \frac{1}{2} \sin \omega_2 t - \frac{1}{4} \sin(2\omega_1 + \omega_2)t$$
$$+ \frac{1}{4} \sin(2\omega_1 - \omega_2)t \tag{2.124}$$

$$\sin \omega_1 t \sin^2 \omega_2 t = \frac{1}{2} \sin \omega_1 t - \frac{1}{4} \sin(2\omega_2 + \omega_1)t$$
$$+ \frac{1}{4} \sin(2\omega_2 - \omega_1)t \tag{2.125}$$

Realizing that "negative" frequencies are the same as positive, it becomes clear that components arise at frequencies ω_1, ω_2 and $|2\omega_1 \pm \omega_2|, |2\omega_2 \pm \omega_1|$. The most problematic of these, in terms of proximity to the true input frequencies, are at $2\omega_1 - \omega_2$ and $2\omega_2 - \omega_1$, because these are inevitably closer to ω_1 and ω_2.

To see this graphically, consider Figure 2.81. With only linear terms (a purely linear amplifier), the two frequencies are the only ones present at the output. The square-law terms introduce other components at twice the frequency of each individual input, as well as the sum of the input frequencies. Since these are much higher in frequency, they may be outside the bandwidth of interest. However the third-order (or cubic) term introduces new frequency components at $2\omega_1 - \omega_2$ and $2\omega_2 - \omega_1$, and if ω_1 is of a similar value to ω_2, the resulting undesired frequency components will be very close to the desired components. Even though the unwanted components are small, they are not zero – we observe distortion in the region of the wanted frequencies comparable in magnitude with the third-order gain term.

So, even though the harmonics of each individual frequency end up being at much higher frequencies, the fact that there is more than one signal present at a time leads to numerous sum and difference frequencies, which may be within the bandwidth of a given system.

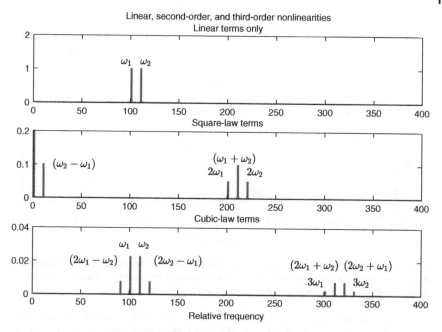

Figure 2.81 Illustrating the effect of a nonlinearity in the amplifier system, resulting in intermodulation terms. The example shown uses $G_1 = 1$, $G_2 = 0.1$, and $G_3 = -0.01$. The frequency scale is arbitrary. Note that the amplitude scales are not equal.

2.5.10 External Noise

Real-world wireless systems suffer from several problems, which must be overcome if the system is to be a reliable form of communication. First, there is the issue of multiple users and interference from other transmitters using the same RF band. Second, unintentional interference from other sources must be taken into account.

To put this in perspective, Figure 2.82 shows a captured signal on the 2.4 GHz wireless band, used for WiFi. Several channels are utilized for wireless complying with the 802.11 family of standards (IEEE, 2012). Here, we see the signals from two physically separate transmitters: one nearby, with a much higher signal strength, and a transmitter which is located further away. The two do not overlap in this figure, but with many wireless transmitters, this is always a possibility if their installations are not coordinated. Furthermore, we see the received signals emanating from, in this case, a microwave oven. The interference has a somewhat broad band, and for this type of emitter, several pronounced peaks are evident. These are not present continually, but come and go.

Wireless systems must be able to cope with these various types of interference. The interference may be transient (a microwave oven, for example) or semiregular (such as another wireless base station). Broadband

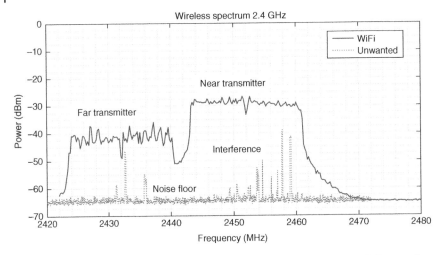

Figure 2.82 Wireless 2.4 GHz channel usage and interference. Two separate WiFi networks may or may not interfere with each other, background interference may exist for short or long periods of time, and background noise is always present.

noise is also present, and this is shown in the *noise floor* of Figure 2.82. Channel-adaptive methods that build upon the ideas already discussed, as well as spread-spectrum and discrete multitone modulation (Chapter 3), are used in high-noise wired and wireless environments to effect high-rate digital transmission. These are employed in combination with various channel coding and error detection methods so as to reduce the impact of transmission errors (Chapter 5).

2.6 Optical Transmission

Along with electrical cabling and wireless transmission, fiber optics forms the third foundation of signal and data transmission. Each of these methods has specific attributes that make it more suited to certain applications. In the case of fiber optics, the main attribute is the ability to deliver very-high-speed data transfer. But along with that advantage comes several other practical problems, such as lack of mobility (for which wireless is preferable) and the difficulty of tapping into an optical fiber (for which electrical cables are more suited). This section introduces some of the key concepts in fiber optics, including the optical fibers themselves, light sources, light detectors, and the measurement of typical optical fiber characteristics.

2.6.1 Principles of Optical Transmission

While the concept of using light as a messenger dates back to early recorded history, the present-day implementation of optical communication using glass optical fibers is generally traced to Kao and Hockham (1966), in that they first analyzed the technical requirements such as attenuation in glass for the specific purpose of lightwave communication. The history of the apparently nonintuitive use of "glass pipes" can in fact be traced back much further (Hecht, 2004, n.d.). In the same way, more recent history shows that the almost parallel invention and subsequent development of the laser (Hecht, 2010) was essential to the practical realization of fiber-optic communications.

Some of the important advantages of using optical radiation in a glass "pipe" (an optical fiber) are:

1) Exceedingly high data rates are achievable (due to the high bandwidth available, much higher than coaxial cable).
2) Immunity from crosstalk (a fiber within a cladding is not susceptible to external EM radiation).
3) Very low loss (thus enabling long distance transmission).

As with any communication method, three things are required: an emitter of power, at some specific frequency (or wavelength); a means to guide the power; and finally a receiver that converts the received power back to electrical voltages. In the case of fiber optics, the emitter is usually a laser diode (LD) (also termed a diode laser), which is able to emit in the infrared (IR) region. The waveguide is a glass fiber that (surprisingly) is quite flexible, and the receiver is a photodiode or similar semiconductor detector (which has some degree of sensitivity at the wavelength of the laser's IR emission). The key problems are:

1) How to produce optical power that can be switched so as to carry information.
2) How to match the emitted wavelength to the waveguide, such that the power is maintained over a long distance.
3) How to detect the received optical signal and convert it to a voltage.

Of course, the modulation of the optical power to enable synchronization of the receiver is another design aspect.

A key issue at this point is that virtually all present fiber-optic systems do not employ visible light. Rather, they employ IR radiation, which has a longer wavelength and is not visible to the human eye. In the visible light spectrum, we have at one end the shorter wavelengths, which are perceived as violet in color. This bends the most in a prism and has a wavelength of the order of 400 nm (NASA, n.d.). Assuming that the speed of propagation is that of light in a vacuum ($v = c \approx 3 \times 10^8$ m s^{-1}), the frequency may be found using $v = f\lambda$ to be approximately 7.5×10^{14} Hz or 750 THz. Shorter-wavelength radiation is not visible to the human eye and is termed *ultraviolet*. At the other end of the spectrum, the longer wavelength of red light bends the least in a prism and is of the order of 650 nm. The corresponding frequency is around 460 THz. Longer-wavelength radiation is not visible to the human eye and is termed *infrared*. Clearly, these frequencies are substantially higher than the radio frequencies encountered earlier.

Within the IR region, there are three subdivisions which are generally referred to as near IR, mid IR, and far IR. Although the definition of the precise boundaries of these regions vary, they are broadly in accordance with the ISO 20473 scheme, which denotes *near IR* as 0.78 − 3 μm, *mid IR* as 3 − 50 μm, and *far IR* as 50 − 1000 μm (ISO, 2009). Current telecommunication systems fall mostly into the 1310 and 1550 nm ranges,[1] and thus are in the near-IR region. The primary reason for this is that glass fibers may be manufactured with very low optical loss (attenuation) in this region. The following discussion is mainly focused on the near-IR bands mentioned. However, most of the underlying principles such as reflection and refraction apply equally well to visible light.

The production of optical fibers with very low loss at a specific wavelength is not all that is required, as mentioned above. The detector must also have a region of overlap in its spectral sensitivity. This concept is illustrated in Figure 2.83, where the emitter and detector only partially overlap. Since the region of overlap is invariably not matched to the region of peak power output of the emitter, or the peak sensitivity of the detector, a reduction in received power is unavoidable.

2.6.2 Optical Sources

One of two related methods for generating the light source in fiber-optic telecommunications are utilized: the Light-Emitting Diode (LED) and the LD (also sometimes termed a Diode Laser). LEDs produce light on the basis of semiconductor action, whereby a PN (*p*-type/*n*-type) junction emits radiation over a range of optical wavelengths, which may be broadly defined as one color for visible LEDs. LDs operate on a related principle, except that the junction of P and N materials forms an optically resonant cavity. This cavity produces coherent or in-phase light, based on the Light Amplification by Stimulated Emission of Radiation (LASER) principle. The historical development of the laser is chronicled in detail in Hecht (2010). A very good reference for some of the more technical aspects touched on here, especially semiconductor lasers and optical fibers, is Paschotta (2008).

An essential aspect of the LD is that the optical cavity is resonant at some particular wavelength, in a way similar to electrical standing waves as discussed in Section 2.4.8. This is illustrated in Figure 2.84. Two reflecting surfaces are required, although one (on the right in the figure) allows a small proportion of the light to exit. The optical cavity in the middle contains the beam itself, which is sandwiched between the energy pump source. In the original gas lasers, this was a very bright flashlamp. In the LD, this is a PN junction doped with appropriate elements. The end result is a standing wave as illustrated at the top of the figure. Emission of photons by one atom may initiate emission of

---

1 Remember that 1 μm = 1000 nm.

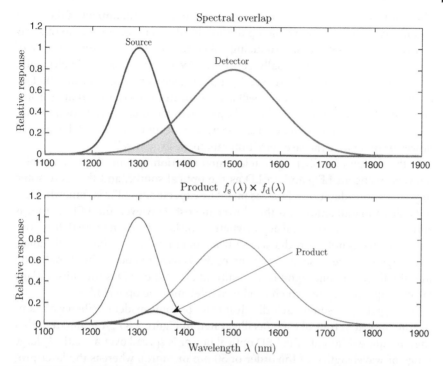

Figure 2.83 Illustrating an optical emitter and detector response overlap. Precise matching is almost never possible. This leads to a smaller electrical signal at the detector output, as well as additional noise due to the wider detector bandwidth.

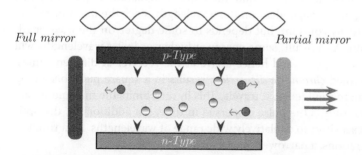

Figure 2.84 Illustrating the basic laser principle. External energy is supplied by the junctions at the top and bottom, stimulating the emission of cascades of photons. The lasing medium has a high gain over a defined optical wavelength. The stimulated radiation thus emitted bounces back and forward within the cavity to form a standing wave, with a fraction released to provide the laser output.

photons by others, hence leading to a cascade effect. Containment of the beam within the cavity serves to build up the amplitude via reflection and in this way forms an optical oscillator, producing coherent radiation. The temperature of the LD is critical and is usually controlled by external means. Additionally, the optical output must be sensed by a photodiode arrangement on one face of the cavity, so as to produce feedback to control the laser current. For low currents, the LD acts as a regular LED, producing incoherent radiation. Above the lasing threshold current, laser operation occurs. This must be limited however; otherwise the laser junction rapidly destroys itself.

In the design of a fiber-optic telecommunication system, there is a choice between using an LED and an LD as the optical source, and there are some important considerations. First, as might be expected, an LED is somewhat less difficult to manufacture and thus lower in cost. However, the LED output is spread over a broad physical angle, whereas the laser is much more tightly constrained. That is not to say that a laser beam does not diverge, but that the angle of divergence is quite small. Focusing optics is used to couple the IR radiation into the fiber, and one significant disadvantage of the LED is the substantially lower proportion of IR, which can be coupled into the optical fiber.

Looking deeper, some other disadvantages become evident. When we examine the optical output of each (as illustrated in Figure 2.85), it is apparent that they are quite different. The LED optical power is spread over a relatively large range of wavelengths (of the order of 50 nm or more), whereas the laser produces quite narrow linewidths. It is clear that, for a given amount of power, more radiation is produced in a narrower spectral range from the laser. A subtle effect results from the propagation of the IR through the fiber, due to refraction or bending of the wave as it passes through the fiber structure. The optical fiber itself has a given refractive index, and this slows the radiation. Importantly, this reduction in speed is dependent on the wavelength, since the index of refraction is not actually constant, but depends to some degree on the wavelength of the radiation. As a result, some energy at higher or lower wavelengths will travel at different speeds and hence arrive at the receiver at different times. This effect, termed *chromatic dispersion*, results in a square pulse becoming successively more rounded as it travels. This in turn limits the maximum data rates, since not only is the pulse dispersed in time, but additionally, the optical receiver is sensitive to a relatively wide range of wavelengths. As a result of these considerations, a narrower spectral width is preferable.

In the laser emission spectrum illustrated in Figure 2.85, a number of distinct laser lines are shown. In fact, this depends on the type of laser employed, and the characteristic shown is indicative of the Fabry–Pérot (FP) laser construction. Another type, the distributed feedback (DFB), is able to produce a single narrow line, but the cost is invariably higher, and power output lower. This is because the DFB laser employs a special optical grating, called a Bragg grating, to reduce the secondary lasing peaks.

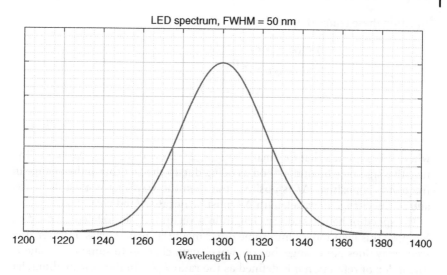

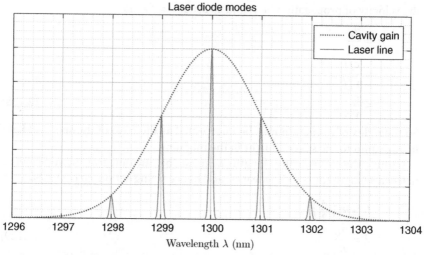

Figure 2.85 Illustration of the optical emission of an LED (top) and Fabry–Pérot (FP) laser diode (bottom). Note the different wavelength scales. The region of 1300 nm shown lies in the infrared spectrum and is not visible to the eye.

Some insight may be gained by calculating the spacing of the individual narrow lines. For a velocity in the laser cavity of v_c, let the center frequency and wavelength be f and λ, respectively. So $v_c = f\lambda$, and for some change in frequency Δf and wavelength $\Delta\lambda$, the fundamental frequency–wavelength product rule still holds, so

$$v_c = (f + \Delta f)(\lambda + \Delta\lambda) \tag{2.126}$$

Equating these (since they are both v),

$$(f + \Delta c)(\lambda + \Delta \lambda) = f\lambda$$

$$\cancel{f\lambda} + f\Delta\lambda + \lambda\Delta f + \cancel{\Delta f \Delta \lambda}^{0} = \cancel{f\lambda}$$

$$f\Delta\lambda \approx -\lambda\Delta f$$

$$\frac{\Delta f}{f} = -\frac{\Delta\lambda}{\lambda} \tag{2.127}$$

where it is assumed that for small changes, $\Delta f\, \Delta\lambda \approx 0$ since $\Delta f\, \Delta\lambda \ll f\lambda$. The negative sign may be understood as being due to the fact that if the change in frequency is positive, Δf is positive, but then the wavelength must decrease, and so the change in wavelength $\Delta\lambda$ must be negative (and vice versa).

The speed of light c and wavelength in a vacuum λ_0 are related by $c = f\lambda_0$. Suppose this light moves into some other medium such as glass or water. The frequency does not change, but the wavelength changes to some new value λ. The index of refraction n is defined as the ratio λ_0/λ. In the new medium, let the velocity of propagation be v, and as usual $v = f\lambda$. So

$$n = \frac{\lambda_0}{\lambda}$$

$$= \frac{c/f}{v/f}$$

$$= \frac{c}{v} \tag{2.128}$$

So the index of refraction may be interpreted as the ratio of the speed of light in a vacuum to the speed of light in the medium.

Within the cavity, the length L must be an integral multiple of a half-wavelength for standing waves, and hence lasing action, to occur. So

$$L = m\left(\frac{\lambda}{2}\right)$$

$$\therefore \lambda = \frac{2L}{m} \tag{2.129}$$

where m is an integer. Using this together with the value of index of refraction $n = c/v$,

$$v = f\lambda$$

$$\therefore \quad \left(\frac{c}{n}\right) = f\left(\frac{2L}{m}\right)$$

$$f = \frac{mc}{2nL} \tag{2.130}$$

Putting m as successive integers, it follows that the difference is

$$\Delta f = \frac{c}{2nL} \tag{2.131}$$

Now utilizing the ratio of frequency and wavelength as in Equation (2.127),

$$\Delta\lambda = -\Delta f \left(\frac{\lambda}{f}\right)$$

$$= -\left(\frac{c}{2nL}\right)\left(\frac{\lambda}{c/\lambda}\right)$$

$$= -\frac{\lambda^2}{2nL} \tag{2.132}$$

The negative sign is really immaterial in this case, since it is the *magnitude* of the change that is of interest. To make a concrete example, consider a laser with a center wavelength 1300 nm, a refractive index within the cavity of $n = 2.8$, and a cavity length 0.3 mm. Then,

$$\Delta\lambda = \frac{\lambda^2}{2nL}$$

$$= \frac{(1300 \times 10^{-9})^2}{2 \times 2.8 \times (0.3 \times 10^{-3})}$$

$$\approx 1.0 \text{ nm}$$

This is the spacing between successive laser lines, as indicated previously in Figure 2.85.

2.6.3 Optical Fiber

We now turn to the question of how light may be made to travel down an optical fiber, even when the fiber is not perfectly straight. Common intuition tells us that this should not be possible. The key is to convince the light to stay within an optical cavity, and initially it was thought that a reflective outer coating might work, much like a mirrored surface on the outside of a glass fiber. However, what works in practice is not reflection, but its counterpart – refraction. Refraction takes place where a light beam moves from one medium with a given refractive index to another medium, with a different refractive index.

Two types of optical fiber are in common use: single mode and multimode. The *mode* refers to the containment method within the fiber. Single-mode fiber is exceedingly small compared with multimode fiber. However, even the latter is quite small by everyday standards. The radiation employed in practical fiber-optic systems lies in the IR region of the optical spectrum, and although it is common to refer to it as "light," that is not to say that it is visible to the human eye. In fact, IR at these wavelengths can be dangerous to the eye, and care must be taken in dealing with optical fiber systems that employ laser light. The IEC 60825 series of standards refer to laser safety in general (IEC, 2014).

Figure 2.86a depicts a step-index multimode glass optical fiber. The key point to note is that the inner core is held within an outer cladding and that the inner

(a) (b)

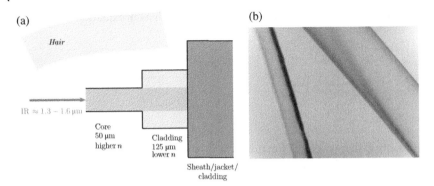

Figure 2.86 Multimode step-index fiber cross section (a), with typical sizes shown. More examples are given at the Fiber Optic Association The Fiber Optic Association (n.d.). The image in (b) shows a single-mode optical fiber with a human hair. The magnification is ×500.

core has a higher index of refraction. In practice, several additional layers of physical protection are included, depending on the intended installation of the fiber (for example, in buildings, underground, or undersea). The core of the single-mode fiber pictured in Figure 2.86b is smaller again, at 10 um.

It is not immediately clear just why the differing refractive indexes could contain light and furthermore why the light could still be contained even if the fiber bends around corners. Recall the definition of refraction, which was used previously in characterizing the operation of semiconductor lasers. When a pencil is placed in a glass of water, it does not appear to be straight. This is due to *refraction*, which occurs because the speed of the light propagation is different in different media. The index of refraction is *higher* in the *denser medium* (water) than in air. This leads naturally to the definition of refractive index in terms of speed of propagation, defined in Equation (2.128), which we repeat here for convenience:

$$n = \frac{c}{v}$$

$$\text{Refractive index} = \frac{\text{Speed of light in vacuum}}{\text{Speed of light in medium}} \qquad (2.133)$$

This is the inverse of what is employed in many other fields, where it is conventional to normalize a quantity with respect to a known standard. That is to say, the formula would have the terms up the other way. For air, n is very close to one. For water, it is close to about $1\left(\frac{1}{3}\right)$. Thus, according to the above equation, the speed of light in water must be *lower*. For completeness, we should also point out that n is not completely independent of the wavelength λ, but varies slightly with the wavelength. This fact is important in analyzing some of the deficiencies of fiber optics. However, for most purposes, n is assumed to be a constant for a given medium.

Refraction is quite important in explaining the principles of operation of fiber optics, so let us consider it for a moment. Figure 2.87 illustrates what occurs to a plane wave traveling in one medium (air, for example) with a lower refractive index (call it n_1) moves into a medium with a higher refractive index n_2. The peaks of the wavefront may be imagined as the bars perpendicular to the direction of travel. At point A, one part of the wave enters the boundary, and at B, another part of the wave enters the boundary at a different point. Since the medium on the lower side of the diagram has a higher index of refraction, the wave must slow down for Equation (2.128) to hold. The angle of approach is θ_1, and the angle of exit is θ_2. Note that these angles are always measured with respect to the normal or perpendicular to the interface and that the light bends toward the normal when entering the medium with higher refractive index.

Consider the triangle formed by the points $AA'B$. The angle at A' is a right angle (90°), and it may be shown that $\angle A'AB$ is equal to θ_1. For a separation L between A and B, we may write the sine ratio as

$$\sin \theta_1 = \frac{A'B}{L} \tag{2.134}$$

In a similar way, for the triangle $AB'B$,

$$\sin \theta_2 = \frac{AB'}{L} \tag{2.135}$$

Rearranging and equating L values in each,

$$\frac{A'B}{\sin \theta_1} = \frac{AB'}{\sin \theta_2} \tag{2.136}$$

Next, remember that velocity is the change in distance over time, or $v = d/t$. In the same time, say, t_o, the distance traveled by each wave end is the same, and so the velocity in each may be expressed as distance over time:

$$v_1 = \frac{A'B}{t_o}$$

$$v_2 = \frac{AB'}{t_o}$$

Finally, the refractive index for each medium tells us that

$$v_1 = \frac{c}{n_1}$$

$$v_2 = \frac{c}{n_2}$$

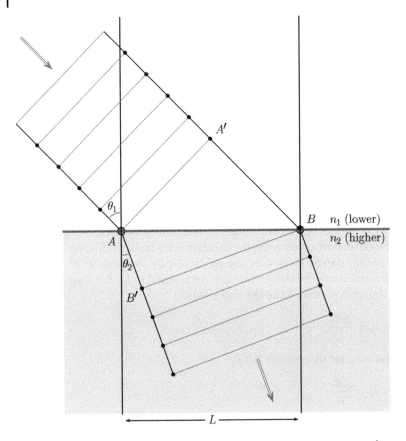

Figure 2.87 Motivating the derivation of Snell's law. The plane wave enters at the top and moves into the medium with a higher refractive index at the boundary *AB*.

Using these expressions,

$$\frac{A'B}{\sin \theta_1} = \frac{AB'}{\sin \theta_2}$$

$$\frac{v_1 t_o}{\sin \theta_1} = \frac{v_2 t_o}{\sin \theta_2}$$

$$\frac{c/n_1}{\sin \theta_1} = \frac{c/n_2}{\sin \theta_2}$$

$$n_1 \sin \theta_1 = n_2 \sin \theta_2 \qquad (2.137)$$

The final equation is termed Snell's law of refraction. In the case of *reflection*, the applicable law is just $\theta_1 = \theta_2$ (where θ_2 is the angle of reflection), but of

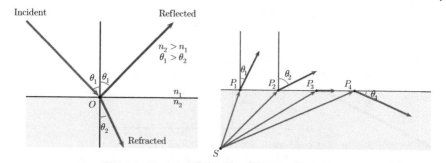

Figure 2.88 Principle of refraction at an interface (left) and total internal reflection (right). This shows that light emanating from a point S may be kept inside the material with higher refractive index, provided the outside material has a lower refractive index, and the angle is shallow enough with respect to the axis of the core.

course in traveling through different media, it is necessary to incorporate the indexes of refraction, and Snell's law does this. Using this equation, it may be deduced that for a given θ_1, if $n_2 > n_1$, then it follows that $\theta_2 < \theta_1$, hence the "bending" of the path toward the normal.

We can apply this concept to the development of an optical fiber, as illustrated in Figure 2.88. The diagram on the left shows both reflection and refraction occurring as a ray is incident on a boundary. But on the right, we move things about such that the ray source S is situated in the medium with the higher refractive index. In this case, the ray is moving from a *higher* refractive index to a lower index, and thus the refracted ray moves *away* from the normal. Moving from P_1 to P_2, the angle $\theta_2 > \theta_1$. Continuing on with this process, θ_2 must increase as the path of the ray from S moves toward being parallel with the medium, until at point P_3 the refracted wave is, in effect, parallel with the boundary. This is termed the *critical angle*. Making the angle from S toward P_4 even more shallow, we see that the ray is in fact reflected back into the medium, and no refraction occurs. This results in *total internal reflection*.

The above ideas – Snell's law and total internal reflection – now help us to see how light may be trapped in a fiber, forming a "light pipe." In Figure 2.89, we have shown a *core* of fiber with a higher refractive index, surrounded by a cladding of slightly lower refractive index. The light source (usually a laser) comes from the left and must be carefully coupled into the fiber. Assuming that the light travels through air, the refractive index is $n_{air} = 1.0002 \approx 1$. As might be imagined, the angle of entry θ_e plays a very important role.

At point A, we have from Snell's law

$$n_{air} \sin \theta_e = n_1 \sin \theta_1 \tag{2.138}$$

Since $n_{air} \approx 1$,

$$\sin \theta_e = n_1 \sin \theta_1 \tag{2.139}$$

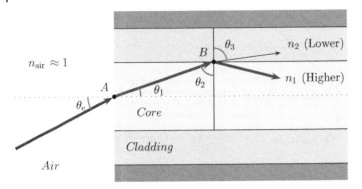

Figure 2.89 Illustrating the light entry angle and numerical aperture for multimode step-index fibers.

At B,

$$n_1 \sin \theta_2 = n_2 \sin \theta_3 \tag{2.140}$$

For total internal reflection to occur as described previously, θ_3 must approach 90° and so $\sin \theta_3 = 1$. Then,

$$n_1 \sin \theta_2 = n_2 \tag{2.141}$$

It is necessary to express this in terms of the external angle of entry of the light, θ_e. Using the internal angles of a triangle, $\theta_1 + \theta_2 = 90°$. Combining,

$$n_1 \sin(90° - \theta_1) = n_2$$
$$\therefore \quad n_1 \cos \theta_1 = n_2 \tag{2.142}$$

Now remembering that $\cos^2\theta + \sin^2\theta = 1$, squaring both sides of the above equation gives

$$n_1^2\cos^2\theta_1 = n_2^2$$
$$n_1^2(1 - \sin^2\theta_1) = n_2^2$$
$$n_1^2 - n_2^2 = n_1^2\sin^2\theta_1 \tag{2.143}$$

Using the equation at point A,

$$\sin \theta_e = n_1 \sin \theta_1$$
$$\therefore \quad \sin^2\theta_e = n_1^2\sin^2\theta_1 \tag{2.144}$$

Equating the equations for $n_1^2\sin^2\theta_1$,

$$n_1^2 - n_2^2 = \sin^2\theta_e$$
$$\therefore \quad \sin \theta_e = \sqrt{n_1^2 - n_2^2} \tag{2.145}$$

This is the *numerical aperture*. In order for light to traverse through the fiber, the greatest angle the incoming light can approach from is θ_e. This in turn depends on the refractive indexes of the core and cladding. Remember, too, that for small angles, $\sin \theta \approx \theta$, and so

$$\theta_e \approx \sqrt{n_1^2 - n_2^2} \tag{2.146}$$

This result also shows us that the index of refraction of the core (n_1) must be greater than the index of refraction of the cladding (n_2); numerically this is so because of the square root in the numerical aperture equation.

Finally, we have the means to understand how light may be guided in a cavity: The inner core material must have a higher refractive index than the outer cladding. The cladding is thus critical to the propagation of the light and not merely for physical protection of the core.

2.6.4 Optical Fiber Losses

One of the most important aspects of deploying fiber cable is the losses encountered. Loss of optical power means that the received signal is smaller than what is initially launched into the cable, and this may degrade the effective bit rate that the link is capable of. Losses result primarily from two sources: the cable itself and any connections made along the cable run.

Early attempts to create optical fiber for commercial use had, as a target, a loss of less than 20 dB km^{-1} (Henry, 1985). This is still an extremely high loss (recall the definition of the decibel – a loss of 20 dB is quite large), However, this barrier was eventually overcome, and losses in optical fiber can be substantially lower than electrical losses in cabling, as shown in the indicative figures contained in Table 2.3. Since the losses in optical fiber are predominantly due to water-related impurities, it is necessary to select an IR band that has minimal absorption (and so minimal loss) at the water absorption wavelengths. Although various wavelength bands may be employed, the most common are in the 1310, 1550, and 1625 nm wavelength ranges.

There are in fact only relatively few narrow regions of optical wavelength that are able to yield acceptable attenuation rates, of the order of 2 dB km^{-1} or less

Table 2.3 Transmission medium broad comparison.

Medium	Frequency range	Attenuation dB km^{-1}
Twisted pair (UTP)	1 kHz – 1 MHz	5–150
Coaxial cable (Coax)	1 MHz – 1 GHz	1–50
Optical fiber	~300 THz	0.2–1

Indicative figures only; see Henry (1985).

(2 dB loss results in around 2/3 of the optical power remaining). Optical loss in a fiber very much depends on the wavelength used. Minute quantities of impurities, particularly water, in the fiber result in substantially increased losses.

Consider a numerical example, with some hypothetical figures. Suppose we have an optical fiber with a quoted a loss of 1 dB km^{-1}, into which an IR laser launches 1 mW of power. We first convert dB into a linear multiplicative gain. It must be remembered that a *loss* causes attenuation or reduction in the signal, and thus this loss is mathematically a fractional gain. A loss must therefore be written as a negative dB gain. Since gain in dB is $10\log_{10}G$,

$$
\begin{aligned}
G_{\text{dB}} &= 10\log_{10}G \\
\therefore -1 &= 10\log_{10}G \\
-0.1 &= \log_{10}G \\
G &= 10^{-0.1} \\
&\approx 0.80
\end{aligned}
\tag{2.147}
$$

So the received power is

$$
\begin{aligned}
P_{\text{rx}} &= G \times P_{\text{tx}} \\
&= 0.8 \times 1 \text{ mW} \\
&= 0.8 \text{ mW} \\
&= 800 \text{ μW}
\end{aligned}
\tag{2.148}
$$

Once we have calculated the loss of a fiber segment, we can extend into multiple interconnected fibers. Each segment in the overall link is joined, either spliced using heat (fusion) splicing or connected using an optical connector. The latter typically has higher loss, as we will see in the next section. Suppose our hypothetical transmission line consists of four short segments as shown in Figure 2.90.

The loss in this example is given as 1 dB m^{-1}. To convert this into a gain multiplier, we again use the standard dB relationship attenuation factor, converting -1 dB into a gain of approximately 0.8. Note that, once again, the decibel gain is negative because it is a loss or attenuation. Over two segments of 1 m, the gain is 0.8×0.8. Over four segments it is $0.8^4 \approx 0.4$.

To do this calculation directly in dB, we can say that the total attenuation is the sum of all the dB attenuations. Thus $4 \times 1 = 4$ dB over the whole 4 m length.

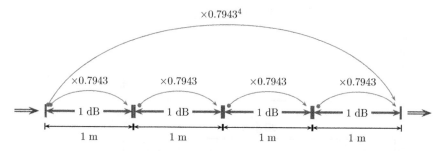

Figure 2.90 Illustrating the calculation of fiber loss over four segments. The numerical gains, which are less than unity, are multiplied. But an equivalent method, arguably easier in practice, is to add the dB figures. The dB figures are understood to be negative, since they represent a loss (<0 dB).

So the gain is

$$-4 = 10\log_{10}G$$
$$-0.4 = \log_{10}G$$
$$G = 10^{-0.4}$$
$$\approx 0.4$$

Thus the dB per length figure is *additive*. This method can be extended to any number of fibers of any length. Additionally, the connector loss between segments of cable may easily be added if working in dB. Figure 2.91 shows both decibel gain (increase in signal level) and attenuation (reduction in signal level).

2.6.5 Optical Transmission Measurements

Having considered how to calculate fiber loss, it is appropriate to now consider how we would actually measure the performance of a fiber connection once installed. In the event that a fiber breaks or a connector fails, it is also desirable to be able to determine where along the fiber length the problem exists. This is especially important since optical fibers are used on long-haul interconnections, often involving buried cabling.

One method that is in widespread use is the Optical Time-Domain Reflectometer (OTDR). Interestingly, this technique requires access to only one end of fiber – not both ends (transmit and receive), as might be expected. As the name implies, the method hinges upon reflection at the ends of the fiber. Such an approach was originally proposed in order to locate faults in the fiber (Ueno and Shimizu, 1976). However, OTDR goes further: It exploits what is otherwise a defect in optical fibers – the very small amount of backreflection, which occurs within the fiber itself.

The amount of backscatter is clearly dependent upon the length of the fiber as well as the attenuation experienced due to loss in the fiber. The loss per unit

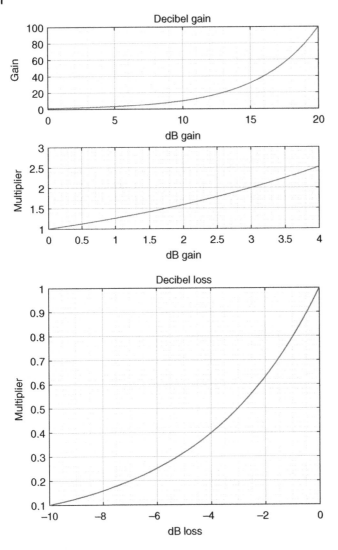

Figure 2.91 Using the decibel scale: top, for a gain >1 (positive dB) and bottom, losses (gain <1, or negative dB). Note the position of 0 dB in each case, as well as 3 dB and −3 dB points.

length is an important parameter in characterizing a fiber. Since the introduction of the idea of using one transmit and receive end (Barnoski and Jensen, 1976; Barnoski et al., 1977; Personick, 1977), the basic concept has evolved to the point where portable handheld instruments are available to perform these measurements. The successful use of such instruments may require some

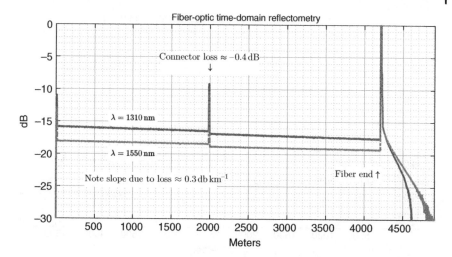

Figure 2.92 Optical time-domain reflectometry test with a long cable and one join. The cable join is visible at around 2 km, and the fiber loss may be calculated from the slope of the overall trend line. Note also the differing loss characteristics for different wavelengths.

degree of expertise in tuning the measurement parameters and interpreting the results.

The general principle of OTDR is as follows. Since the backscatter itself is quite small, a short pulse of the order of tens of nanoseconds is introduced into one end of the cable. The returned signal is recorded, along with a precise timing measurement. Large reflections identify catastrophic faults such as breaks in the cable, and smaller reflections occur at connector interfaces, since the matching from one fiber end to another is invariably imperfect. If we approximate the refractive index as $n = 1.5$, then the propagation speed within the fiber equates to a round-trip time for a pulse of close to $10~\mu s~km^{-1}$. To achieve resolution at this level, the returned signal must be sampled at even shorter intervals.

The very small backscatter from within the fiber itself is also useful to determine the fiber loss. Since this signal is quite small, it is normal to use multiple separated pulses and then to average the result. The idea behind this is that the average of a repeatable signal builds up in magnitude, but the average of the random noise amplitude generally decays in proportion to $1/\sqrt{N}$, where N is the number of sweeps averaged. There is a tradeoff with many of the parameters, including launched power, laser pulse width, and resolution.

Figure 2.92 shows the result of an OTDR experiment using two joined fibers of approximately 2 km length. The horizontal scale is calibrated in distance from the OTDR laser source. There is normally a significant loss in launching the power into the cable. We can see that the first fiber segment is 2 km long

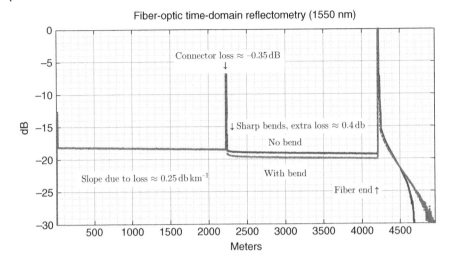

Figure 2.93 Optical time-domain reflectometry test at $\lambda = 1550$ nm with sharp fiber bends introduced. The additional loss thus introduced should ideally be avoided in practice.

and it is joined to the second fiber segment with a connector whose loss is 0.4 dB. Note the slope of the initial segment of the line. This shows the attenuation in decibels per kilometer (dB km^{-1}). A large backreflection is visible at the end of the second fiber, after around 2.2 km. Finally, the end of the fiber shows a very large drop-off where no light is reflected and only noise is present.

Figure 2.93 shows the result of a different cable configuration, with several tight bends introduced at the midpoint coupling in order to illustrate the effect of bends sharper than the recommended bend radius. An additional fixed loss of about 0.4 dB is introduced into the link due to these bends. Clearly this is undesirable, and proper routing of fiber cables is necessary to minimize the chance of introducing such losses. Any diversions must be done so as not to exceed the minimum recommended bend radius.

2.7 Chapter Summary

This chapter covered some of the key means by which information may be sent and received over a distance, using electrical wiring, wireless radio transmission, and optically using fibers. The following are the key elements covered in this chapter:

- Digital pulse transmission: line coding, synchronization, and scrambling.
- Principles of RF: change in frequency by mixing; the definition of RF bands and their use in practice.

- Transmission lines: reflections and standing waves.
- Propagation of RF and basic principles of antennas.
- Optical communications: light sources and detection, principles of optical fibers, and design and test of fiber links.

Problems

2.1 A wireless Local Area Network (LAN) uses the 2.4 GHz band. What is the corresponding wavelength?

2.2 Plot the Hamming window equation as given in Equation (2.15). Verify that it gives a smooth "taper" function, starting at $0.54 - 0.46 = 0.08$ and ending at 0.08, and a peak of $h = 1$.

2.3 Power levels and voltage magnitudes may be determined from a spectrum analyzer display.
a) In Figure 2.7, the sine wave amplitude is 200 mV peak to peak. Verify that this corresponds to the power level of -10.30 dBm shown on the figure.
b) Determine the absolute and relative voltages of the harmonics in Figure 2.8, and compare to the power levels that were measured as -8.20, -17.73, -22.18, and -25.10 dBm.

2.4 The linear feedback shift register code in Section 2.4.4 used a feedback register of 1001 binary. Using this code, determine what the feedback would be if the feedback register was 1011 binary instead. What are the implications of this?

2.5 Using the tapered frequency response for the raised-cosine filter (Section 2.4.2), evaluate the integrals for Equations (2.27) and (2.28) to give the time-domain impulse response for the raised cosine filter.

2.6 A coaxial cable with unknown termination impedance is subjected to a pulse from a matched signal source. The pulse is from 0 to 2 V when measured open circuit at the source itself. The waveform monitored at the signal source is shown in Figure 2.94.
a) Determine the line length, assuming that the propagation velocity is 2×10^8 m s^{-1} for this particular cable.
b) Determine the termination impedance as a proportion of the characteristic impedance.

2.7 The waveforms shown in Figure 2.95 are each the result of launching a voltage pulse as shown into a coaxial cable.

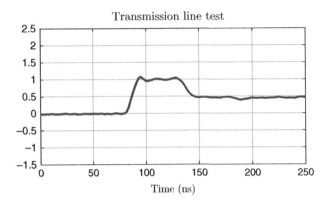

Figure 2.94 Transmission line with a square pulse input.

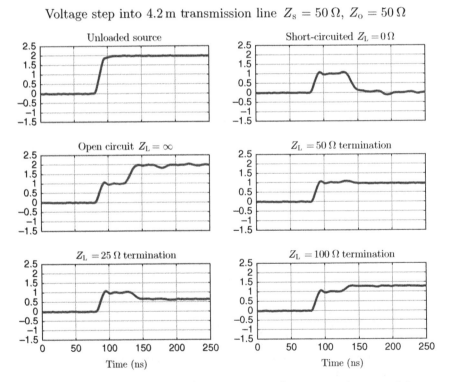

Figure 2.95 Experimental waveforms for investigating reflection on a 4.2 m transmission line.

a) Explain each of the impedance cases.
b) Using the given terminating impedances, determine the reflection coefficient in each case, calculate the expected voltage levels, and compare to the waveforms shown.
c) Show how to mathematically determine the length of the cable using measurements from the graphs.

2.8   The binomial approximation is used in determining some approximations for antennas. This states that $(1 + x)^\alpha \approx 1 + \alpha x$ for $x \lesssim 0.5$. Plot these functions using MATLAB to determine if this is a reasonable approximation.

2.9   A 1 GHz radio signal is transmitted from a stationary transmitter toward a receiver moving at 100 km h^{-1}. Calculate the approximate frequency shift.

2.10   The isotropic radiators in a phased array as shown in Figure 2.69 are fed by the same current magnitudes with equal phases and have a separation of $d = \lambda$. What is the shape of the pattern produced?

2.11   Given an incoming RF signal ω_{RF} and a local oscillator at frequency ω_{LO}, mathematically derive the output frequencies from an ideal mixer. Explain the significance of having two output frequencies. What image frequency would also produce a spurious IF signal?

2.12   An FM system is designed for an intermediate frequency (IF) of 10.7 MHz. Suppose we wish to tune stations over the FM band from 88 to 108 MHz.
a) At the lower end of the band, what would the local oscillator (LO) frequency need to be in order to let the difference frequency through, for tracking above the RF? For tracking below the RF?
b) At the upper end of the band, what would the local oscillator (LO) frequency need to be in order to let the difference frequency through, for tracking above the RF? For tracking below the RF?
c) Explain why the range of frequencies is not as dramatic as with an AM receiver operating in the $540 - 1600$ kHz band with an IF of 455 kHz.

2.13   Section 2.5.8 showed the local oscillator (LO) as having a frequency below that of the radio frequency (RF). This is termed low-side injection. It is also possible to have the LO frequency higher than the RF, which is termed high-side injection.
a) Draw a diagram similar to Figure 2.73, which shows high-side injection.

b) For the same IF of 10 MHz, what would the required LO frequency be in order to receive an RF signal of 100 MHz?

c) For this LO frequency, and a required IF of 10 MHz, what image frequency might be fed through?

2.14 The 2.4 GHz wireless channels 1–13 are separated by 5 MHz and have a width of 20 MHz (IEEE, 2012). Channel 1 has a center frequency of 2412 MHz, and it follows that channel 13 is centered at 2472 MHz. What two channels are in use in Figure 2.82?

2.15 The optical time-domain reflectometer (OTDR) is very useful in characterizing a fiber link. It does this using short laser pulses and then measuring the amount of light reflected at terminations and backscattered from within the fiber.

a) Show that a 5% reflection is the equivalent to a loss of about -13 dB.

b) Show that the round-trip time (RTT) of a light pulse in a fiber is about $10 \ \mu s \ km^{-1}$, assuming a refractive index of $n = 1.5$.

3

Modulation and Demodulation

3.1   Chapter Objectives

On completion of this chapter, the reader should:

1) Be able to define many of the common types of modulation and their variants.
2) Be able to explain the spectral effects of modulation and why certain types of modulation are used in differing situations.
3) Be able to draw block diagrams of modulators and demodulators, both analog and digital, and derive the mathematical expressions for their form where appropriate.
4) Be able to explain, using mathematical notation, the operation of modulators and demodulators.

3.2   Introduction

If a signal is to be transmitted over some distance, be it an analog signal such as audio or digital data, then it must somehow be transformed or *modulated* so as to conform to the required characteristics of the transmission medium. This might be a wired connection, or a wireless/radio link, or even an optical transmission system. The transmission medium is, in effect, the carrier of the information, and it is necessary to modulate, or change, the characteristics of that carrier waveform in order to carry the signal information. The reverse operation, demodulation, must occur at the receiver. It is necessary to take the received signal and transform it back into the original. In practice, exact reversal may not be possible due to nonlinear operations in the system or external noise.

As might be expected, there is no one single method of modulation that suits all types of situations. The main aspect is the frequency of operation of the

Communication Systems Principles Using MATLAB®, First Edition. John W. Leis.
© 2018 John Wiley & Sons, Inc. Published 2018 by John Wiley & Sons, Inc.
Companion website: www.wiley.com/go/Leis/communications-principles-using-matlab

carrier – for example, a radio frequency (RF) carrier may operate at 160 MHz, yet the signal to be transmitted may be audio of 16 kHz notional maximum frequency – a factor of 10 000 difference. There are several methods by which the 16 kHz signal could be superimposed on the 160 MHz carrier signal, each with advantages and disadvantages. Where sharing must occur, such as in radio systems with limited radio bandwidth, the possibility of interfering with other users must be taken into account. With digital transmission systems, the data rate is often (though not always) a critical parameter. This chapter examines modulation for various transmission systems, as well as the inverse operation, that of demodulation.

To ground the concept of modulation, consider the equation of simple sine waveform that we wish to transmit:

$$m(t) = A_{\mathrm{m}} \sin(\omega_{\mathrm{m}} t + \varphi_{\mathrm{m}}) \tag{3.1}$$

This is one single tone, but real signals such as speech and music are made up of many such waves in combination. The receiver typically wishes to recover the way in which the amplitude A_{m} of the sine wave of frequency ω_{m} varies over time, not just for one waveform but for many simultaneously. For a signal in the audio range, a frequency of 1000 Hz might be representative. Thus $\omega_{\mathrm{m}} = 2\pi \times 1000 \approx 6280$ rad s^{-1}. If the radio carrier is 100 MHz, then the carrier frequency $\omega_{\mathrm{c}} = 2\pi \times 100 \times 10^6$ rad s^{-1}, a vastly different frequency. So the question is this: How to superimpose the lower frequency on the higher one, and how to reverse this? The RF carrier may be represented by

$$x_{\mathrm{c}}(t) = A_{\mathrm{c}} \sin(\omega_{\mathrm{c}} t + \varphi_{\mathrm{c}}) \tag{3.2}$$

So we essentially have three options: to change the amplitude A_{c}, the frequency ω_{c}, or the phase φ_{c} of the carrier in response to the input waveform $m(t)$. Modulation is concerned with superimposing the signal $m(t)$ to be transmitted onto the carrier $x_{\mathrm{c}}(t)$ to form some new signal $x_{\mathrm{m}}(t)$. Demodulation is the process of recovering $m(t)$, or at least a close approximation to $m(t)$, from the noisy or imperfect received signal.

3.3 Useful Preliminaries

The understanding of modulation systems relies heavily on the equations that represent the various signals present. These signals are combined in various ways, and the analysis largely comes down to sine and cosine functions from trigonometry. This section reviews the mathematical concepts that will be used in the development of modulation and demodulation block diagrams.

Figure 3.1 Lengths and angles for trigonometry. Angle θ is shown to be less than 90°, but this need not be the case, and the concept can be generalized to any angle.

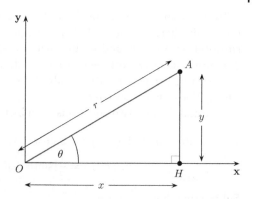

3.3.1 Trigonometry

The well-known trigonometric ratios sine, cosine, and tangent are illustrated in Figure 3.1. The relationship between the angles and lengths in triangle $\triangle OAH$ is fixed. For a triangle with angle θ, x-axis length x, y-axis length y, and distance r from the origin, the definitions of sine and cosine give

$$\sin \theta = \frac{y}{r}$$
$$\cos \theta = \frac{x}{r}$$

from which it follows that

$$y = r \sin \theta$$
$$x = r \cos \theta$$

Now we know $x^2 + y^2 = r^2$ (Pythagoras' theorem for triangles), so using the above for x and y,

$$x^2 + y^2 = r^2$$
$$r^2\cos^2\theta + r^2\sin^2\theta = r^2$$
$$\cos^2\theta + \sin^2\theta = 1$$

Now suppose the angle is actually made up of not one angle, but the sum of two others. This compound angle is of the form $\theta = \alpha \pm \beta$. Some useful results for this compound angle are as follows:

$$\sin(\alpha + \beta) = \sin \alpha \cos \beta + \cos \alpha \sin \beta \tag{3.3}$$

$$\sin(\alpha - \beta) = \sin \alpha \cos \beta - \cos \alpha \sin \beta \tag{3.4}$$

$$\cos(\alpha + \beta) = \cos \alpha \cos \beta - \sin \alpha \sin \beta \tag{3.5}$$

$$\cos(\alpha - \beta) = \cos \alpha \cos \beta + \sin \alpha \sin \beta \tag{3.6}$$

The first, $\sin(\alpha + \beta)$, can be derived from a geometric construction, which we do not give here. The second identity $\sin(\alpha - \beta)$ may be found from the first by substituting $\beta \to -\beta$, and noting that $\sin(-\beta) = -\sin\beta$ and $\cos(-\beta) = \cos\beta$. The third identity $\cos(\alpha + \beta)$ may be derived by replacing $\alpha \to \alpha + (\pi/2)$ in the first expansion. Thus $\sin[(\pi/2) + (\alpha + \beta)]$ becomes $\cos(\alpha + \beta)$, and $\sin[(\pi/2) + \alpha]$ becomes $\cos\alpha$.

If we add the first two equations term by term, we obtain

$$\sin(\alpha + \beta) + \sin(\alpha - \beta) = 2\sin\alpha\cos\beta$$

$$\therefore \quad \sin\alpha\cos\beta = \frac{1}{2}[\sin(\alpha + \beta) + \sin(\alpha - \beta)] \tag{3.7}$$

Subtracting them, we have

$$\cos\alpha\sin\beta = \frac{1}{2}[\sin(\alpha + \beta) - \sin(\alpha - \beta)] \tag{3.8}$$

Using the second set of identities,

$$\cos\alpha\cos\beta = \frac{1}{2}[\cos(\alpha + \beta) + \cos(\alpha - \beta)] \tag{3.9}$$

and finally

$$\sin\alpha\sin\beta = \frac{1}{2}[\cos(\alpha - \beta) - \cos(\alpha + \beta)] \tag{3.10}$$

Substituting $\alpha = \beta = \theta$ gives some other commonly used relations:

$$\sin 2\theta = 2\sin\theta\cos\theta \tag{3.11}$$

$$\cos 2\theta = \cos^2\theta - \sin^2\theta \tag{3.12}$$

$$\sin^2\theta = \frac{1}{2}(1 - \cos 2\theta) \tag{3.13}$$

$$\cos^2\theta = \frac{1}{2}(1 + \cos 2\theta) \tag{3.14}$$

These trigonometric formulas are useful in analyzing and designing modulators and demodulators. In dealing with waveforms at a frequency ω radians per second (or f cycles per second, where $\omega = 2\pi f$), the representation becomes $\sin\omega t$ or $\cos\omega t$. All that's necessary is to replace the fixed angles such as θ, α, or β by the argument ωt. This is reasonable, since the product ωt with appropriate units becomes ω (rad s^{-1}) $\times t$ (s) $= \omega$ (rad), which is the correct angular measure. Table 3.1 summarizes these results.

Table 3.1 Summary of useful trigonometric formulas.

$$\sin(\alpha + \beta) = \sin\alpha\cos\beta + \cos\alpha\sin\beta$$
$$\sin(\alpha - \beta) = \sin\alpha\cos\beta - \cos\alpha\sin\beta$$
$$\cos\alpha + \beta) = \cos\alpha\cos\beta - \sin\alpha\sin\beta$$
$$\cos(\alpha - \beta) = \cos\alpha\cos\beta + \sin\alpha\sin\beta$$
$$\sin\alpha\sin\beta = [\cos(\alpha - \beta) - \cos(\alpha + \beta)]/2$$
$$\cos\alpha\cos\beta = [\cos(\alpha - \beta) + \cos(\alpha + \beta)]/2$$
$$\sin\alpha\cos\beta = [\sin(\alpha + \beta) + \sin(\alpha - \beta)]/2$$
$$\cos\alpha\sin\beta = [\sin(\alpha + \beta) - \sin(\alpha - \beta)]/2$$
$$\sin 2\theta = 2\sin\theta\cos\theta$$
$$\cos 2\theta = \cos^2\theta - \sin^2\theta$$
$$\sin^2\theta = (1 - \cos 2\theta)/2$$
$$\cos^2\theta = (1 + \cos 2\theta)/2$$

3.3.2 Complex Numbers

We introduce here the notion of *complex numbers*[1] in order to facilitate some of the later theoretical developments, especially the Fourier transform (Section 3.9.7). For our purposes, the complex number, consisting of a real part plus an imaginary part, is a useful extension to trigonometry. But there are more advantages to the complex notation, such as being able to succinctly represent phase shift, and less complicated derivations when multiplication of trigonometric quantities is required.

We use the complex operator j, though the symbol ι may be employed in other fields. The complex operator $j = \sqrt{-1}$ is used to separate the conventional or *real* part of a complex quantity from the j or *imaginary* part. This may seem an arbitrary and unnecessary definition at first. But consider that we learn counting integers $1, 2, 3, \ldots$ initially, then the symbol for zero. After that, negative integers may be introduced to solve certain problems like $4-6$. Following this, integral fractions such as $1/2$ or $2/3$ may be defined, with certain rules for their addition and multiplication. Finally, real numbers such as 2.63 or -3.98 may be used, with the notion of place value arithmetic holding.

To achieve some mathematical solutions, we need to learn and use certain constructs, such as that a negative number multiplied by a positive number is negative, but a negative number multiplied by another negative number yields a positive quantity. The complex number is not used in everyday dealings, but if we wish to solve problems such as $z^2 = -1$, then it is necessary – and again, certain rules apply. Naturally, such a notation follows established rules of algebra – with the new extension that $j^2 = -1$ or, equivalently, $j = \sqrt{-1}$.

---

1 This section may be omitted if not studying the theoretical aspects of the Fourier transform.

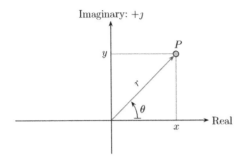

Figure 3.2 A point in the complex plane defines the cosine magnitude (real part) and the sine magnitude (the j or complex part). So $x + j\,y$ is equivalent to $re^{j\theta}$.

To introduce the key concepts, we begin with a geometric representation. A point on the complex plane may be represented as[2]

$$P = x + j\,y \tag{3.15}$$

as illustrated in Figure 3.2. The rectangular notation $x + jy$ may be augmented by polar notation, using magnitude and counterclockwise angle. The length r and angle θ in the illustration are related to the length and angle, by using geometry and trigonometry, respectively:

$$r = \sqrt{x^2 + y^2} \tag{3.16}$$

$$\theta = \arctan\left(\frac{y}{x}\right) \tag{3.17}$$

As a further step, the so-called Euler representation of a complex number is

$$re^{j\theta} = r(\cos\theta + j\sin\theta) \tag{3.18}$$

We can see that this embodies the projection of the real part ($r\cos\theta$) and the imaginary part ($r\sin\theta$) but written using an exponential notation.

Combining these concepts, we find that $1j$ is equivalent to $1e^{j\pi/2}$. This may be demonstrated by substitution into the formula above, giving $1e^{j\pi/2} = \cos(\pi/2) + j\sin(\pi/2) = 1j$. Furthermore, it is seen to be consistent with the geometric representation. The quantity $re^{j\theta}$ when multiplied by j yields

$$re^{j\theta} \times 1e^{j\pi/2} = re^{j(\theta + \pi/2)} \tag{3.19}$$

This shows that multiplication by j effects a counterclockwise rotation.

---

2 The j may be written either after the quantity (for example, $5j$ or πj) or before (for example, $j5$ or $j\pi$).

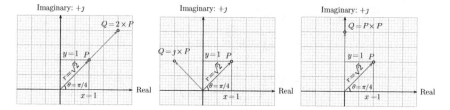

Figure 3.3 Multiplying the point $P = (1 + j1)$ by another complex number. From left to right: $P \times (2 + j0) = (2 + j2)$, $P \times (0 + j1) = (-1 + j1)$, and $P \times (1 + j1) = (0 + j2)$.

Some specific examples are shown in Figure 3.3. Let $P = (1 + j1)$. Taking each in turn, we have

$$(1 + j1) \times (2 + j0) = 2(1 + j1) + j0(1 + j1)$$
$$= (2 + j2)$$
$$(1 + j1) \times (0 + j1) = 0(1 + j1) + j1(1 + j1)$$
$$= j1 + j^2$$
$$= (-1 + j1)$$
$$(1 + j1) \times (1 + j1) = 1(1 + j1) + j1(1 + j1)$$
$$= (1 + j1) + (j1 + j^2)$$
$$= (1 + j1) + (j1 - 1)$$
$$= (0 + j2)$$

In the first case, the complex number P is multiplied by a constant, which changes the length but leaves the angle unchanged. In the second case, the complex number P is multiplied by j, which leaves the length unchanged but rotates the angle counterclockwise by 90°. In the final case, the complex number P is multiplied by $(1 + j1)$, which is a combination of the previous cases. The length is changed, and the angle is rotated. The product of the lengths is $\sqrt{2} \times \sqrt{2} = 2$, and the angle is $\pi/4 + \pi/4 = \pi/2$.

Noting that $P = (1 + j1) = \sqrt{2}e^{j\pi/4}$, we can rework these three examples using polar notation as follows:

$$\sqrt{2}e^{j\pi/4} \times 2e^{j0} = 2\sqrt{2}e^{j\pi/4}$$
$$= 2\sqrt{2}\left(\cos\frac{\pi}{4} + j\sin\frac{\pi}{4}\right)$$
$$= 2\sqrt{2}\left(\frac{1}{\sqrt{2}} + j\frac{1}{\sqrt{2}}\right)$$
$$= 2(1 + j1)$$
$$= (2 + j2)$$

$$\sqrt{2}e^{j\pi/4} \times 1e^{j\pi/2} = \sqrt{2}e^{j3\pi/4}$$

$$= \sqrt{2}\left(\cos\frac{3\pi}{4} + j\sin\frac{3\pi}{4}\right)$$

$$= \sqrt{2}\left(-\frac{1}{\sqrt{2}} + j\frac{1}{\sqrt{2}}\right)$$

$$= (-1 + j1)$$

$$\sqrt{2}e^{j\pi/4} \times \sqrt{2}e^{j\pi/4} = (\sqrt{2})^2 e^{j2\pi/4}$$

$$= 2e^{j\pi/2}$$

$$= (0 + j2)$$

3.4 The Need for Modulation

Chapter 2 introduced and analyzed the concept of up/downconversion, which is used to translate a signal from a lower frequency to a much higher frequency, and back again. As well as being translated in frequency, the transmitted signal also needs to have the original modulating signal superimposed upon it. This process is termed *modulation*, with *demodulation* occurring at the receiver. Ideally, these are the perfect inverses of each other: what one does, the other undoes. Modulation and up/downconversion are distinct operations required for different reasons but share one important concept: that of multiplying two signals to produce another, which is translated in frequency.

Upconversion is when a lower frequency, usually an *intermediate frequency* (IF), is translated to a radio frequency or RF. Downconversion is the reverse and occurs in the receiver. The purpose of this up/downconversion is so that most of the signal operations – especially modulation – can occur at an IF. This is done because it is easier and less expensive to build circuits and processing systems that operate at lower frequencies. Converting the very high frequencies to lower frequencies as soon as possible has direct benefits in terms of performance and cost.

A signal to be modulated $m(t)$ and carrier $x_c(t)$ is depicted in Figure 3.4. Suppose the modulation is a cosine wave of the form

$$m(t) = A_m \cos \omega_m t \tag{3.20}$$

and the carrier (of much higher frequency) is

$$x_c(t) = A_c \cos \omega_c t \tag{3.21}$$

These could be multiplied together to give

$$m(t)\, x_c(t) = A_m A_c \cos \omega_c t \cos \omega_m t$$

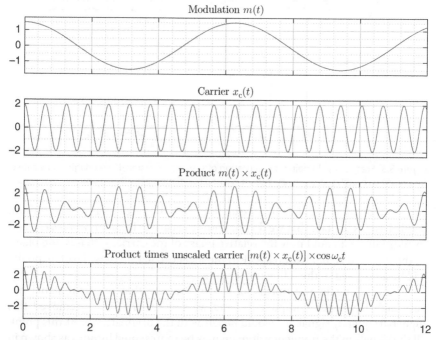

Figure 3.4 Converting a lower-frequency signal $m(t)$ up in frequency using multiplication by a much higher signal $x_c(t)$, and back down again, also via multiplication. The final result (lower panel) may be filtered to remove the high-frequency component, effectively leaving just the envelope, which is essentially the original $m(t)$ waveform.

Using the $\cos\alpha\cos\beta$ expansion, this may be simplified to

$$m(t)\,x_c(t) = \frac{1}{2}A_m A_c[\cos(\omega_c + \cos\omega_m)t + \cos(\omega_c - \cos\omega_m)t]$$

$$= \frac{1}{2}A_m A_c \cos(\omega_c \pm \omega_m)t \qquad (3.22)$$

It is important to note that the *product* of cosines has now become the *sum* and *difference* of cosines, with new frequencies $\omega_c + \omega_m$ and $\omega_c - \omega_m$. We refer to these as the *sum and difference frequencies*.

Figure 3.4 shows the modulating signal $m(t)$, the carrier $x_c(t)$, and their product. It may be seen that the lower-frequency signal effectively changes the envelope of the higher-frequency signal, and this corresponds to the operation of multiplication. The *modulated* signal is what is transmitted.

The *demodulation* of this particular type of modulated signal at the receiver may be attempted by multiplying by the carrier waveform by a local oscillator waveform with the same frequency as the carrier. We assume that we know the

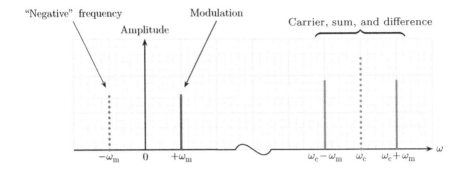

Figure 3.5 Frequency domain representation of signal conversion. If we imagine a negative frequency to match the given ω_m, then it is just a translation of both $+\omega_m$ and $-\omega_m$ by ω_c.

carrier frequency, but of course we do not know the carrier amplitude. A more subtle issue is the fact that we do not know the carrier frequency *precisely*, and in addition we do not know the relative phase of the carrier – that is, the phase of the received signal with respect to the local oscillator. These last few points will be addressed in the following sections.

The time-domain waveforms as shown indicate that the final signal has very high frequency components, and we can see that if we removed this higher-frequency signal, we would be left with the original $m(t)$. At this point, all the signals in the frequency domain may be represented by lines as shown in Figure 3.5. The signals shown are the modulation at frequency ω_m, the carrier at a much higher frequency ω_c, and the two signals at frequencies $\omega_c \pm \omega_m$. These are termed as the *sidebands*. We can think of the upconversion as first creating a "negative frequency" centered on zero, followed by translation of the positive and negative frequencies up by an amount equal to the carrier frequency.

This visualization will be very helpful in understanding the more complicated modulation schemes, discussed in later sections. Several questions now arise. First, is this simple multiplication the best scheme to use in all situations? What do we mean by "best" in this context? And what about the unanswered questions regarding (re)generating the local oscillator?

3.5 Amplitude Modulation

Assuming we stay with the basic multiplication scheme for transmission, we then need some way to obtain the local carrier. This is rather difficult, and in the very earliest schemes, the carrier itself was just transmitted alongside the sidebands at $\omega_c \pm \omega_m$. This simple type of modulation is termed *Amplitude Modulation* (AM). Consider the carrier as a sinusoidal:

$$x_c(t) = A_c \cos \omega_c t \tag{3.23}$$

We can achieve modulation by changing the amplitude of the carrier waveform over time. AM was the first type of modulation investigated historically and still finds widespread use. It is also used in conjunction with other more advanced types of modulation as will be seen later in the chapter. At its simplest, on–off keying of an old-fashioned telegraph signal may be considered to be AM, since that is either modulation on ($A_c = A_m$) or modulation off ($A_c = 0$). Note that we could just as easily have used a sine function – the end result will differ a little mathematically, but the conclusions are the same.

Figure 3.6 illustrates a straightforward approach to AM to begin with, together with the waveforms we obtain. The modulation $m(t)$ to be transmitted may in principle be any signal, but for the purposes of analysis, it is usual to just use a pure tone sinusoidal signal.

Mathematically, the resultant modulated waveform is then

$$x_{AM}(t) = m(t) \cos \omega_c t + A_c \cos \omega_c t \tag{3.24}$$

The carrier is significantly higher in frequency than the modulation – in fact so high, that it would not normally be visible on this scale. However, for the purposes of illustration, it is customary to "slow down" the carrier so that a few cycles may be seen. After the operation of multiplication of the carrier by the modulation, and then further addition of some of the carrier waveform, the modulated waveform looks as shown. In effect, it simply changes the amplitude of the carrier in response to the amplitude of the modulation. This waveform may be analyzed by first defining the *modulation index* as

$$\mu = \frac{A_m}{A_c} \tag{3.25}$$

(a)

(b)

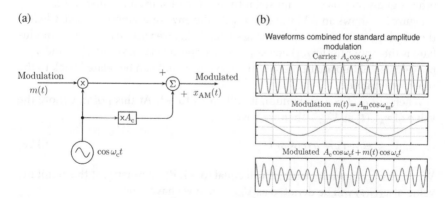

Figure 3.6 Generating an AM waveform using multiplication and addition. (a) Generating an AM signal. (b) The waveforms at each stage.

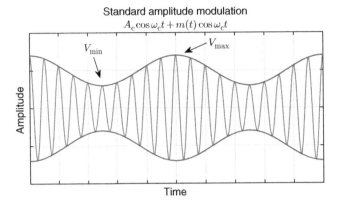

Figure 3.7 AM modulation parameter calculation, showing the AM waveform with its envelope superimposed.

Using a fixed-frequency modulation signal,

$$m(t) = A_{\mathrm{m}} \cos \omega_{\mathrm{m}} t \tag{3.26}$$

the AM signal may then be rearranged to

$$
\begin{aligned}
x_{\mathrm{AM}}(t) &= A_{\mathrm{c}} \cos \omega_{\mathrm{c}} t + A_{\mathrm{m}} \cos \omega_{\mathrm{m}} t \, \cos \omega_{\mathrm{c}} t \\
&= A_{\mathrm{c}} (1 + \mu \cos \omega_{\mathrm{m}} t) \cos \omega_{\mathrm{c}} t
\end{aligned} \tag{3.27}
$$

Note that it doesn't matter whether we use sine or cosine for the carrier, since just a phase shift is required. Also, in general the modulation signal $m(t)$ will be some more complex form such as speech, music, or digital data, but all these other signals may be decomposed into the sum of sine and cosine waves.

Figure 3.7 shows an AM waveform with the envelope superimposed. Clearly, the envelope reflects the modulating signal. What we wish to recover (demodulate) is the upper (or lower) envelope. The amplitudes marked at V_{max} and V_{min} provide useful information about the waveform and can be related back to the modulation index.

When $\cos \omega_{\mathrm{m}} t$ is a maximum, it will equal to $+1$. At this point, denote the value of $x_{\mathrm{AM}}(t)$ as V_{max}. Then we have

$$V_{\mathrm{max}} = A_{\mathrm{c}} (1 + \mu) \cos \omega_{\mathrm{c}} t \tag{3.28}$$

When $\cos \omega_{\mathrm{m}} t$ is a minimum, it will equal to -1. By symmetry, at this point the value of $x_{\mathrm{AM}}(t)$ will be denoted as V_{min}. Then we have

$$V_{\mathrm{min}} = A_{\mathrm{c}} (1 - \mu) \cos \omega_{\mathrm{c}} t \tag{3.29}$$

Dividing these two equations at the peak of the carrier (when $\cos \omega_c t = 1$) and solving for μ, we find that

$$\mu = \frac{V_{max} - V_{min}}{V_{max} + V_{min}} \qquad (3.30)$$

Thus, it is possible to determine the modulation index from the waveform measurements. Furthermore, examination of the figure shows that the carrier amplitude is really just the average of the maximum and minimum:

$$A_c = \frac{V_{max} + V_{min}}{2} \qquad (3.31)$$

and the modulation amplitude is the average of the *difference*

$$A_m = \frac{V_{max} - V_{min}}{2} \qquad (3.32)$$

3.5.1 Frequency Components

The process of AM changes the amplitude of the carrier. The carrier itself is a single, pure tone – but what frequency components are present?

Substituting a single-tone modulation $m(t) = A_m \cos \omega_m t$ into the AM generation equation, we have

$$x_{AM}(t) = A_m \cos \omega_m t \, \cos \omega_c t + A_c \cos \omega_c t \qquad (3.33)$$

The first term is the product of two sinusoids. It is not immediately obvious what frequency components this would produce. However, using the trigonometrical expansion for $\cos \alpha \cos \beta$ followed by the substitutions $\alpha \rightarrow \omega_c t$ and $\beta \rightarrow \omega_m t$, we end up with[3]

$$\cos \omega_c t \cos \omega_m t = \frac{1}{2}[\cos(\omega_m t + \omega_c t) + \cos(\omega_m t - \omega_c t)]$$

This sum/difference of cosines is the form we require, and so the AM waveform is

$$x_{AM}(t) = A_c \cos \omega_c t + \frac{A_m}{2} \cos(\omega_c \pm \omega_m)t \qquad (3.34)$$

Evidently, the spectrum for a fixed amplitude modulating signal $m(t)$ at frequency ω_m results in a frequency component at ω_c with amplitude A_c, as well as at $\omega_c \pm \omega_m$ with amplitude $A_m/2$. The former is the result of adding the carrier, and the latter is, indirectly, the result of the multiplication of carrier and modulation. Since $\mu = A_m/A_c$, the amplitude $A_m/2$ may be rewritten as $\mu A_c/2$ – that is, it is proportional to the modulation index μ.

---

3 Remember that $\cos(-\theta) = \cos \theta$.

To generate an AM waveform, we can use the following MATLAB code:

```
% time
N = 2*1024;
Tmax = 10;
dt = Tmax/(N-1);
t = 0:dt:Tmax;

% carrier
Ac = 2;
fc = 4;
wc = 2*pi*fc;
xc = cos(wc*t);

% modulation
Am = 0.5;
fm = 0.5;
wm = 2*pi*fm;
xm = cos(wm*t);

% AM generation
mu = Am/Ac;
xam = Am*xm.*xc + Ac*xc;

plot(t, xam);
xlabel('time s');
ylabel('amplitude');
```

This type of code layout will be useful to illustrate a number of principles in this chapter, so it is worth taking the time to examine it. We generate $N = 2 \times 1024$ points for a "smooth" plot, with an arbitrary time maximum of $T_{max} = 10$. The actual value is a matter for scaling – for example, it could be in microseconds. Likewise, the carrier frequency $f_c = 4$ Hz could be scaled accordingly. If we scale the time axis in 10^{-6} units of time, then the frequency is scaled by 10^{+6}, and so $f_c = 4$ would correspond to 4 MHz.

The sidebands are separated from the carrier by an amount equal to the modulating frequency. Figure 3.8 shows time waveforms with their corresponding frequency spectra. As a result of the mathematical analysis, we now know that the frequency components will consist of the carrier of amplitude A_c, with sidebands either side of the carrier, at frequencies $\omega_c \pm \omega_m$ of amplitude $A_m/2 = \mu A_c/2$. The frequencies come directly from our analysis, where we had cosine terms of the form $\cos(\omega_c \pm \omega_m)t$. The frequency spectrum shows the *amplitude*, so that even if they were negative, the magnitude would be positive.

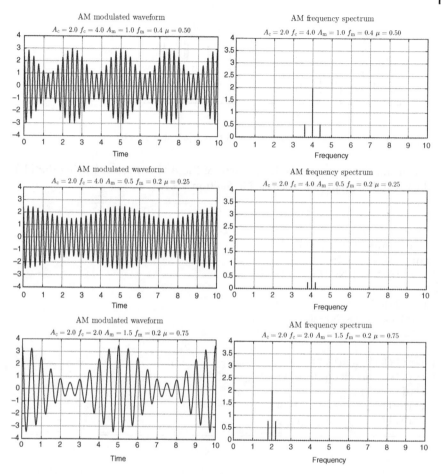

Figure 3.8 AM modulation showing time waveforms (left) and corresponding frequency spectra (right).

The two sidebands mean that AM uses a bandwidth effectively twice the modulating frequency. This in turn implies that the bandwidth required is larger than it ought to be and has implications when we have multiple RF channels with different AM signals. Figure 3.9 illustrates this in the frequency domain. Each of the radio channels must be strictly limited in their bandwidths as illustrated, and this in turn places a restriction on the highest frequency that may be modulated onto each channel.

We can find and plot the frequency components as follows. The FFT operation shown below converts the time waveform into its corresponding frequency

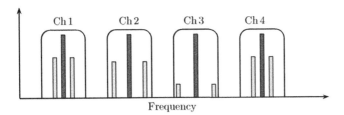

Figure 3.9 AM signal bandwidth and its effect on adjacent channels.

spectrum. For now, we use it to illustrate the AM concepts, leaving the FFT details and theory to be discussed in Section 3.9.7.

```
% frequency
df = (1/dt);
fam = abs(fft(xam));
fam = fam/N*2;
f = [0:N-1]/N*df;
K = 100;
k = 1:K;
maxfreq = (K/N)*df;

% plot to frequency maxfreq using bars
bar(f(k), fam(k));
axis([0 maxfreq 0 4]);
grid('on');
xlabel('frequency Hz');
ylabel('amplitude');
```

3.5.2 Power Analysis

As well as frequency and related bandwidth considerations, the issue of power is important. More power implies the need for larger capacity subsystems such as the output electronics and antenna. But more power consumption implies higher cost and shorter battery life for portable transmitters. Even if we do use more power, we want to be sure that the power is actually performing a useful function. The previous discussion on the spectrum of AM indicates that a substantial amount of power is used just in transmitting the carrier, and this does not actually help transmit the modulation signal itself.

To analyze the power and efficiency of AM, first recall that for a peak amplitude A, the RMS value of a waveform is $V_{\text{RMS}} = A/\sqrt{2}$. We found earlier that the carrier amplitude in AM is A_c and the sidebands each have an amplitude of $A_m/2 = \mu A_c/2$. The total AM signal power is thus

$$P_{\text{total}} = P_{\text{carrier}} + 2 \times P_{\text{sideband}}$$

$$= A_c^2 + 2 \times \left(\frac{\mu A_c}{2}\right)^2$$

$$= P_{\text{carrier}}\left(1 + \frac{\mu^2}{2}\right) \tag{3.35}$$

It is reasonable to define the efficiency as the power in the sidebands (which actually transmit "information") divided by the total power consumed and is thus

$$\eta = \frac{P_{\text{sidebands}}}{P_{\text{total}}}$$

$$= \frac{\mu^2}{2 + \mu^2} \tag{3.36}$$

It follows that when $\mu = 0$, the efficiency is zero. In that case, the power is completely in the carrier (and there is no modulation, which is not useful at all). However, when $\mu = 1$, the efficiency is $1/3$. This shows that the efficiency of AM is not especially good, and as noted earlier, a large proportion of the power is used in simply transmitting the carrier. Thus, we reach the conclusion that AM is not a very efficient scheme in terms of power efficiency, as much of the power is wasted in transmitting the carrier.

3.5.3   AM Demodulation

Once we have the modulated signal, we need to solve the receiver's problem: demodulation. To state this more concisely, we want to obtain $m(t)$, or at least an approximation to it, from only the received signal $x_{\text{AM}}(t)$. Essentially, the problem may be considered as recovering the upper (or lower) envelope of the modulated waveform, as may be observed from the waveforms shown so far. A simple rectification of the waveform, followed by a lowpass filter, may suffice. This is shown in Figure 3.10. The diode detector is, in effect, keeping only the positive half of the waveform and clamping the negative half to zero. A related approach is shown in block-diagram form in Figure 3.11, where we now square the samples. The squaring operation may be performed by nonlinear devices and thus is a low-cost alternative. Of course, in a digital sampled-data system, it is simple to compute the square of successive samples.

The waveforms resulting from the squaring and filtering operation are illustrated in Figure 3.12. In effect, the higher-frequency (RF or *radio frequency*) components are filtered out to leave the *audio frequency* (AF), which approximates the original waveform.

To analyze this approach, we just square the waveform to give

$$x_{\text{AM}}^2(t) = A_c^2 \cos^2 \omega_c t (1 + \mu \sin \omega_m t)^2 \tag{3.37}$$

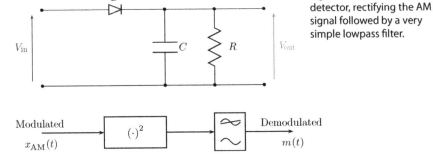

Figure 3.10 A diode detector, rectifying the AM signal followed by a very simple lowpass filter.

Figure 3.11 AM demodulation via squaring and first-order filtering.

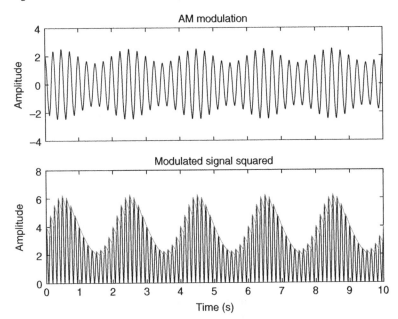

Figure 3.12 AM demodulation as squaring of the input, the envelope of the peaks is shown superimposed.

which may be expanded to yield

$$x_{AM}^2(t) = \frac{A_c^2}{2}\left[\left(1 + \frac{\mu^2}{2}\right) + \left(1 + \frac{\mu^2}{2}\right)\cos 2\omega_c t\right.$$

$$+ \frac{\mu^2}{4}\cos(2\omega_c \pm 2\omega_m)t$$

$$+ \mu\cos(2\omega_c \pm \omega_m)t$$

$$\left. + 2\mu\cos\omega_m t + \frac{\mu^2}{2}\cos 2\omega_m t\right] \tag{3.38}$$

Clearly, there are constant terms, a number of high-frequency components present, and other terms. Those in the band around the modulating frequency are only the ω_m term (the original frequency) and the $2\omega_m$ term (twice the original frequency). The $\cos \omega_m t$ term is what is desired, but all the others are not, and thus create distortion. After lowpass filtering to remove the high-frequency components, and a DC block to remove the constant terms, we are left with the demodulated signal as

$$x_{AM}(t) = A_m A_c \cos \omega_m t + \left(\frac{\mu A_c}{2}\right)^2 \cos 2\omega_m t \qquad (3.39)$$

Since $\mu < 1$, then $\mu^2 \ll 1$ so the distortion introduced by the $\cos 2\omega_m t$ term is somewhat less than the desired component. What remains is a term that is proportional to the original signal, $A_m \cos \omega_m t$.

Another approach is synchronous demodulation, also called coherent demodulation. In this method, a signal of the same frequency as the carrier has to be available, and it is used to demodulate the waveform. A key advantage of synchronous demodulation (as opposed to asynchronous demodulation, where the carrier is not available) is that it generally gives better performance – the distortion components are reduced, and also it is less sensitive to received noise. Synchronous demodulation requires a local oscillator, which increases the complexity of the system. It is considered further in Section 3.8, where it is shown that it is useful for other modulation schemes, not just AM.

3.5.4 Variations on AM

It was demonstrated that a considerable amount of power is employed in just transmitting the carrier in AM. In addition, the bandwidth is effectively twice the modulating (input) signal bandwidth, because we have two sidebands – the upper and the lower.

If we omitted the carrier altogether, the result is *double-sideband* AM (DSB-AM) (usually just referred to as DSB). This is simply a matter of just multiplying the modulating signal by the carrier – much as was done when upconversion and downconversion to/from radio frequencies was introduced. The double sideband modulated signal is

$$x_{DSB}(t) = A_c m(t) \cos \omega_c t \qquad (3.40)$$

This is illustrated in Figure 3.13. Taking the case of a modulation $m(t) = A_m \cos \omega_m t$, then we have

$$x_{DSB}(t) = A_c A_m \cos \omega_m t \, \cos \omega_c t \qquad (3.41)$$

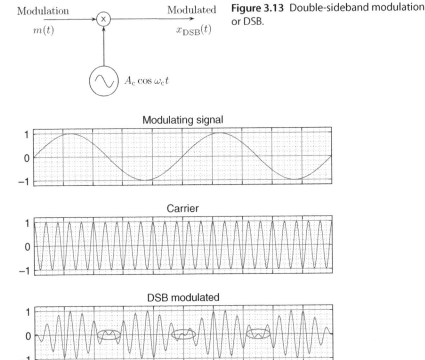

Figure 3.13 Double-sideband modulation or DSB.

Figure 3.14 The phase reversal of the modulated DSB waveform produces cancelation of the carrier.

Using the $\cos \alpha \cos \beta$ expansions, substituting $\alpha \to \omega_m t$ and $\beta \to \omega_c t$ results in

$$x_{\text{DSB}}(t) = A_c A_m \cos \omega_m t \, \cos \omega_c t \tag{3.42}$$

$$= \frac{A_c A_m}{2} \cos(\omega_c \pm \omega_m)t \tag{3.43}$$

There are two frequency components present at $\omega_c \pm \omega_m$, but no carrier. This leads to an interesting observation: since mathematically we found that there was no carrier present, why can we still see it in the modulated signal? The answer lies in the phase change. Figure 3.14 shows the modulated signal – note in particular the circled portions of the waveform. At these points, corresponding to the zero crossings of the modulating signal, the phase of the modulated signal is reversed. So calculating or measuring the frequency components over any length of time results in an average of zero, since the alternating portions of the carrier cancel each other.

If a locally generated carrier is available, then synchronous demodulation is possible for DSB, as illustrated in Figure 3.15. It is important that the local

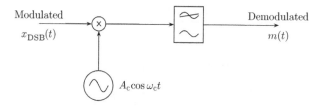

Figure 3.15 Synchronous DSB demodulation. Matching of the local oscillator phase to the received signal phase is critical.

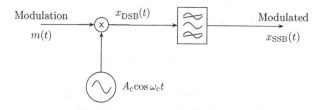

Figure 3.16 SSB modulation using bandpass filtering.

carrier oscillator precisely matches the phase of the received signal, otherwise incorrect demodulation results.

DSB thus does not explicitly transmit the carrier, and so saves power. However, it still needs the same bandwidth as conventional AM. The next step is to remove one of the sidebands, leaving only the other sideband. If this is possible, then the bandwidth required would only be that of the original baseband signal, not twice the original bandwidth. Single sideband (SSB) achieves this – it is really DSB with one of the sidebands removed. Some methods in use for generating SSB signals include simple bandpass filtering, the Hartley modulator, and the Weaver modulator.

Bandpass filtering is conceptually the simplest to understand, but in practice the most difficult to implement. It consists of a double-sideband modulator, followed by a bandpass filter to select the required upper (USB) or lower sideband (LSB) (Figure 3.16). The primary disadvantage with this approach is the need to create precise bandpass filters that operate at very high frequencies.

This difficulty leads us to consider some alternatives. These other methods – the Hartley modulator or phasing method and the Weaver modulator – both rely on shifting signals in phase to achieve the required modulation. Both of these methods are well suited to digital implementation with sampled data. They use phase shifts extensively, so it may be worthwhile to revisit Section 1.3, or at least use it as a reference when working through this section.

First, we examine the phasing method or Hartley modulator after Hartley (1923). The signal path is shown in Figure 3.17, where we see that the carrier is required in both upper and lower modulator branches, with a phase shift

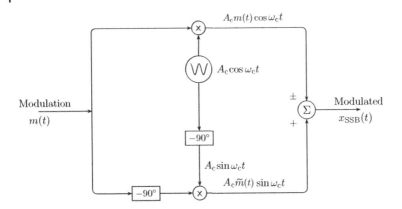

Figure 3.17 SSB modulation using phasing, also called a Hartley modulator.

of 90°. The incoming modulation must also be shifted by 90°. The add/subtract at the output stage produces either the USB (if the signals are subtracted) or the LSB (added). This method is sometimes called the phasing method of SSB generation, because it uses the signals and phase-shifted versions of the signals rather than filtering.

For a sinusoidal modulation test case, we can start with

$$m(t) = A_m \cos \omega_m t \tag{3.44}$$

In the diagram, the carrier is a cosine in the upper branch $x_c(t) = A_c \cos \omega_c t$, with cosine replaced by sine on the lower branch. So on the upper branch,

$$\begin{aligned} x_u(t) &= m(t)A_c \cos \omega_c t \\ &= A_m A_c \cos \omega_c t \cos \omega_m t \end{aligned} \tag{3.45}$$

On the lower branch, the modulation $\cos \omega_m t$ is converted into $\sin \omega_m t$ by the phase delay of $-90°$. Similarly, the cosine carrier $\cos \omega_c t$ becomes $\sin \omega_c t$ due to the phase delay of $-90°$. So the net output on the lower branch is

$$x_l(t) = A_m A_c \sin \omega_c t \sin \omega_m t \tag{3.46}$$

The output could be formed by the sum or difference

$$\begin{aligned} x_{SSB}(t) &= A_m A_c (\cos \omega_c t \cos \omega_m t \pm \sin \omega_c t \sin \omega_m t) \\ &= A_m A_c \cos(\omega_c \mp \omega_m)t \end{aligned} \tag{3.47}$$

Either the $+$ case is selected for the upper sideband or the $-$ for the lower sideband. Note that this method requires the basic operations of signal multiplication, lowpass filtering, and phase shifting. The first two are relatively straightforward, but phase shifting can present difficulties. For the carrier phase shift, the range of frequencies is quite narrow (ideally zero for a fixed carrier) and this

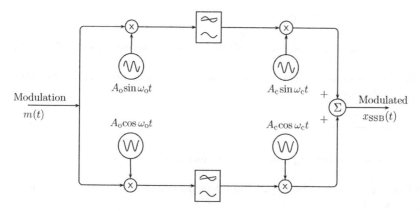

Figure 3.18 SSB modulation using Weaver's method.

does not present great difficulties. Indeed, a phase shift of a fixed frequency is just a delay. However, the modulator requires a −90° phase delay for the incoming modulation, which will have a finite bandwidth. Precise phase delays are difficult to achieve over a range of frequencies with analog components, although a constant phase delay over a range of frequencies may be achieved using a digital filtering technique known as the Hilbert transform.

The Hartley method requires phase shifts at various points. It can be difficult to achieve a constant phase shift across a range of frequencies, especially when the frequency is high. An alternative, the Weaver modulator (Weaver, 1956), does away with the phase shift requirements, at least for the wide-bandwidth modulating signal. *Quadrature* or 90° phase-shifted carrier signals are still required, but this is considerably easier than phase-shifting a wideband signal. This method is shown diagrammatically in Figure 3.18.

As shown in the diagram, quadrature oscillators at the carrier frequency (the $A_c \sin \omega_c t$ and $A_c \cos \omega_c t$) are required, together with quadrature oscillators $A_o \sin \omega_o t$ and $A_o \cos \omega_o t$ at a new frequency ω_o. Considering the upper path, we have

$$A_m A_o \cos \omega_m t \sin \omega_o t = \frac{A_o A_m}{2}[\sin(\omega_m + \omega_o)t - \sin(\omega_m - \omega_o)t] \quad (3.48)$$

This is lowpass filtered to leave the lower frequency ($\omega_m - \omega_o$) component, followed by multiplication by the carrier,

$$-\frac{A_m A_o A_c}{2} \sin(\omega_m - \omega_o)t \sin \omega_c t$$

$$= -\frac{A_m A_o A_c}{4}[\cos(\omega_m - \omega_o - \omega_c)t - \cos(\omega_m - \omega_o + \omega_c)t]$$

Similarly for the lower path,

$$A_m A_o \cos\omega_m t \cos\omega_o t = \frac{A_m A_o}{2}[\cos(\omega_m + \omega_o)t + \cos(\omega_m - \omega_o)t] \quad (3.49)$$

With a lowpass filter to leave the $(\omega_m - \omega_o)$ component, followed by multiplication by the carrier, we have

$$\frac{A_m A_o A_c}{2} \cos(\omega_m - \omega_o)t \cos\omega_c t$$

$$= \frac{A_m A_o A_c}{4}[\cos(\omega_m - \omega_o - \omega_c)t + \cos(\omega_m - \omega_o + \omega_c)t]$$

Adding the output of the upper and lower branches, the $\cos(\omega_m - \omega_o - \omega_c)t$ terms cancel, leaving

$$x_{SSB}(t) = \frac{A_m A_o A_c}{2} \cos(\omega_m - \omega_o + \omega_c)t \quad (3.50)$$

Note that this is frequency-shifted compared with conventional SSB, where we would have $(\omega_c + \omega_m)$. If we regroup and write as

$$x_{SSB}(t) = \frac{A_m A_o A_c}{2} \cos[(\omega_c - \omega_o) + \omega_m]t \quad (3.51)$$

then it can be seen that the "carrier" is effectively $(\omega_c - \omega_o)$, rather than ω_c as we usually have. Setting the oscillator frequency ω_o to be half the bandwidth $\omega_b/2$ is an effective solution, which just moves the effective carrier frequency down by $\omega_b/2$.

Turning now to demodulation of an SSB signal, a simple mixing approach could be tried again using a local oscillator at frequency ω_c. Taking the single-tone USB case, the modulated signal is

$$x_{USB}(t) = A_m A_c \cos[(\omega_c + \omega_m)t] \quad (3.52)$$

The demodulated signal would be

$$\hat{x}(t) = A_m A_c \cos[(\omega_c + \omega_m)t] \cos\omega_c t$$

$$= \frac{A_m A_c}{2}[\cos(2\omega_c + \omega_m)t + \cos\omega_m t] \quad (3.53)$$

This might appear to be all that is necessary, since the $2\omega_c$ term could be filtered to leave the audio signal. However in practice, the carrier frequency may not be known precisely at the receiver. To understand and model this, let the offset be $\delta\omega_c$. In that case, the demodulated signal would be

$$\hat{x}(t) = A_m A_c \cos[(\omega_c + \omega_m)t] \cos(\omega_c + \delta\omega_c)t$$

$$= \frac{A_m A_c}{2}[\cos(2\omega_c + \omega_m + \delta\omega_c)t + \cos(\omega_m - \delta\omega_c)t] \quad (3.54)$$

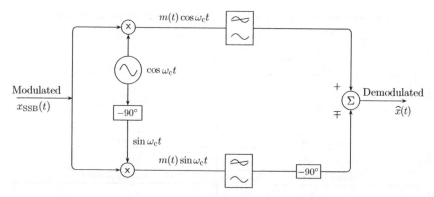

Figure 3.19 SSB demodulation using the Hartley phasing approach.

So the recovered frequency would be off by $\delta\omega_c$, and the recovered tone would be shifted in proportion to the error in the carrier signal. Clearly, this is highly undesirable.

A phasing approach, similar to that employed in modulation, works better. This is depicted in Figure 3.19. If the input is the USB,

$$x_{\text{USB}}(t) = A_m A_c \cos[(\omega_c + \omega_m)t] \tag{3.55}$$

then the upper branch of the demodulator is

$$
\begin{aligned}
x_u(t) &= A_m A_c \cos(\omega_c + \omega_m)t \cos\omega_c t \\
&= \frac{A_m A_c}{2}[\cos(2\omega_c + \omega_m)t + \cos(\omega_m t)]
\end{aligned}
\tag{3.56}
$$

Lowpass filtering of this signal leaves

$$x_u(t) = \frac{A_m A_c}{2}\cos\omega_m t \tag{3.57}$$

In the lower branch of the demodulator,

$$
\begin{aligned}
x_l(t) &= A_m A_c \cos(\omega_c + \omega_m)t \sin\omega_c t \\
&= \frac{A_m A_c}{2}[\sin\omega_m t + \sin(2\omega_c + \omega_m)t]
\end{aligned}
\tag{3.58}
$$

After lowpass filtering, this becomes

$$x_l(t) = \frac{A_m A_c}{2}\sin\omega_m t \tag{3.59}$$

and finally with a delay of $-90°$,

$$\tilde{x}_l(t) = -\frac{A_m A_c}{2}\cos\omega_m t \tag{3.60}$$

If $x_u(t)$ and $\tilde{x}_l(t)$ are added, we have zero net result. But if these signals are subtracted, we have $\cos\omega_m t$, which is the original signal. So we conclude that a

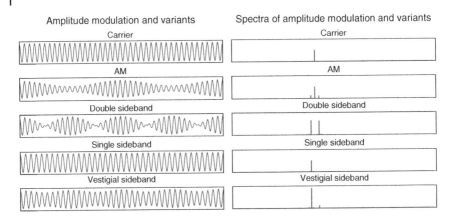

Figure 3.20 The waveforms and spectra of AM modulation variants – AM, DSB, SSB, and VSB.

USB signal could be demodulated. If the input is the LSB, then using a similar approach we find that subtraction yields zero, but addition yields the original tone. So, this structure could be used to demodulate either LSB or USB, just by selecting addition or subtraction at the final junction.

In all of the above, it is important to remember that there are various gain constants in the system along the way: the RF amplifiers produce gain and thus increase the signal amplitude, whereas the propagation path results in a loss of signal amplitude. So there are many constants lumped in together in the path from transmitter to receiver.

Figure 3.20 shows the waveform and spectra of the AM modulation variants discussed so far, as well as vestigial sideband (VSB). VSB is a compromise of sorts between having only one sideband, and having only a small amount of the other sideband present. In conventional AM, the effect of the modulation index is visible, as compared with DSB where the signal amplitude crosses zero at a rate equal to twice the modulating frequency. It is not possible to determine the modulation type from observing the SSB waveform alone. Turning to the frequency spectra, AM consists of the carrier and some sideband information, DSB has the carrier suppressed, SSB has only one sideband, and VSB has most power in one sideband but some residual or vestige of the other sideband. Note that the vertical power axes are not the same, so as to show the presence of the small second sideband in VSB.

3.6 Frequency and Phase Modulation

AM conveys the original signal by means of change in the amplitude of the carrier. Any noise within the transmitter or receiver, or in the transmission path, will affect the amplitude of the modulated signal. This change in amplitude will, to an AM demodulator, be indistinguishable from the original modulation – in other words, noise looks like a wanted signal. Thus, AM is somewhat susceptible to noise.

In Frequency Modulation (FM), the amplitude of the modulated signal waveform does not change. FM demodulation does not depend on the amplitude of the received, modulated signal – and therein lies the inherent advantage of FM. This section explains FM, as well as a closely related technique, *Phase Modulation* (PM). The early development of FM due to the work of Armstrong (1936) was controversial, as it promised reduced noise susceptibility as compared with simpler AM systems, which were all that was available at the time.

3.6.1 FM and PM Concepts

For alternatives to AM, which is sensitive to noise and not very power efficient, it may be helpful to return to the original proposition for modulation. That is, we wish to transmit a signal $m(t)$ by means of a change in the carrier wave $x_c(t)$:

$$x_c(t) = A_c \cos(\omega_c t + \varphi_c) \tag{3.61}$$

So far, in AM only the amplitude A_c of the carrier was changed, and this was made to vary with the modulating signal $m(t)$. But from the equation, it is clear that we have other parameters to manipulate: the frequency ω_c and phase φ_c. Since frequency and phase are related, it is not surprising that the modulation schemes resulting from a change in frequency or phase are also related.

Consider Figure 3.21, which shows how we might conceptualize the generation of a *time* waveform as stepping through the *phase* angle. As we incrementally step through phase angles, a "lookup" of a sine wave tells us the corresponding amplitude. When we reach the end of this sinewave lookup table, we simply revert to the start again, since the waveform is repetitive. The *frequency* of the wave is governed by *how fast* we step through the phase angles. Thus, the *rate of change of the phase* is actually the *frequency*. In reverse, given a frequency (in radians per second) of a waveform, and a time δt, we can work out how many radians were stepped through in that time period. Thus, accumulating or *integrating (summing up) frequency* over time tells us the *phase*.

It might be worth noting that technically *all* of the argument of the cosine function – the $(\omega_c t + \varphi_c)$ part of the expression – is a phase angle. The exact notation has been discussed for some time (see, for example, van der Pol, 1946); however conventional usage in telecommunications is to call φ the phase angle and to denote it as positive or negative according to the problem being discussed. It is also necessary sometimes to refer to the *instantaneous frequency*, since if we keep changing the phase φ_c, then the actual frequency is changed either side of ω_c.

Conceptually, changing the frequency of the carrier is probably easiest to understand. Consider Figure 3.22, which shows a (co)sinusoidal signal for the modulation. Comparing the frequency modulated waveform to the carrier, it

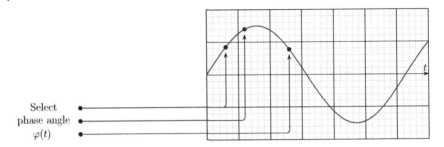

Figure 3.21 Generating a time waveform viewed as stepping through a phase angle.

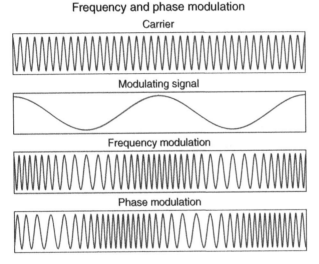

Figure 3.22 Comparison of frequency modulation and phase modulation for a sinusoidal modulation signal. The cosine modulating signal covers a range of amplitudes from positive to negative. Note the phase difference between FM and PM.

should be clear that the frequency of the modulated signal is highest when the modulating signal amplitude is largest, and the frequency is lowest when the modulating amplitude is lowest. The frequency matches that of the carrier when the modulating voltage is zero. This is as it should be: the modulation is just nudging the carrier oscillator up or down. A second example, that of Figure 3.23, helps to clarify this. Now we have a sawtooth waveform that ramps up, and the frequency of the modulated signal ramps up correspondingly. When the modulation suddenly falls back to the starting point, the frequency of the modulated signal returns back to its lowest value.

PM is a little more subtle. In the first case, with a cosine modulation, it appears that the PM is similar in nature but time shifted. As the modulation voltage falls,

Figure 3.23 Comparison of frequency modulation and phase modulation for a sawtooth modulation signal. The sawtooth (ramp) modulating signal starts at zero and ramps up to a maximum value, then falls back to zero. Note the gradual frequency increase in FM, and the abrupt phase change in PM.

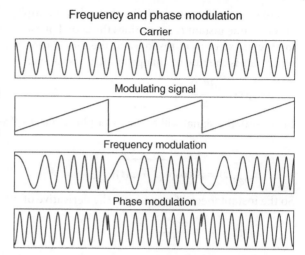

the phase angle added to the ωt argument is decreasing. This manifests itself as an apparent decrease in the frequency of the modulated signal. Similarly, when the modulating voltage is rising, the ever-increasing phase makes the $(\omega t + \varphi)$ term appear as an increasing frequency. Looking at the sawtooth modulation, when the modulating voltage is increasing, the phase angle is slowly increasing up to the maximum. When the modulating voltage drops down to the minimum, there is an abrupt change in the phase of the modulated signal.

3.6.2 FM and PM Analysis

These observations may be explained mathematically by remembering that frequency (radians per second) is the rate change of phase (in radians), or

$$\omega(t) = \frac{d\,\varphi(t)}{dt} \tag{3.62}$$

The phase may not in fact be fixed, but time varying. Thus, FM and PM are closely related and are sometimes grouped under the heading of *angle modulation*. It is helpful to first consider a generalized angle-modulated signal as

$$x_{\text{angle}}(t) = A \cos \theta(t) \tag{3.63}$$

where $\theta(t)$ is the time-varying angle and is comprised of a specific frequency multiplied by time, plus a phase:

$$x_{\text{angle}}(t) = A \cos[\overbrace{\omega_c t + \varphi(t)}^{\theta(t)}] \tag{3.64}$$

For PM, an input signal $m(t)$ would result in

$$x_{\text{PM}}(t) = A \cos[\omega_c t + k_p m(t)] \tag{3.65}$$

where k_p is a constant multiplier that, when multiplied by the modulation $m(t)$ at some time instant t, determines the actual phase angle. So, PM is just changing the phase term in response to the modulation voltage. The instantaneous frequency is the rate of change of the phase angle:

$$\omega_i(t) = \frac{d\theta(t)}{dt} \tag{3.66}$$

In general, the signal will be a carrier plus a phase offset:

$$x_{\text{angle}}(t) = A \cos[\overbrace{\omega_c t + \varphi(t)}^{\theta(t)}] \tag{3.67}$$

So the instantaneous frequency is the derivative of $\theta(t)$ with respect to t,

$$\omega_i(t) = \omega_c + \frac{d\varphi(t)}{dt} \tag{3.68}$$

We could write this as a frequency deviation away from the carrier

$$\omega_i(t) - \omega_c = \frac{d\varphi(t)}{dt} \tag{3.69}$$

This tells us that PM is actually composed of a frequency whose rate of change is proportional to the phase angle, which was in turn due to the modulating voltage. Thus, as the *rate of change* of the modulating voltage increases, the *frequency* of the *phase-modulated* (PM) waveform increases, and vice versa.

FM, on the other hand, varies the *instantaneous* frequency in response to the modulation. The instantaneous frequency is comprised of the fixed carrier, plus an amount proportional to the modulating voltage $m(t)$:

$$\omega_i(t) = \omega_c + k_f m(t) \tag{3.70}$$

Note that the modulating voltage could be positive or negative at any instant. Now, the angle $\theta(t)$ is the cumulative sum, or integral, of the instantaneous frequencies, so

$$\begin{aligned} \theta(t) &= \int \omega_i(t) dt \\ &= \int [\omega_c + k_f m(t)] \, dt \\ &= \omega_c t + k_f \int m(t) dt \end{aligned} \tag{3.71}$$

So the final equation for an FM signal for modulation $m(t)$ is

$$x_{\text{FM}}(t) = A \cos\left[\omega_c t + k_f \int_0^t x_m(\tau) d\tau\right] \tag{3.72}$$

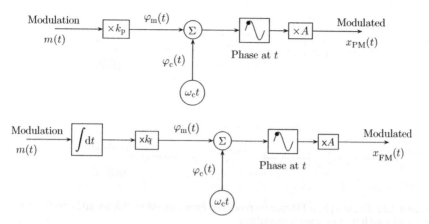

Figure 3.24 Phase modulation conceptual diagram (top). The phase angle is determined from the "prototype" sine wave, with the specific point (or phase) being determined by the current oscillator position added to the scaled modulation signal. Frequency modulation (bottom) is similar, but with the phase angle determined not by the instantaneous value but the cumulative value of the input.

where τ is a dummy variable of integration.[4] The variable τ disappears when we evaluate the integral using the limits of 0 and t, leaving a function of t. In other words, we may state that the phase angle of an FM signal depends on the time-accumulated modulating voltage.

3.6.3 Generation of FM and PM Signals

To see how PM may be generated, consider Figure 3.24. Here we have the carrier frequency multiplied by time to arrive at the current point in terms of phase angle. This is then added to the scaled modulation signal and used as an input to the sinusoidal function. The phase angle thus generated is used to index the amplitude and that instant in the sine wave table of values, as indicated by the dot in the waveform. Of course, this is just a mathematical function, $\sin\theta$, where $\theta(t)$ is actually a function of time. FM, in a similar way, is also shown in Figure 3.24. Here we have the carrier frequency multiplied by time to arrive at the current point in terms of phase angle. Now, however, the modulation amplitude is cumulatively summed, as indicated by the integral box, to arrive at a phase angle offset. The phase angle offset is thus continually increasing and should therefore be thought of as generating a particular frequency. This is then added to the phase as determined by the carrier, and the sinusoidal function table is again used to determine the amplitude at that point.

Thus, for FM, the *cumulative* modulation voltage gives rise to a *frequency that is offset with respect to the carrier*. If the modulation voltage is positive, the

---

4 Note that some authors use $2\pi k_f$ instead of k_f for the constant multiplier.

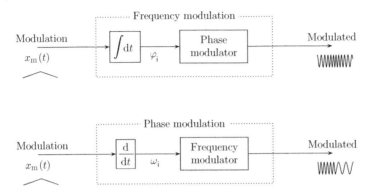

Figure 3.25 Showing how FM may be produced from a phase modulator and how PM may be produced from a frequency modulator.

cumulative phase advance will be a successively increasing function, giving rise to a frequency increase over and above the carrier frequency. If the modulation voltage is held constant but is negative, the cumulative phase advance will be a decreasing function. Since this is added to the carrier phase angle (due to the ωt term), the net effect will be a successively reduced phase offset, which appears as a reduced ω – that is, a reduced frequency. It is important to remember that this cumulative function does not increase forever, because the modulating waveform $m(t)$ typically goes up and down, with a long-term average of zero.

Thus, the link between FM and PM is one of integration of the signal for FM. As shown in Figure 3.25, we can generate FM using a PM modulator, by first integrating the signal. Conversely, we can generate PM using an FM modulator by first differentiating the signal. The modulation voltage input can of course be anything; a linear ramp is shown in the diagram, because it is instructive to compare what happens in each case. The waveforms of Figure 3.26 show the resultant waveforms for two types of modulation input: a ramp up/down and a positive/negative step. The choice of these two modulating waveforms is significant. In going from the ramp to the step (left to right in the diagram), we differentiate the modulation. Conversely, in going from right to left, we integrate the modulation. Note how the PM of the ramp is identical to the FM of the step waveform.

3.6.4 The Spectrum of Frequency Modulation

FM effectively varies the frequency of the carrier, and does not keep the same carrier as embedded in AM. So as might be expected, there are varying frequency components in the frequency modulated signal. Intuitively, the spectral components would be harmonically related to the modulation frequency ω_m

Relationship between frequency and phase modulation

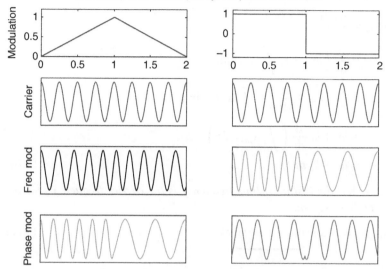

Figure 3.26 FM and PM modulation waveform comparison. In going from ramp to the step (left to right), we differentiate the modulation; in going from right to left, we integrate the modulation. Phase modulation of the ramp is identical to the frequency modulation of the step waveform.

and would be centered on the carrier frequency ω_c. This is true, but whether those frequency components are actually present or not depends on the frequency deviation, as will be shown. In fact, under some circumstances, it is possible to make the carrier disappear altogether and leave only the sidebands.

The FM equation that we have developed is

$$x_{FM}(t) = A \cos \left(\omega_c t + k_f \int_0^t x_m(\tau)\mathrm{d}\tau \right) \tag{3.73}$$

We can look at it in terms of phase angle $\theta(t)$. The phase angle is

$$\theta(t) = \omega_c t + k_f \int_0^t x_m(\tau)\mathrm{d}\tau \tag{3.74}$$

Since the instantaneous frequency $\omega_i(t)$ is the rate change of phase

$$\omega_i(t) = \frac{\mathrm{d}\theta(t)}{\mathrm{d}t} \tag{3.75}$$

the instantaneous frequency may be expanded as

$$
\begin{aligned}
\omega_i(t) &= \frac{d\theta(t)}{dt} \\
&= \frac{d}{dt}\left(\omega_c t + k_f \int_0^t x_m(\tau)d\tau\right) \\
&= \frac{d}{dt}(\omega_c t) + \frac{d}{dt}\left(k_f \int_0^t x_m(\tau)d\tau\right) \\
&= \omega_c + k_f \frac{d}{dt}\left(\int_0^t x_m(\tau)d\tau\right) \\
&= \omega_c + k_f\, m(t) \tag{3.76}
\end{aligned}
$$

This shows that *the instantaneous frequency is the carrier frequency plus (or minus) an offset frequency*:

$$
x_{FM}(t) = A\cos\{\underbrace{[\omega_c + k_f m(t)]\, t}\} \tag{3.77}
$$
$$
\omega_i(t)
$$

For a test signal of $m(t) = A_m \cos\omega_m t$, the FM signal according to the above definitions would be

$$
\begin{aligned}
x_{FM}(t) &= A\cos\left(\omega_c t + k_f \int_0^t x_m(\tau)d\tau\right) \\
&= A\cos\left(\omega_c t + \frac{k_f A_m}{\omega_m}\sin\omega_m t\right) \tag{3.78}
\end{aligned}
$$

The *change* in frequency is the multiplier $\Delta\omega = k_f A_m$. That is, the deviation depends on the constant k_f, and the amplitude of the modulating signal A_m. Note that we've assumed that ω_m, the modulating frequency, is constant. This is valid because at present, we are considering a single-frequency test signal only. By defining the FM modulation index β as

$$
\beta = \frac{k_f A_m}{\omega_m} \tag{3.79}
$$

the FM signal for pure-sinusoidal modulation simplifies to

$$
x_{FM}(t) = A\cos(\omega_c t + \beta\sin\omega_m t) \tag{3.80}
$$

The modulation index β may also be written as the relative change in frequency:

$$
\beta = \frac{\Delta\omega}{\omega_m} \tag{3.81}
$$

The concept of FM modulation index defined in this way parallels that of the AM modulation index μ. Note that if we had analyzed PM instead, we would have $k_p A_m$. Again, this modulation index is valid only for a pure tone modulating signal input. Furthermore, the FM modulation index may also be written

$$\beta = \frac{\Delta\omega}{\omega_m} = \frac{2\pi\Delta f}{2\pi f_m} = \frac{\Delta f}{f_m} \qquad (3.82)$$

Thus β is the frequency deviation ratio (DR) for single-tone modulation, whether in Hz or rad s^{-1}.

The modulation index is also called the deviation ratio (DR); however, the term DR applies more generally, not just to a specific input type. For commercial FM broadcasting, the deviation is $\Delta f = 75$ kHz with $f_{max} = 15$ kHz, so the DR=75/15 = 5. A DR of one or greater is termed wideband FM, whereas a DR of less than one is called narrowband FM.

We can now write the FM signal for pure sinusoidal cosine modulation in a simpler form

$$x_{FM}(t) = A\cos(\omega_c t + \beta\sin\omega_m t) \qquad (3.83)$$

When FM was first invented, there was considerable debate as to its merits over AM, and it was shown in Carson (1922) that an FM signal with sinusoidal input and modulation index β could be written as a series of the form

$$x_{FM}(t) = A\sum_{n=-\infty}^{n=+\infty} J_n(\beta)\cos(\omega_c + n\,\omega_m)t \qquad (3.84)$$

where $J_n(\beta)$ is the mathematical Bessel function. We can write out the first few terms to see the pattern in the series. Note that n is positive and negative and starts with zero, ± 1, ± 2, and continues on for higher harmonics. The series terms are

$$\begin{aligned}
x_{FM}(t) = {} & J_0(\beta)A\cos\omega_c t \\
& + J_1(\beta)A\cos(\omega_c + \omega_m)t + J_{-1}(\beta)A\cos(\omega_c - \omega_m)t \\
& + J_2(\beta)A\cos(\omega_c + 2\omega_m)t + J_{-2}(\beta)A\cos(\omega_c - 2\omega_m)t \\
& + J_3(\beta)A\cos(\omega_c + 3\omega_m)t + J_{-3}(\beta)A\cos(\omega_c - 3\omega_m)t \\
& + \cdots
\end{aligned} \qquad (3.85)$$

Each ω_m is multiplied by n, and so with $n = 0$ we have the carrier frequency ω_c alone, since the actual frequency is $\omega_c + n\omega_m$. With $n = 1$ we have the component at ω_m above ω_c, or $\omega_c + \omega_m$; with $n = -1$ we have a term at $\omega_c - \omega_m$. Each term thus represents a sine wave at frequency $\omega_c + n\omega_m$, whose amplitude is weighted by the corresponding Bessel J coefficient. The subscript n refers to the component number, and the argument β refers to the modulation index for the particular situation that we are given with the test signal. So, the weighting of each cosine term is $J_n(\beta)$.

This may be seen graphically in Figure 3.27. On the left is the time-domain waveform, with the modulation parameters as given. Note that these are for

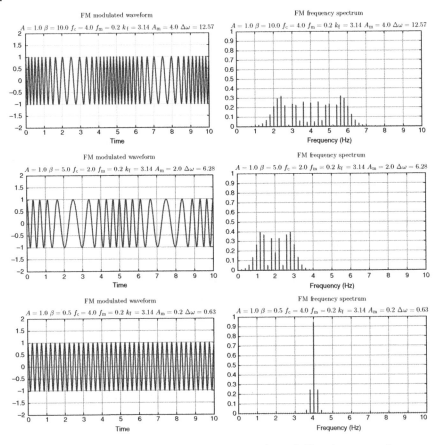

Figure 3.27 Frequency modulation showing time waveforms (left) and corresponding frequency spectra (right).

a test signal case of a single-tone sinusoidal modulation. To the right of each time waveform, the corresponding frequency components are shown. These frequency plots are given in terms of Hz frequency, so it is necessary to perform the calculation $\omega = 2\pi f$ to translate to radian frequency. In the first case, the carrier frequency is $f_c = 4$ Hz, and we can see the frequency components are spread either side of this, corresponding to $f_c + nf_m$, where $f_m = 0.2$ Hz in this case. The amplitude of each frequency component is determined by the corresponding Bessel coefficient $J_n(\beta)$.

The derivation of this is explained in Section 3.6.5, with the end result shown in Table 3.2. Note that the table only shows positive n values, since the magnitude spectrum is symmetrical. Normally, we are only interested in the amount of each harmonic, not whether it is sine or negative sine (or cosine, for

Table 3.2 A Bessel table for determining sideband amplitudes in frequency modulation.

| | | | | | Bessel functions $J_n(\beta)$ | | | | | |
| | | | | | Modulation index β | | | | | |
β	J_0	J_1	J_2	J_3	J_4	J_5	J_6	J_7	J_8	J_9	J_{10}
0	1.00	—	—	—	—	—	—	—	—	—	—
0.25	0.98	0.12	—	—	—	—	—	—	—	—	—
0.5	0.94	0.24	0.03	—	—	—	—	—	—	—	—
1	0.77	0.44	0.12	0.02	—	—	—	—	—	—	—
1.5	0.51	0.56	0.23	0.06	0.01	—	—	—	—	—	—
2	0.22	0.58	0.35	0.13	0.03	—	—	—	—	—	—
2.4	0.00	0.52	0.43	0.20	0.06	0.02	—	—	—	—	—
2.5	−0.05	0.50	0.45	0.22	0.07	0.02	—	—	—	—	—
3	−0.26	0.34	0.49	0.31	0.13	0.04	0.01	—	—	—	—
4	−0.40	−0.07	0.36	0.43	0.28	0.13	0.05	0.01	—	—	—
5	−0.18	−0.33	0.05	0.36	0.39	0.26	0.13	0.05	0.02	—	—
6	0.15	−0.28	−0.24	0.11	0.36	0.36	0.25	0.13	0.06	0.02	—
7	0.30	0.0	−0.30	−0.17	0.16	0.35	0.34	0.23	0.13	0.06	0.02
8	0.17	0.23	−0.11	−0.29	−0.10	0.19	0.34	0.32	0.22	0.13	0.06
9	−0.09	0.24	0.15	−0.18	−0.26	−0.06	0.20	0.33	0.31	0.21	0.13
10	−0.25	0.04	0.26	0.06	−0.22	−0.23	−0.01	0.22	0.32	0.29	0.21
12	0.05	−0.22	−0.08	0.19	0.18	−0.07	−0.24	−0.17	0.05	0.23	0.30
15	−0.01	0.20	0.04	−0.19	−0.12	0.13	0.21	0.03	−0.17	−0.22	−0.09

that matter). The case of $J_0(0)$ is as we would expect: for a modulation index of $\beta = 0$, we actually have no modulation, and the only frequency component present must, by definition, be the carrier. This is the $n = 0$ component, and there are no others. Small values (< 0.01) of the J coefficient are shown as a dash and may be considered to be negligible.

The Bessel coefficients may be calculated using several methods, one of which is to solve

$$J_n(\beta) = \frac{1}{\pi} \int_0^\pi \cos(\beta \sin t - nt)\mathrm{d}t \qquad (3.86)$$

MATLAB incorporates the function besselj() to determine these coefficients. To calculate $J_n(\beta)$ for $n = 3$ and $\beta = 5$, we can use this function directly.

```
n = 3;
beta = 5;
besselj(n, beta)
ans =
    0.3648
```

We can use MATLAB or Table 3.2 to validate the previous spectral plots. For example, the third plot given has $\beta = 0.5$, so we use the row corresponding to this β value to read off the J values. These correspond to the harmonic amplitudes (0.94, 0.24, and 0.03). A curious case arises when $\beta = 2.4$ – there is no carrier component. That is, $J_0(2.4) = 0$. That is quite different to AM, where the carrier is always present.

When using these values, it is important to understand that the sideband components are symmetrical either side of the carrier (or at least, where the carrier should be). Another important property of the Bessel functions in this context is the fact that

$$\sum_{n=-\infty}^{\infty} J_n^2(\beta) = 1 \tag{3.87}$$

This means that the power sum is normalized to unity. So if the relative power is increased or decreased, it is just a matter of scaling the Bessel values accordingly. We can demonstrate the symmetry and the above summation of J^2 as follows:

```
n = -6:6;
beta = 2.5;
bc = besselj(n, beta)
    0.0042   -0.0195   0.0738   -0.2166   0.4461   -0.4971
   -0.0484
    0.4971    0.4461   0.2166    0.0738   0.0195    0.0042
sum(bc.^2)
ans =
    1.0000
```

Figure 3.28 shows the spectrum analyzer plot of an FM signal with the following parameters:

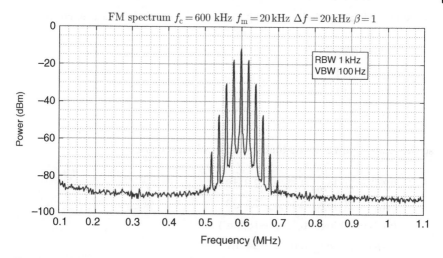

Figure 3.28 Measured spectrum for FM, $\beta = 1$.

Parameter name	Symbol	Value
Carrier frequency	f_c	600 kHz
Carrier amplitude	A_c	200 mVpp
Modulating frequency	f_m	20 kHz
Frequency deviation	Δf	20 kHz

From these values, the modulation index is calculated as

$$\frac{\Delta f}{f_m} = 1$$

The unmodulated carrier RMS amplitude is

$$V_{rms} = \frac{V_{pp}/2}{\sqrt{2}}$$

The unmodulated carrier power is

$$10\log_{10}\left(\frac{V_{rms}^2/50}{1 \times 10^{-3}}\right) = -10 \text{ dBm}$$

The sidebands are scaled according to the Bessel coefficient values. For $\beta = 1$,

$$J_0(\beta) = 0.77$$
$$J_1(\beta) = 0.44$$
$$J_2(\beta) = 0.12$$
$$J_3(\beta) = 0.02$$

so the relative power levels that should be observed are

$$P_0(\beta) = 20 \log_{10} 0.77 = -2.3 \text{ dB}$$
$$P_1(\beta) = 20 \log_{10} 0.44 = -7.1 \text{ dB}$$
$$P_2(\beta) = 20 \log_{10} 0.12 = -18.8 \text{ dB}$$
$$P_3(\beta) = 20 \log_{10} 0.02 = -34 \text{ dB}$$

These are relative to the unmodulated carrier power, which was found to be -10 dBm. The measured figures on the graph are -11.9, -17.9, -30.5, and -47 dBm. The first few components are in good agreement, but as the power diminishes, it becomes increasingly difficult to measure the power accurately. After all, -47 dBm is an exceedingly small value (about 20 nW).

Figure 3.29 shows the spectrum analyzer plot of an FM signal with the following parameters:

Parameter name	Symbol	Value
Carrier frequency	f_c	600 kHz
Carrier amplitude	A_c	200 mVpp
Modulating frequency	f_m	20 kHz
Frequency deviation	Δf	48 kHz

The only change is for Δf, so that $\Delta f / f_m = 2.4$, and for $\beta = 2.4$ the corresponding Bessel values are

$$J_0(\beta) = 0$$
$$J_1(\beta) = 0.52$$
$$J_2(\beta) = 0.43$$
$$J_3(\beta) = 0.20$$

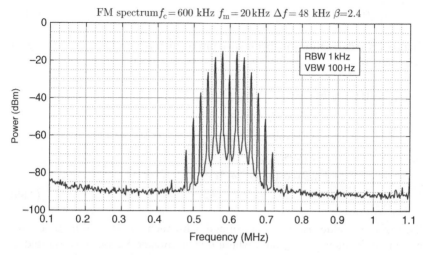

Figure 3.29 Measured spectrum for FM, $\beta = 2.4$.

so the relative power levels that should be observed are

$$P_0(\beta) = 20 \log_{10} 0.0 = \text{undefined}$$
$$P_1(\beta) = 20 \log_{10} 0.52 = -5.7 \text{ dB}$$
$$P_2(\beta) = 20 \log_{10} 0.43 = -7.3 \text{ dB}$$
$$P_3(\beta) = 20 \log_{10} 0.20 = -14 \text{ dB}$$

These are again relative to the unmodulated carrier power, which was found to be -10 dBm. The measured figures on the graph are $-27, -15, -18$, and -26 dBm. Once again, there is in general good agreement, but less so as the power diminishes.

3.6.5 Why Do the Bessel Coefficients Give the Spectrum of FM?

As illustrated in Section 3.6.4, the Bessel coefficients give the magnitude of the FM spectrum. It is instructive to learn how this theory comes about, and in doing so, a useful general principle involving multiplication of waveforms is revealed.

Recall that the FM spectrum consists of a component (possibly zero) at the carrier frequency, and other components spaced at integral multiples of the modulation frequency, away from the carrier. This is similar to (but not quite the same as) the Fourier series (Section 2.3.1), where the fundamental and harmonics at multiples of the fundamental frequency are present. We can rewrite the single-tone FM modulation signal with amplitude $A = 1$ (since it is just a scaling constant) as

$$x_{FM}(t) = \cos(\omega_c t + \beta \sin \omega_m t) \tag{3.88}$$

The goal is to determine the spectrum component magnitudes if the same signal is written as

$$
\begin{aligned}
x_{FM}(t) = &\, J_0(\beta) \cos \omega_c t \\
&+ J_1(\beta) \cos(\omega_c + \omega_m)t \\
&+ J_{-1}(\beta) \cos(\omega_c - \omega_m)t \\
&+ J_2(\beta) \cos(\omega_c + 2\omega_m)t \\
&+ J_{-2}(\beta) \cos(\omega_c - 2\omega_m)t \\
&+ \cdots
\end{aligned} \tag{3.89}
$$

Each of the J values is to be determined, with each corresponding to a component at frequency $\omega_c \pm k\omega_m$, where k is an integer. Suppose we wanted to determine the component $J_2(\beta)$, which corresponds to the frequency $\omega_c + 2\omega_m$. This is one specific case, but it will reveal a method that can be used for all the components. We multiply each side of the expansion by the sinusoidal term we wish to extract, in this case $\cos(\omega_c + 2\omega_m)t$, and then integrate the result over one period τ_m of the modulation waveform.

$$
\begin{aligned}
\int_0^{\tau_m} x_{FM}(t) \times \cos(\omega_c + 2\omega_m)t\, dt \;=\; &\int_0^{\tau_m} \cos(\omega_c t + \beta \sin \omega_m t) \times \cos(\omega_c + 2\omega_m)t\, dt \\
=\; & J_0(\beta) \int_0^{\tau_m} \cos \omega_c t \times \cos(\omega_c + 2\omega_m)t\, dt \quad \longrightarrow 0 \\
& + J_1(\beta) \int_0^{\tau_m} \cos(\omega_c + \omega_m)t \times \cos(\omega_c + 2\omega_m)t\, dt \quad \longrightarrow 0 \\
& + J_{-1}(\beta) \int_0^{\tau_m} \cos(\omega_c - \omega_m)t \times \cos(\omega_c + 2\omega_m)t\, dt \quad \longrightarrow 0 \\
& \qquad\qquad\qquad\qquad\qquad \text{not zero} \\
& + J_2(\beta) \int_0^{\tau_m} \cos(\omega_c + 2\omega_m)t \times \cos(\omega_c + 2\omega_m)t\, dt \\
& + J_{-2}(\beta) \int_0^{\tau_m} \cos(\omega_c - 2\omega_m)t \times \cos(\omega_c + 2\omega_m)t\, dt \quad \longrightarrow 0 \\
& + \cdots
\end{aligned} \tag{3.90}
$$

The result of this helps us in extracting the component required, since all but one of the integrations will be shown to equal zero. A numerical demonstration of this is shown in the code below, for the case of harmonic component $n = 2$ and modulation index $\beta = 10$.

```
% determining the integrals for computing the FM spectrum

N = 1000;
beta = 10;

n = 2;
taum = 1;
wm = 2*pi/taum;
wc = 10*wm;

t = linspace(0, taum, N);
dt = t(2) - t(1);

% the FM signal
xfm = cos(wc*t + beta*sin(wm*t));

% the modulation
xm = cos(wm*t);

% carrier signal,
% carrier plus modulation frequency,
% carrier plus twice modulation frequency

xc = cos(wc*t);
xh1 = cos(wc*t + wm*t);
xh2 = cos(wc*t + 2*wm*t);

Integral11 = dt*sum(xh1.*xh1);
Integral12 = dt*sum(xh1.*xh2);

Integral21 = dt*sum(xh2.*xh1);
Integral22 = dt*sum(xh2.*xh2);

fprintf(1, 'Product-Integral terms:\n');
fprintf(1, 'Int 11 = %f    Int 12 = %f    Int 21 = %f
        % Int 22 = %f\n', ...
    Integral11, Integral21, Integral21, Integral22);
```

The result shows that terms multiplied by other terms and integrated are zero, and the only terms that remain will be when a component is multiplied by itself.

```
Product-Integral terms:
Int 11 = 0.5010    Int 12 = 0.0010    Int 21 = 0.0010
                                      Int 21 = 0.5010
```

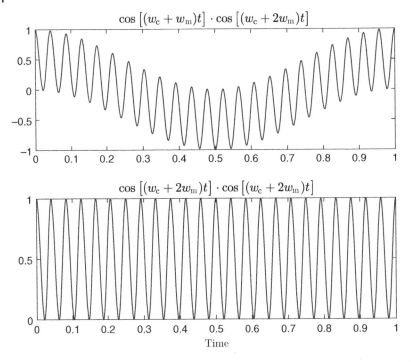

Figure 3.30 Harmonic multiplications for deriving the FM spectrum. The upper panel shows two different frequencies multiplied, with an average of zero. The lower panel shows two identical frequencies multiplied, with an average of 0.5.

This is illustrated in Figure 3.30, where it may be observed that the average of the upper function (different frequency) is zero, whereas the lower function (same frequency) is not zero.

Next, we can expand the left-hand side of the $x_{FM}(t)$ Equation (3.88) using the $\cos \alpha \cos \beta$ expansion:

$$\int_0^{\tau_m} \cos(\omega_c t + \beta \sin \omega_m t) \times \cos(\omega_c + 2\omega_m)t \; dt$$

$$\overbrace{= \frac{1}{2} \int_0^{\tau_m} \cos(2\omega_c t + \beta \sin \omega_m t + 2\omega_m t) \; dt}^{\text{Term 1=0}}$$

Term 2: forms Bessel integral for $n=2$

$$\overbrace{+ \frac{1}{2} \int_0^{\tau_m} \cos(\beta \sin \omega_m t - 2\omega_m t) \; dt} \qquad (3.91)$$

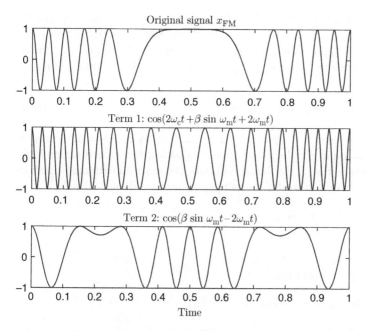

Figure 3.31 The expansion of the FM equation (top) yields two terms: term 1 and term 2. By symmetry, it may be observed that term 1 has an average of zero, whereas term 2 does not.

These two terms are shown in Figure 3.31. The integral for term 1 is zero. The integral for term 2 is evidently of a similar form to the Bessel function $J_2(\beta)$. These may be calculated using the previous code as a starting point and adding

```
fprintf(1, 'FM Expansion terms:\n');
term1 = cos(2*wc*t + beta*sin(wm*t) + wm*t);
term2 = cos(beta*sin(wm*t) - n*wm*t);
IntegralTerm1 = dt*sum(term1);
IntegralTerm2 = dt*sum(term2);
fprintf(1, 'Term 1 = %f   Term 2 = %f\n', IntegralTerm1,
                                           IntegralTerm2);
```

that results in

```
FM Expansion terms:
Term 1 = 0.000998   Term 2 = 0.255631
```

Term 1 is practically zero (the small value is due to rounding errors). Term 2, however, has a finite and nonnegligible value. Once again, this may be seen with reference to the function plots, shown in Figure 3.31.

The second term is where the Bessel function $J_2(\beta)$ comes into play. Finally, equating the simplified version of Equation (3.91) with the constant resulting from Equation (3.90), we have a method of determining the frequency component magnitude using the Bessel integral. For the general case, this is as shown before

$$J_n(\beta) = \frac{1}{\pi} \int_0^\pi \cos(\beta \sin t - nt) \, dt \tag{3.92}$$

Generalizing the result for all possible harmonics, it may be seen that the amplitude of each harmonic may be found by just evaluating the Bessel function $J_n(\beta)$. A numerical evaluation of the Bessel function, as compared with the (superior) built-in function, demonstrates the validity of these results:

```
t = linspace(0, pi, N);
dt = t(2) - t(1);

Jarg = cos(beta*sin(t) - n*t);
Jcalc = (1/pi)*sum(Jarg*dt);

JMatlab = besselj(n, beta);
disp('Compare Bessel evaluations');
fprintf(1, 'Calculated %f, MATLAB built-in %f\n', Jcalc,
                                                  JMatlab)
```

The corresponding output for $n = 2$ and $\beta = 10$ is

```
Compare Bessel evaluations
Calculated 0.255631, MATLAB built-in 0.254630
```

Of course, the above code may be modified for other harmonics by changing n to another integer and for another modulation index by changing β. Although the derivation is involved, it provides a very useful result: that we can determine each component n for a given modulation index β by simply evaluating the Bessel function $J_n(\beta)$.

3.6.6 FM Demodulation

Demodulation of FM requires recovery of the original modulating signal $m(t)$, given the frequency modulated signal $x_{FM}(t)$. PM is similar, so if we can solve either FM demodulation or PM demodulation, we can solve the other. Many methods for FM demodulation have been employed, including digital means (Farrell et al., 2005).

Recall that FM is really just changing the *instantaneous frequency* of the signal about the carrier in response to the modulation amplitude. Thus, tracking the received signal frequency is conceptually what is needed. This tracking is a little difficult to achieve in practice, but it can be done using methods to be outlined in Section 3.7.

To address the problem, imagine a frequency-selective filter that produced a higher average output for higher frequencies and a lower average output for lower frequencies. In other words, a higher *frequency* is converted back into an increase in voltage (and the converse – a lower frequency to a lower voltage). To set out on a solution path, it is reasonable to start with the FM signal equation for single-tone modulation:

$$x_{FM}(t) = A \cos(\omega_c t + \beta \sin \omega_m t) \tag{3.93}$$

The problem is to recover the original modulation signal $m(t)$. In this simplified single-tone case, we should be able to recover the sinusoidal signal $m(t) = A_m \cos \omega_m t$.

What would happen if we took the derivative of the FM signal described by Equation (3.93)? This might not be an obvious step, but it will lead the way to creation of a frequency-selective discriminator. Setting $u = \omega_c t + \beta \sin \omega_m t$ and using the chain rule of calculus $dx/dt = (dx/du)(du/dt)$, we arrive at

$$\frac{d\, x_{FM}(t)}{dt} = -A \sin(\omega_c t + \beta \omega_m \sin \omega_m t) \times (\omega_c + \beta \omega_m \cos \omega_m t) \tag{3.94}$$

$$= -A\omega_c \overbrace{\sin(\omega_c t + \beta \omega_m \sin \omega_m t)}^{\text{carrier frequency around } \omega_c}$$

$$- \underbrace{(A\beta \omega_m \cos \omega_m t)}_{\text{constant} \times m(t)} \underbrace{\sin(\omega_c t + \beta \omega_m \sin \omega_m t)}_{\text{carrier frequency around } \omega_c} \tag{3.95}$$

This looks formidable, but the various terms of this equation are not dissimilar to an AM signal: it is really just the carrier, plus the carrier times the modulation. The "carrier" in this case would ideally be $\sin \omega_c t$, but it appears in the above as $\sin(\omega_c t + \beta \omega_m \sin \omega_m t)$. Since $\omega_c \gg \omega_m$, we can safely say that, as an approximation, we can ignore the extra part involving ω_m wherever it occurs in conjunction with ω_c. Also, using $\beta = k_f A_m / \omega_m$, the term $A\beta \omega_m \cos \omega_m t$ may be simplified to $Ak_f A_m \cos \omega_m t$. So as an approximation,

$$\frac{d\, x_{FM}(t)}{dt} \approx - \overbrace{A\omega_c\, \sin \omega_c t}^{\text{Scaled carrier at } \omega_c} - \overbrace{(Ak_f \sin \omega_c t)}^{\text{Scaled carrier at } \omega_c} \overbrace{(A_m \cos \omega_m t)}^{m(t)} \tag{3.96}$$

The result is, in effect, an *amplitude-modulated* signal. In other words, the FM signal has been converted into an AM one, and we know how to demodulate that already. One significant disadvantage to keep in mind, though, is that differentiating a signal (which is really finding its rate change) is usually not a good idea. This is because noise will be present in any real system, and the amplified rate change of noise will occur along with the desired signal, and thus more noise may be introduced.

To confirm our understanding, the following MATLAB code shows how to create an FM waveform and then finds the rate of change. This is simply the difference between successive calculated points on the waveform. Figure 3.32 shows the results of running this code.

```
% waveform parameters
N = 2000;
Tmax = 20;
dt = Tmax/(N-1);
t = 0:dt:Tmax;
fs = 1/dt;

% carrier
fc = 3;
wc = 2*pi*fc;
xc = cos(wc*t);

% modulating signal
fm = 0.2;
wm = 2*pi*fm;
Am = 1;
xm = Am*cos(wm*t);

% FM modulation parameters
A = 2;
kf = 10;

% integral of xm
xmi = cumsum(xm)*dt;

% combine with carrier to produce FM
xfm = A*cos(wc*t + kf*xmi);

% first stage of FM demodulation - differentiation to
% produce AM
dxfm = diff(xfm)/dt;

% plot the signals
subplot(4,1,1); plot(xm);
subplot(4,1,2); plot(xc);
subplot(4,1,3); plot(xfm);
subplot(4,1,4); plot(dxfm);
```

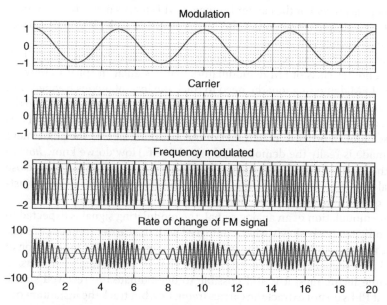

Figure 3.32 Differentiating an FM signal reveals another signal that is amplitude modulated. The timescale is arbitrary, depending on the frequencies of the waveforms concerned. The signals are xc (carrier), xm (modulation), xfm (modulated), and finally the rate-of-change dxfm.

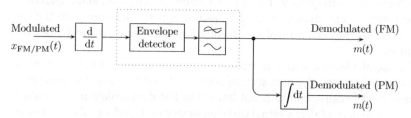

Figure 3.33 Asynchronous FM demodulation. The dotted part is effectively an AM demodulator. A preceding section (not shown) would limit the amplitude of the incoming signal, so as to reduce any spurious noise amplitude spikes.

The integral of the frequency changes gives us the phase change. So, if the signal is PM rather than FM, we can add an integrator to the output of the FM demodulator. The approach is shown diagrammatically in Figure 3.33.

What if we had the carrier signal available? AM incorporates the carrier in the transmitted signal, so recovery of the carrier is not excessively difficult. FM does not explicitly incorporate the carrier into the transmitted signal. In fact, the carrier may be zero at various times, as shown by the previous analysis using the Bessel functions. When $\beta = 2.4$, we see that $J_0(\beta) = 0$, indicating that no component is present at the carrier frequency. This makes demodulation more difficult, and simple extraction of the carrier is not feasible. It is necessary to track the frequency somehow.

A frequency-tracking device, commonly called a Phase-Locked Loop (PLL), leads to another class of FM demodulators that is similar in spirit to how we described FM demodulation: the need to track the instantaneous frequency of the incoming signal. The key difference is that a local signal is generated, and the system adjusts, in real time, the frequency (or more precisely, the phase) of this local oscillator in order to match that of the incoming signal. The adjustment that is made is really the demodulated signal itself. How do we know *how* to adjust the local oscillator? A feedback loop is necessary to compare how close the local oscillator is to the incoming signal. The closeness gives us the adjustment required and also the direction of adjustment (up or down).

For the modulation of an analog signal, the modulating signal is expected to change only relatively slowly when compared with the carrier frequency. Thus, it ought to be possible to track the instantaneous frequency. For modulation of a digital data sequence, the frequency would change more rapidly – but using these same techniques, it is still possible to demodulate the received signal. Note that PLLs do not extract the carrier frequency, but track the instantaneous frequency. As we have found, phase is a concept closely related to frequency, and shortly we will demonstrate how tracking the phase, rather than the frequency, achieves our goal – hence the P in PLL.

To summarize, there two main classes of FM demodulator: discriminators, which transform frequency changes into amplitude changes; and coherent or synchronous methods, which need a signal that is synchronous with the carrier. Tracking the carrier involves a feedback loop to follow the instantaneous frequency.

The use of a feedback loop to track the carrier frequency is quite important in its own right, and the next section is devoted to developing the concept. It is important because its use is not limited to FM demodulation – when it is necessary to demodulate a signal carrying binary or digital signals, it becomes essential to know the carrier frequency and phase. This is so that the receiver logic can ascertain the correct point at which to make a decision as to whether a particular amplitude, at a certain time, represents a binary 1 or 0.

3.7 Phase Tracking and Synchronization

A large class of demodulation methods rely on having the carrier waveform available to the receiver. These are termed *synchronous demodulation* approaches, and unless the carrier is explicitly transmitted (as in AM), then it must be regenerated somehow at the receiver. These synchronous

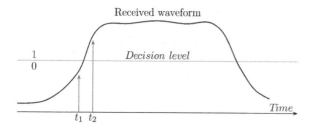

Figure 3.34 Determining the correct time to sample a waveform is critical. In this example, a higher value is interpreted as a binary 1, and a lower value as a binary 0. As illustrated, incorrect timing could lead to the wrong decision and hence an incorrect binary value.

demodulation methods generally lead to better quality of the reproduced signal $m(t)$ when the modulating signal is analog (such as voice or music).

Even more importantly, when the modulating signal is digital (a binary bit stream), then it becomes absolutely essential to have precise timing information available at the receiver, so that the correct bit value may be recovered. That is, if the bit 0/1 decision is made at the wrong time, then the wrong bit value may be assumed at the receiver, as illustrated in Figure 3.34. The principal means for achieving this synchronization from the received waveform alone is known as a PLL. There are many variations on the basic PLL concept and the blocks used within the PLL. This section aims to explain the basic PLL concept, and a variation known as a Costas loop that has found widespread use in digital or binary demodulation.

Consider a local oscillator, which is to be synchronized with the transmitted signal oscillator. Only a modulated version of the waveform is available at the receiver, but consider for the moment the simpler problem where the received signal is a pure sinusoid. The local oscillator may be oscillating at approximately (but not precisely) the correct frequency. Additionally, because of component tolerances – and perhaps the movement of either the transmitter or receiver – the timing or phase may not be exact. We also do not know the amplitude, but as it turns out this is relatively unimportant. The aim is to "tune" the phase of the oscillator to that of the received signal. If this is done continually, the frequency is then implicitly tracked, since the frequency is the rate of change of phase. That is, it may be necessary to adjust the phase increments so as to achieve a signal that is earlier or later with respect to the received signal.

We can do this using a local oscillator that is "nudged" by the incoming signal, so as to attain the correct synchronization. Consider a child's swing in a playground. If it is swinging back and forth, we can increase the amplitude of the swing by pushing at just the right time. If we want to make the swing go faster or slower (that is, change the oscillation frequency), then we can do so by applying pushes a little earlier or later to the peak of the existing cycle. We can't change the frequency instantaneously, but after a few cycles we can move the

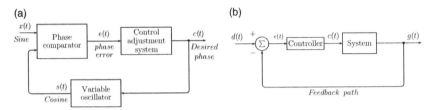

Figure 3.35 A phase-locked loop, which may be considered as a type of control system. The phase comparator determines how close the waveforms are and guides the oscillator via the controller to either increase or decrease its frequency so as to more closely align the timing (or phase) with the incoming waveform. (a) The phase-locked loop (b) A generic control system.

oscillation to a frequency of our choosing (faster or slower). If the swing itself is the local oscillator, the incoming wave that we want to synchronize with corresponds to the application of pushes to the swing. The timing has to be just right, since the natural resonant frequency will depend on the mass of the child and the length of the ropes attached to the swing.

Figure 3.35a shows how we can do this using a feedback loop. The variable oscillator is set to a particular frequency, and this is compared with the incoming waveform. Any difference is then used as an error signal, indicating whether the frequency needs to be tuned up or down. Such a feedback control system is used in many engineering systems, and a generic block diagram is depicted in Figure 3.35b. Here, we imagine that the phase error is generated by a simple subtraction: the desired phase minus the actual phase. The "system" is really the oscillator itself, and the job of the "controller" is to quickly adjust the oscillator drive signal up or down. This needs to be done as rapidly as possible, but with no error in the longer-term when the incoming signal is stable. That is to say the error signal $e(t)$ ought to be zero in the steady-state operating condition.

Thus, there are three main elements: a phase detector, which generates an error signal according to the averaged phase difference of two waveforms; a variable oscillator whose frequency can be controlled; and a control adjustment system that acts on the error to effect the desired frequency change (or phase increment). The oscillator may be an analog component design, or it may be a digital oscillator such as a direct digital synthesizer (Section 1.6).

Let the input signal be $\sin \omega t$ and the oscillator reference signal be $\cos \omega t$. Note that this means the reference is always 90° in advance of the input. As a result, the averaged product of sine times cosine is always zero, which we need in order for the zero-error condition in the steady state. If a signal with synchronized timing is required for further demodulation, then it is just a matter of taking the cosine signal and delaying it to produce the sine. Note that if both were sine signals, the product would *not* be zero. Signals such as a sine–cosine pair, whose averaged product is zero, are termed *orthogonal signals* and are discussed further in Section 3.9.3.

(a)

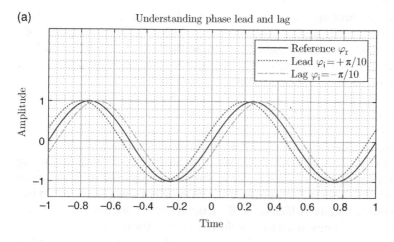

(b)

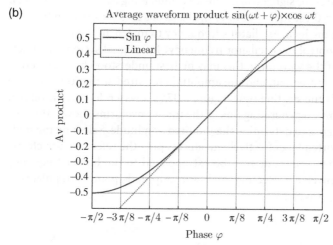

Figure 3.36 Waveforms with a phase difference (a) and determining the phase difference by averaging over a few cycles the product of the input and local oscillator (b).

Imagine that, initially, the sine and cosine signals are exactly phase locked. The error is zero, since the average product is zero. Now if the *input* phase changes by φ, then the average product is $\sin(\omega t + \varphi) \cos \omega t$, which many be expanded to $[\sin(2\omega t + \varphi) + \sin \varphi]/2$. Lowpass filtering should ideally remove the 2ω frequency component, leaving $(1/2) \sin \varphi$. For a positive φ, this is a positive number, indicating that the reference needs to be advanced a little. If it is a negative number, we need to delay or retard the reference a little. Each case is illustrated in Figure 3.36a.

What is important here is that the error signal is related to the phase change. It is not a linear proportion, which would be of the form $K\varphi$, but rather $K \sin \varphi$.

However, for small angles, $\sin \varphi \approx \varphi$, and this is the usual case when we are relatively close to the desired synchronized or "locked" condition. Figure 3.36b illustrates this approximate linearity. It is important to understand that the horizontal axis is *not* time in this figure, but rather phase angle, and the output indicated on the vertical axis is the averaged product.

We can turn this around, and imagine that the input signal phase is fixed, but the *local oscillator* is a little out – after all, we do not know the precise phase angle to start with. In that case, the input is $\sin \omega t$, but the oscillator is generating a waveform described as $\cos(\omega t + \varphi)$. The product is then $\sin \omega t \cos(\omega t + \varphi)$, which expands to $[\sin(2\omega t + \varphi) - \sin \varphi]/2$. So once again, lowpass filtering (effectively, averaging) over a number of samples is employed, which gives an average of $-(1/2) \sin \varphi$. In this case, if the phase φ is a small positive number, then the product signal (which is, in effect, the error) is a negative number. This makes sense, as we now need to slow down or retard the oscillator a little. And if φ is negative, the error signal will be positive, indicating the need to speed up the oscillator. From these ideas, we have the basis of a feedback system that can continually adjust the local oscillator's frequency to match an input.

So how do we control the oscillator frequency? In a digital system that takes explicit samples of the waveform, this may be best understood by considering a fixed small time increment, and the amplitude required at that time relative to the amplitude at the starting time. Figure 3.37a shows a reference waveform and a starting point. In generating one of the three waveforms shown, the goal at the very next step is to determine the required amplitude. Selecting the middle point in the dotted box effectively means keeping the reference waveform, selecting the higher amplitude may be extrapolated to the higher-frequency waveform, and conversely selecting the lower-amplitude point extrapolates to a lower-frequency waveform.

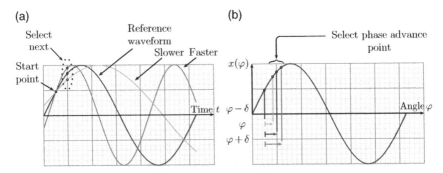

Figure 3.37 To derive the amplitude at the next step, and thus the overall waveform, the amplitude must be selected according to the fixed step φ plus or minus a small difference δ. Accordingly, this yields a faster or slower waveform. (a) Selecting the next amplitude at each step (b) Next step amplitude from phase advance/retard.

The amplitude itself may be read off the plot of a sine versus angle plot as in Figure 3.37b. The reference point (shown as φ) indicates the corresponding amplitude at the next phase-step increment. A little less, say, $\varphi - \delta$, yields a lower amplitude on the curve, whereas a little more, say, $\varphi + \delta$, yields a slightly higher amplitude on the curve. So for a fixed time increment, we can read off the next amplitude required from the phase graph. Repeating this process at each time step thus creates a continuous waveform. The Numerically Controlled Oscillator (NCO) is stepped at each sample according to this principle, so that the *rate of change* per step forces the frequencies to align.

Finally, we need the control part of the loop. Recall that this is to ensure that the error is in fact forced to zero. It could, in the simplest case, be just a constant multiplier – the greater the phase error, the greater the phase step required for each point on the wave. Too great a step, though, can cause the frequency to rapidly rise up. Likewise, too small a step may mean that the local oscillator waveform is too slow to catch up to changes in the input.

Since it is necessary to ensure that the error signal is forced to zero, it is better to incorporate some element of integration, or cumulative summation of the error. This enables tracking of a constantly varying input frequency. The sum over time (or integral) of the error must be zero, and since it is located within a feedback loop, the error must eventually be forced to a steady-state value of zero, with the system stabilizing.

The complete loop is then as shown in Figure 3.38. The constant α is just a multiplier, and in simple terms controls how fast the loop reaches steady-state synchronization. The value of β controls how much of the cumulative error is introduced – while the control signal $r(t)$ is nonzero, the integrator output will ramp up or down, thus tending to increase or decrease the oscillator drive signal. A constant multiplier K is also included for convenience. Tuning K, α, and β then controls the response of the PLL to changes in the input frequency and/or phase.

This structure works well, but we can improve it a little more by using two oscillators that are 90° out of phase (said to be in *quadrature*), as in Figure 3.39. This type of structure is termed a Costas loop, and although originally proposed for analog signal demodulation (Costas, 1956), it has found widespread applicability in digital demodulation. Examination of Figure 3.39 shows that a second arm has been added, which is symmetrical with the phase detector-oscillator of the conventional PLL. Note that there is a phase difference between the two arms. The in-phase or I branch employs a sine signal, whereas the quadrature phase or Q branch employs a cosine signal (as did the basic PLL). The NCO is stepped according to the principles outlined earlier. Finally, the arctan function is shown in the diagram, since sine and cosine components are available (recall that $\tan \theta = \sin \theta / \cos \theta$).

The following code illustrates the operation of a simple Costas loop. It employs a simple sample-averaging process for the lowpass filter, and many enhancements could be made to the basic outline. A great deal of work often

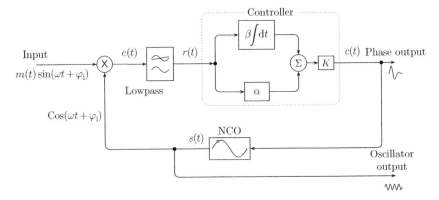

Figure 3.38 The PLL is comprised of phase detector (multiplier plus averaging filter), tunable controller, and numerically controlled oscillator, in a feedback-loop configuration.

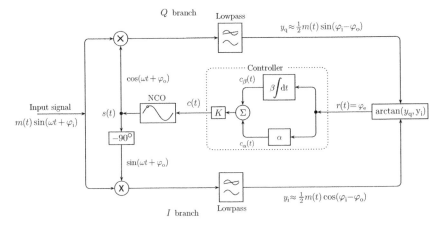

Figure 3.39 The Costas loop extends the basic PLL approach to employ quadrature signals in two separate branches, utilizing the combined phase error of each to drive the oscillator.

goes into the design of the so-called loop filter and associated parameter choices, since this dramatically affects the overall system performance in a given application. Also, it is possible to make the loop become unstable with an inappropriate choice of parameters, which is clearly undesirable.

```
N = 22000;      % total steps
M = 400;        % samples to average for lowpass filter

% phase angle step for simulation (radians per sample)
w = 2*pi/100;
```

```
% NCO frequency (phase step) exactly matches input
wosc = w;

% PLL loop parameters
K = 0.001;
calpha = 1;
cbeta = 0.001;        % faster to reach target, but overshoots
%cbeta = 0.0001;      % slower but does not overshoot

nw = 0;
nwsave = [];
nwosc = 0;
nwoscsave = [];

xMsave = [];

ph = 0;

ca = 0;
cb = 0;
cbprev = 0;

% select phase or frequency change
%TestChangePhase = false;
TestChangePhase = true;

TestChangeFreq = false;
%TestChangeFreq = true;

for n = 1:N

    if TestChangePhase
        % phase change test
        if( n == 8000 )
            ph = 2;
        end
    end

    if TestChangeFreq
        % frequency change test
        if( n == 8000 )
            delw = w*0.02;

            % effect change in frequency
            w = w + delw;
        end
```

```
    end

    xin(n) = sin(nw + ph);

    % oscillator waveforms - sine and cosine with calculated
    % phase shift
    xsin(n) = sin(nwosc);
    xcos(n) = cos(nwosc);

    % averaging M samples for "lowpass" filtering of product
    % of waveforms
    m = n:-1:n-M+1;
    m = m( m > 0 );

    yI = mean(xin(m).*xsin(m));
    yQ = mean(xin(m).*xcos(m));

    if( n < M )
        dw(n) = 0;
    else
        % phase estimator
        xM = atan2(yQ, yI);
        xMsave = [xMsave xM];

        % control algorithm
        ca = calpha*xM;
        cb = cbeta*xM + cbprev;
        yM = ca + cb;

        cbprev = cb;

        % final constant multiplier K
        dw(n) = K*yM;
    end

    nw = nw + w;
    nwosc = nwosc + wosc + dw(n);
end

figure(1);
plot(dw);
title('phase step');

figure(2);
plot(xMsave);
title('control signal');
```

For the purpose of analyzing the Costas loop, we define the in-phase branch as the one that multiplies the incoming signal by a sine function, since the input was assumed to be a sine. Likewise, the quadrature-phase branch refers to the multiplication by the cosine function. In the core of the phase-locked Costas loop, the phase of the sine and cosine oscillators is stepped by an amount dw at each iteration, with the filtered I and Q products yI and yQ computed. The phase estimation is done using the arctangent function, followed by the control loop with tunable parameters α and β.

Two cases are possible in practice: a change in phase of the input (with the frequency remaining constant), or a change in frequency of the input. Figure 3.40 illustrates the case where a step change of phase of the input waveform occurs. The phase step shown is the value in addition to the default oscillator step. Both the phase error and the control signal derived from the phase error are

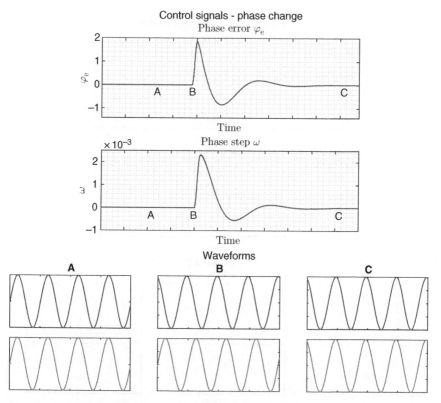

Figure 3.40 PLL response to change in phase. The phase error is shown, together with the control signal derived from it. The waveforms show the input sinusoid and the PLL oscillator sinusoid at the indicated time instants – before (A), during (B), and after (C) the phase change.

shown. In the lower panel of Figure 3.40, careful examination of the input signal (upper) and oscillator sine signal (lower), especially at their start and end points, shows how the phase is identical (A), quite different (B), and restored (C). The restoration to the in-phase condition occurs due to the action of the control loop, which acts to drive the error signal to zero.

Figure 3.41 shows the case where a step change of frequency of the input waveform occurs. In this case, the phase error is again forced to zero; however, the phase step value is permanently increased to reflect the increase in frequency. In other words, the oscillator must continually increment by a slightly increased phase value so as to keep up with the higher-input frequency. Of course, a lower-frequency signal may also be tracked in a similar fashion, with the oscillator phase increment reducing by an appropriate amount.

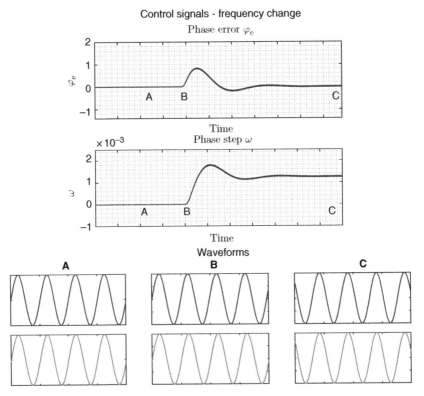

Figure 3.41 PLL response to change in frequency. This should be compared with the previous figure. Note that the phase increment is permanently increased, so as to track the increased input frequency. At time B, the frequency of the input waveform is greater than the oscillator shown below it; however, the PLL action restores the frequency (and phase) match at C.

3.8   Demodulation Using *IQ* Methods

The previous sections explored various means to demodulate AM, FM, and PM signals. Naturally, demodulation depends upon what is practically feasible, and in the past some types of operation have been preferred over others. For example, a phase delay over a certain frequency band may be difficult to achieve using analog electronics. The use of digital sampling and processing opens up a number of possibilities in this regard. In particular, phase shifting and quadrature signal generation are somewhat easier in the digital or sampled domain. The group of methods referred to as *IQ demodulation* – for in-phase/quadrature-phase – are more suited to Digital Signal Processor (DSP) implementation.

If we define an in-phase signal *I* as the cosine signal, and the quadrature *Q* or 90° delayed version as a sine, we have the situation illustrated in Figure 3.42. Here, we see a sinusoidal signal and its delayed version, as well as a cosine and its delayed version. It is apparent that by taking the cosine signal as a reference (in-phase), the sine signal becomes the quadrature or delayed signal. The waveforms may be represented as a point on the plane as shown on the right-hand side of Figure 3.42, where the horizonal (or "**x** axis") is the "cosine axis" and the vertical ("**y** axis") is the "sine axis."

To understand how this facilitates demodulation, it is helpful to recall the trigonometric expansions for sine and cosine products from Section 3.3.1. In particular, Table 3.1 will be useful. The block diagram of Figure 3.43 shows in general terms how *IQ* demodulation may be employed. It is simply a multiplication of the modulated input by both sine and cosine, followed by lowpass filtering. As we will show, further processing of the *I*(*t*) and *Q*(*t*) signals provides a demodulated signal, where the algorithm used in the post-processing is chosen according to the modulation type.

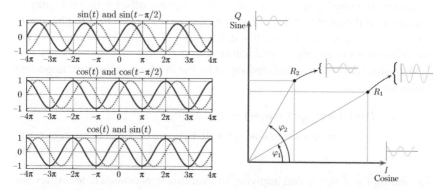

Figure 3.42 Illustrating quadrature signals: time domain (left) and *IQ* plane (right). The magnitude *R* and phase *φ* are represented using *I* as cos *ωt* on the horizontal axis, and *Q* as sin *ωt* on the vertical axis.

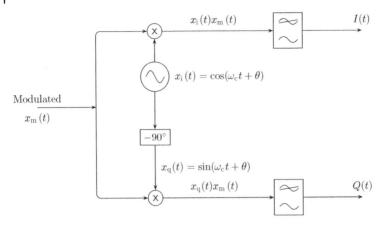

Figure 3.43 Demodulation with quadrature signals: *I* is the cosine component, and *Q* is the sine component.

3.8.1 Demodulation of AM Using *IQ* Signals

An AM signal with cosine carrier and sine modulation $A_m \sin \omega_m t$ is

$$x_{AM}(t) = A_m \sin \omega_m t \, \cos \omega_c t + A_c \cos \omega_c t \tag{3.97}$$

Adopting the convention of defining the in-phase multiplier $x_i(t)$ and quadrature-phase multiplier $x_q(t)$,

$$x_i(t) = \cos(\omega_c t + \theta) \tag{3.98}$$

$$x_q(t) = \sin(\omega_c t + \theta) \tag{3.99}$$

To handle the most general case, we include a phase offset θ above. It may be the case that the local signal $x_i(t)$ has been derived to be precisely in phase with the carrier $x_c(t)$, in which case $\theta = 0$. Including θ in the definition, however, provides for a completely general solution.

Multiplication of the modulated signal by the in-phase or cosine component yields the *I* signal

$$
\begin{aligned}
x_{AM}(t)x_i(t) &= x_{AM}(t) \, \cos(\omega_c t + \theta) \\
&= A_c \cos \omega_c t \cos(\omega_c t + \theta) + A_m \sin \omega_m t \, \cos \omega_c t \, \cos(\omega_c t + \theta)
\end{aligned} \tag{3.100}
$$

Expanding the *I* signal and applying the various trigonometric identities, we find that although a large number of component terms are generated, we may express the result as

$$x_{AM}(t)x_i(t) = [DC] + [2\omega_c \text{ components}] + \left[\frac{A_m}{2} \sin \omega_m t \cos \theta\right] \quad (3.101)$$

The DC or constant value represents a constant offset and may be removed by the next stage of processing. The higher-frequency components at $2\omega_c$ and $2\omega_c \pm \omega_m$ may also be removed using a lowpass filter, since $\omega_c \gg \omega_m$. This leaves

$$I(t) = \frac{A_m}{2} \sin \omega_m t \cos \theta \quad (3.102)$$

If the local in-phase oscillator is exactly in phase with the incoming carrier, then $\theta = 0$, and our demodulation is complete. However, if this is not the case, then we may proceed further by calculating the quadrature component as

$$x_{AM}(t)x_q(t) = x_{AM}(t) \sin(\omega_c t + \theta)$$

Following a similar expansion and simplification process to that of $I(t)$, the result is

$$Q(t) = \frac{A_m}{2} \sin \omega_m t \sin \theta \quad (3.103)$$

Thus the demodulated signal may be obtained as

$$m(t) = \sqrt{I^2(t) + Q^2(t)} \quad (3.104)$$

To confirm our understanding, the following MATLAB code shows how to create an AM waveform and then demodulate it using the IQ signal approach.

```
N = 2000;
Tmax = 20;
dt = Tmax/(N−1);
t = 0:dt:Tmax;
fs = 1/dt;

% carrier
fc = 3;
wc = 2*pi*fc;
xc = cos(wc*t);

% modulation
fm = 0.2;
wm = 2*pi*fm;

xm = cos(wm*t);

Ac = 2;
mu = 0.2;
Am = mu*Ac;
```

```
% amplitude modulation equation
xam = Am*xm.*xc + Ac*xc;

% AM Demodulation — theta is arbitrary
theta = pi/3;
xc = cos(wc*t + theta);
xs = sin(wc*t + theta);

I = xam.*xc;
Q = xam.*xs;
xd = sqrt(I.^2 + Q.^2);
```

Figure 3.44 shows the resulting signals xc (carrier), xm (modulation), xam (modulated), and finally xd. The latter is positive, since $\sqrt{I^2(t) + Q^2(t)}$ is positive. The high-frequency component at twice the carrier frequency ($2\omega_c$) is clearly visible. As found in the mathematical derivation, this waveform requires subsequent lowpass filtering and offset removal as a final stage.

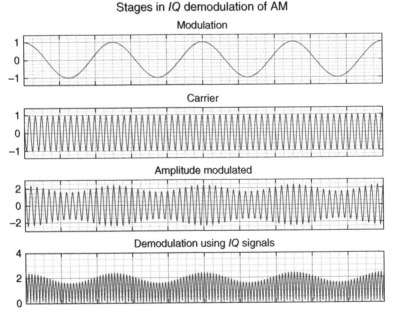

Stages in *IQ* demodulation of AM

Figure 3.44 Waveforms for *IQ* demodulation of AM. Lowpass filtering of the output waveform (lower panel) would remove the double carrier frequency component. Removal of the constant offset is also required. The final output waveform should then correspond to the modulating input (top panel).

3.8.2 Demodulation of PM Using *IQ* Signals

To investigate the demodulation of phase modulated (PM) signals, consider a PM signal with cosine carrier and sine modulation. The carrier is $\cos \omega_c t$, and once again we define

$$x_i(t) = \cos(\omega_c t + \theta) \tag{3.105}$$

$$x_q(t) = \sin(\omega_c t + \theta) \tag{3.106}$$

The PM signal for modulation $m(t)$ is

$$x_{PM}(t) = A \cos[\omega_c t + k_p m(t)] \tag{3.107}$$

For a single-tone test signal, we may utilize a sinusoid modulation of the form

$$m(t) = A_m \sin \omega_m t \tag{3.108}$$

Thus the PM signal for single-tone modulation becomes

$$x_{PM}(t) = A \cos(\omega_c t + k_p A_m \sin \omega_m t) \tag{3.109}$$

Multiplying by the in-phase carrier, we have the I component product

$$
\begin{aligned}
x_{PM}(t)x_i(t) &= x_{PM}(t) \, \cos(\omega_c t + \theta) \\
&= A \cos(\omega_c t + k_p A_m \sin \omega_m t) \cos(\omega_c t + \theta) \\
&= \frac{A}{2}[\cos(2\omega_c t + k_p A_m \sin \omega_m t + \theta) + \cos(k_p A_m \sin \omega_m t - \theta)] \\
&= \left[\frac{A}{2} \cos(k_p A_m \sin \omega_m t - \theta)\right] + [2\omega_c \text{ components}]
\end{aligned} \tag{3.110}
$$

Lowpass filtering leaves

$$I(t) = \frac{A}{2} \cos(k_p A_m \sin \omega_m t - \theta) \tag{3.111}$$

Similarly, the Q component becomes

$$
\begin{aligned}
x_{PM}(t)x_q(t) &= x_{PM}(t) \, \sin(\omega_c t + \theta) \\
&= A \cos(\omega_c t + k_p A_m \sin \omega_m t) \sin(\omega_c t + \theta) \\
&= \frac{A}{2}[\sin(2\omega_c t + k_p A_m \sin \omega_m t + \theta) + \sin(\theta - k_p A_m \sin \omega_m t)] \\
&= \frac{A}{2}[\sin(2\omega_c t + k_p A_m \sin \omega_m t + \theta) - \sin(k_p A_m \sin \omega_m t - \theta)] \\
&= \left[-\frac{A}{2} \sin(k_p A_m \sin \omega_m t - \theta)\right] + [2\omega_c \text{ components}]
\end{aligned}
$$

$$\tag{3.112}$$

Again, lowpass filtering leaves

$$Q(t) = -\frac{A}{2} \sin(k_p A_m \sin \omega_m t - \theta) \tag{3.113}$$

Once again, we need an algorithmic trick to determine the original test signal from the I and Q signals. Taking the arctangent of (Q/I) reveals that PM demodulation may be accomplished using

$$
\begin{aligned}
\arctan\left(\frac{Q(t)}{I(t)}\right) &= \arctan\left[\frac{-(A/2)\sin(k_p A_m \sin \omega_m t - \theta)}{(A/2)\cos(k_p A_m \sin \omega_m t - \theta)}\right] \\
&= -\arctan[\tan(k_p A_m \sin \omega_m t - \theta)] \\
&= -k_p A_m \sin \omega_m t + \theta
\end{aligned}
\tag{3.114}
$$

This is the original modulation, scaled and with an offset. To confirm our understanding, the following MATLAB code shows how to create a PM waveform and then demodulate it using IQ signals. Figure 3.45 shows the resulting signals xc (carrier), xm (modulation), xpm (modulated), and finally xd.

```
N = 2000;
Tmax = 20;
dt = Tmax/(N-1);
t = 0:dt:Tmax;
fs = 1/dt;

% carrier
fc = 3;
wc = 2*pi*fc;
xc = cos(wc*t);

% PM modulation
fm = 0.2;
wm = 2*pi*fm;
Am = 1;
xm = Am*sin(wm*t);

kp = 10;
A = 2;

% phase modulation equation
xpm = A*cos(wc*t + kp*xm);

% PM Demodulation
theta = pi/3;
xc = cos(wc*t + theta);
xs = sin(wc*t + theta);

I = xpm.*xc;
Q = xpm.*xs;
d = -1*atan2(Q, I);
xd = unwrap(d);
```

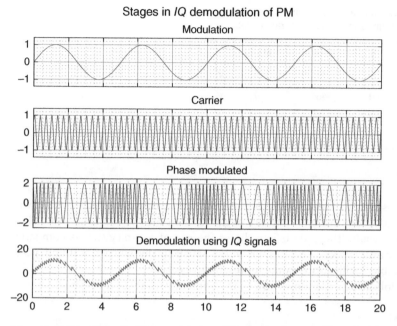

Figure 3.45 Waveforms for *IQ* demodulation of PM. Further lowpass filtering of the output waveform (lower panel) would smooth the demodulated signal. Note the correspondence to the input modulating signal (top).

Notice that subsequent to the `atan2` stage, which calculates the arctangent of Q/I, it is necessary to "unwrap" the phase angle using the MATLAB function `unwrap`. This is because the arctangent is calculated over the range $-\pi$ to $+\pi$, and this range does *not* correspond to a smooth modulating signal. Consider an example in degrees: suppose one output point was calculated as $175°$, and a subsequent point as $8°$ more, or $(175 + 8) = +183°$. The `atan2` function would return the equivalent (in radians) in the range $\pm180°$, which is $-177°$. Now, a "smooth" modulating signal would not jump from $+175$ to -177. Clearly, the equivalent angle of $+183$ is what is required. This is precisely what the `unwrap` function does. Note that it requires not only the present sample, but also the previous sample, in order to compensate for the jumps in value. This is the reason the local carrier phase offset θ is unimportant: the demodulation using this approach is calculated as the phase *difference*, and since θ is assumed constant, it does not affect the result. Finally, as found in the mathematical derivation, this waveform requires subsequent lowpass filtering and offset removal.

Since the arctangent function is used extensively in *IQ* demodulation, it is worth pointing out that there are two common types of arctangent function. The standard arctangent function calculates $\arctan(y/x)$, but this yields an incorrect result (or at least, one that was unexpected) in certain circumstances. If x and y are both positive, there is no problem. But if either is negative, it

Table 3.3 Comparing `atan` and `atan2` functions. The latter gives a true four-quadrant result.

x	y	atan	atan 2	
+1	+1	+45	+45	✓
+1	−1	−45	−45	✓
−1	+1	−45	+135	✗
−1	−1	+45	−135	✗

is impossible to know which one carried the positive sign and which was negative. Additionally, if both are negative, then they would cancel to yield a positive result. The code below shows the use of `atan(y/x)` as compared with `atan2(y,x)`. The former works satisfactorily when $x > 0$ and $y > 0$ only. Table 3.3 illustrates some representative cases.

The following code shows how to experiment with these functions.

```
x = 1;
y = 1;
at = atan(y/x)*180/pi;
at2 = atan2(y,x)*180/pi;
fprintf(1, 'x=%d y=%d  atan=%d, atan2=%d degrees\n', x, y,
                                                      % at, at2);
```

3.8.3 Demodulation of FM Using *IQ* Signals

For the case of FM demodulation, we again start with the definition of the modulated signal. For FM, this is

$$x_{FM}(t) = A \cos \left[\omega_c t + k_f \int_0^t m(\tau) \, d\tau \right] \tag{3.115}$$

For a single-tone (co)sinusoid modulation

$$m(t) = A_m \cos \omega_m t \tag{3.116}$$

the FM signal becomes

$$x_{FM}(t) = A \sin(\omega_c t + \beta \sin \omega_m t) \tag{3.117}$$

with

$$\beta = \frac{k_f A_m}{\omega_m} \tag{3.118}$$

As before, we assume that the carrier is $\cos \omega_c t$ with an unknown phase offset θ, and so

$$x_i(t) = \cos(\omega_c t + \theta) \tag{3.119}$$

$$x_q(t) = \sin(\omega_c t + \theta) \tag{3.120}$$

The *I* component is the incoming FM modulated signal multiplied by the local $x_i(t)$ signal

$$
\begin{aligned}
x_{FM}(t)x_i(t) &= x_{FM}(t)\cos(\omega_c t + \theta) \\
&= A\cos(\omega_c t + \beta \sin \omega_m t)\cos(\omega_c t + \theta) \\
&= \frac{A}{2}[\sin(2\omega_c t + \beta \sin \omega_m t + \theta) + \cos(\beta \sin \omega_m t - \theta)] \\
&= \left[\frac{A}{2}\cos(\beta \sin \omega_m t - \theta)\right] + [2\omega_c \text{ components}] \tag{3.121}
\end{aligned}
$$

After lowpass filtering

$$I(t) = \frac{A}{2}\cos(\beta \sin \omega_m t - \theta) \tag{3.122}$$

The *Q* component is

$$
\begin{aligned}
x_{FM}(t)x_q(t) &= x_{FM}(t)\sin(\omega_c t + \theta) \\
&= A\cos(\omega_c t + \beta \sin \omega_m t)\sin(\omega_c t + \theta) \\
&= \frac{A}{2}[\sin(2\omega_c t + \beta \sin \omega_m t + \theta) - \sin(\beta \sin \omega_m t - \theta)] \\
&= \left[\frac{-A}{2}\sin(\beta \sin \omega_m t - \theta)\right] + [2\omega_c \text{ components}] \tag{3.123}
\end{aligned}
$$

After lowpass filtering

$$Q(t) = -\frac{A}{2}\sin(\beta \sin \omega_m t - \theta) \tag{3.124}$$

Similar to PM, we take the arctangent of (Q/I)

$$
\begin{aligned}
\arctan\left[\frac{Q(t)}{I(t)}\right] &= \arctan\left[-\frac{(A/2)\sin(\beta \sin \omega_m t - \theta)}{(A/2)\cos(\beta \sin \omega_m t - \theta)}\right] \\
&= -\arctan[\tan(\beta \sin \omega_m t - \theta)] \\
&= -\beta \sin \omega_m t + \theta \tag{3.125}
\end{aligned}
$$

This is not quite the original modulation (which, in this example, was a cosine). Recalling that FM incorporates integration as part of its definition, we take the derivative

$$
\begin{aligned}
\frac{d}{dt}\left\{\arctan\left[\frac{Q(t)}{I(t)}\right]\right\} &= -\frac{d}{dt}[\beta \sin \omega_m t + \theta] \\
&= -k_f A_m \cos \omega_m t \tag{3.126}
\end{aligned}
$$

This is the original modulation, inverted and scaled.

To confirm our understanding, the following MATLAB code shows how to create an FM waveform and then demodulate it using *IQ* signals.

```
N = 2000;
Tmax = 20;
dt = Tmax/(N-1);
t = 0:dt:Tmax;
fs = 1/dt;

% carrier
fc = 3;
wc = 2*pi*fc;
xc = cos(wc*t);

% FM modulation
fm = 0.2;
wm = 2*pi*fm;
Am = 1;
xm = Am*cos(wm*t);

A = 2;
kf = 10;

% integral of xm
xmi = cumsum(xm)*dt;

% frequency modulation equation
xfm = A*cos(wc*t + kf*xmi);

% FM Demodulation
theta = pi/3;
xc = cos(wc*t + theta);
xs = sin(wc*t + theta);

I = xfm.*xc;
Q = xfm.*xs;

d = -1*atan2(Q, I);
xd = unwrap(d);
```

Figure 3.46 shows the resulting signals xc (carrier), xm (modulation), xfm (modulated), and finally xd.

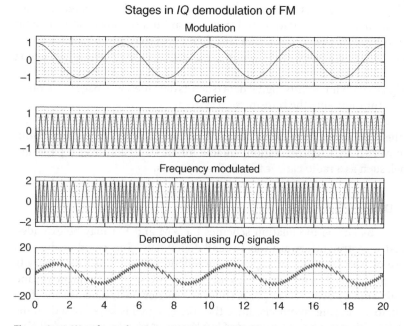

Figure 3.46 Waveforms for *IQ* demodulation of FM. Filtering is required for the output waveform (lower panel), followed by differentiation – at which point it should correspond to the input modulating wave (top panel).

Similar to PM, the FM case requires phase unwrapping. Not shown in the figure is the final lowpass filtering, which in this case must be followed by differentiation as per the derivation. Differentiation of the lower plot may be seen to give a waveform corresponding to the original modulating signal.

The above methods of PM and FM demodulation required an arctangent function to complete the operation. This may be calculated if the signals are sampled, but is not so easy if the signals remain in their original analog form. Hence, *IQ* demodulation is more suited for use in sampled-data systems. The arctan calculation is clearly important, and some researchers have investigated fast and efficient means of calculating this function (for example, Frerking, 2003; Lyons, 2011, section 13.22).

3.9 Modulation for Digital Transmission

All the aforementioned modulation schemes may be used to modulate analog signals. Only one analog modulation signal – typically speech, music, or image intensity – was assumed.

Analog signals such as audio and video, however, may be encoded or quantized to a binary representation and then transmitted serially (that is, one bit after another). At the receiver, they are then converted into an analog form. Although more complex, there are many advantages to transmitting analog information in digital form. Such an approach permits the combination of digitized signals with inherently digital data (such as data files, web pages, and other content), thus unifying the transmission system.

This section introduces digital modulation schemes for passband modulation. The requirement is to take a binary stream of 1 and 0 data (the *bitstream*) and convert it to a representation suitable for transmission over a passband channel, such as a radio carrier for wireless systems.

3.9.1 Digital Modulation

The previous analog modulation schemes for AM and FM may be extended in a relatively straightforward way to send digital data. This could be done, in the case of AM, by using two specific modulation levels – one for binary 0, and one for binary 1. Similarly, FM could employ two specific frequencies, and PM could employ two specific phase shifts of a carrier. Such schemes are often referred to as "keying" methods, since the analog signal is keyed based on the digital data. Thus, we have *Amplitude-Shift Keying* (ASK), *Frequency-Shift Keying* (FSK), and *Phase-Shift Keying* (PSK).

One of the important requirements of a digital transmission system is to maximize the available bitrate for a given bandwidth. So the basic ASK, FSK, and PSK schemes may be extended to multiple bits at a time, simply by changing the number of passband modulation possibilities. For example, to transmit 2 bits at a time using ASK, a total of four amplitude levels would be required. In general, for B bits at a time, 2^B distinct representations would be required. This could be extended to higher rates, but the problem is that the ability to differentiate each case decreases when real channels subject to noise and other imperfections are employed. Consider the case of sending 8 bits at a time, using a total of 256 amplitude levels. If the level of noise exceeds half the spacing between distinct amplitude levels, then the assumed closest amplitude level may decode to an incorrect bit pattern. Similar arguments hold for phase and frequency changes.

It makes sense, at least intuitively, to combine the various fundamental parameters – amplitude, frequency, and phase – to achieve the goal of higher bit rate for digital transmission. The combination (usually encoding several bits) is termed a *symbol*, where each symbol represents a number of bits transmitted during the *symbol interval*. Most commonly, a combination of amplitude and phase changes may be employed to increase the number of distinct symbols able to be represented. This is because it is usually simpler to lock onto a fixed carrier frequency. Additionally, using several frequencies together, each with their own amplitude and phase changes, may also be

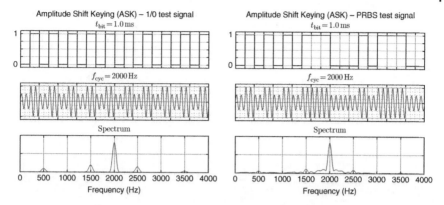

Figure 3.47 Amplitude shift keying in theory, with an alternating 1/0 input signal (left) and PRBS or pseudo-random binary sequence (right) to represent a more realistic transmission scenario.

used to create what are, in effect, parallel channels. Such frequency bands are termed *subchannels*.

Figure 3.47 illustrates ASK. Here, we simply use a fixed carrier at two possible amplitude levels. A special case (termed On–Off Keying) is where one of the levels is zero. This has the advantage that less power may be used; however, it also has the disadvantage that synchronization may be lost during about half of the transmission time. Figure 3.47 shows the frequency spectrum for a pseudo-random binary sequence (PRBS), and it may be observed that it is a special case of AM as examined earlier. For the general case of an arbitrary bit-stream, the carrier remains present, but the power is spread over the sidebands.

Instead of changing the amplitude, the frequency may be changed, resulting in FSK. This is illustrated in Figure 3.48. Here, two separate tones or frequencies are transmitted. As expected, the frequency content contains those two tones, but in addition various sidebands are present, in a way similar to continuous FM transmission. The spacing of the sidebands is determined by the inverse of the bit rate.

The final variant is PSK, as Figure 3.49 illustrates. As may be expected due to the similarities between frequency and PM, the frequency spectrum of PSK is not dissimilar to FSK, with the primary difference being that only one frequency is employed for the carrier. In the output spectrum, multiple frequencies are produced due to the phase transitions. In a real PSK system, it is desirable to reduce the bandwidth of a transmission, and this in turn implies the need to smooth out the phase discontinuities.

Figure 3.50 shows the measured frequency spectra of these signals. Note that the measured signals agree with the theoretical predictions in terms of center frequencies and harmonics and the presence of the noise floor. The RBW and VBW of the spectrum analyzer (Section 2.3.3) must be adjusted to obtain the

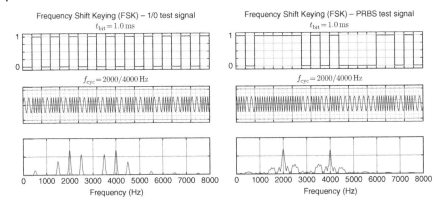

Figure 3.48 Frequency shift keying in theory, with an alternating 1/0 input signal (left) and PRBS (right) to represent a more realistic transmission scenario.

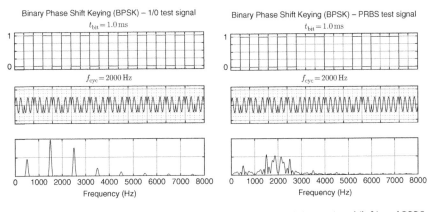

Figure 3.49 Phase shift keying in theory, with an alternating 1/0 input signal (left) and PRBS (right) to represent a more realistic transmission scenario.

necessary resolution. As is usual for spectrum measuring instruments, the vertical axis is calibrated in terms of dBm (Section 1.7.2) to indicate power rather than voltage.

3.9.2 Recovering Digital Signals

One key part of a receiver's processing of a digital data stream is to extract the original signal from the inevitably noise-corrupted received signal. This may be done in one of two common ways: either a *matched filter* or a *correlate–integrate* structure. This section considers the workings of each of these and compares their operation. They are broadly similar in their concept,

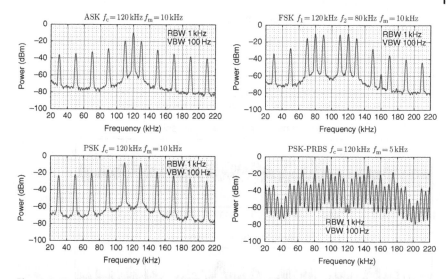

Figure 3.50 Measured spectra for ASK and FSK (top) and PSK (lower). Each shows the spectrum for a 1/0 alternating input sequence. The PSK case shows in addition the spectrum resulting from a pseudorandom binary sequence (PRBS) bitstream. Note that the power is measured in dBm.

but different in their formulation in practice. Understanding the difference between the two is critical, so this section considers both at the same time.

Figure 3.51 illustrates the problem. The pulse waveform that is received is shown in the top panel. This waveshape is the result of the pulse shaping occurring at the transmitter as well as the effects of the channel. In this case, the waveform could represent a binary bitstream of $+1, -1, -1, +1, -1$. Note that the pulse spacing is 200 samples in this example and that the maximum value does not occur at the start of each sample interval, but at some time later.

The assumption is that we know the expected pulse shape due to prior knowledge of the transmitter pulse shaping and/or probing of the channel. Noise is typically added to the waveform along the way, and white Gaussian noise is shown in the middle plot. The addition of the transmitted signal and the noise is shown in the bottom plot. It should be clear that simply detecting the maximum value of amplitude (either positive or negative) may not be a good strategy, since the noise will often produce spurious peaks. The goal, then, is to remove as much of the noise as possible.

One structure that may achieve this is a multiply-and-integrate sequence as illustrated diagrammatically in Figure 3.52. It is assumed that it is possible to generate a continuous, repeating, noise-free (clean) waveform corresponding to the expected pulse shape, and we need to choose whether a positive or negative pulse was transmitted. Multiplying the incoming waveform by the locally generated pulse train produces a series of products, which are then summed

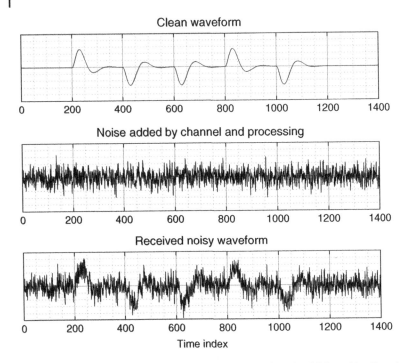

Figure 3.51 The "clean" version of a digital pulse signal (top), additive white Gaussian noise (middle), and the received signal (bottom).

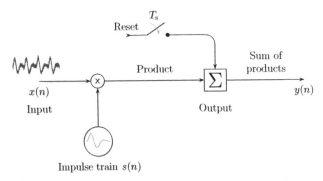

Figure 3.52 Multiplying the incoming wave and integrating the sum over one symbol period.

(integrated) over one symbol interval T_s. The integrator is reset after each symbol interval; at the end of the symbol interval, the resulting time-averaged value is a good indication of the original symbol amplitude transmitted. Of course, amplitudes change over the transmission channel, and so an absolute threshold

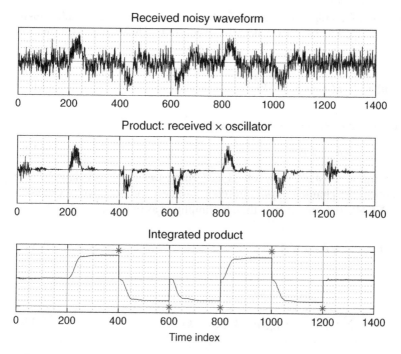

Figure 3.53 Waveforms obtained by the multiply-integrate structure. The stars indicate the sampling point at the end of each symbol interval. After this interval, the multiply-integrate operation is restarted.

comparison is not possible. This is also the reason why pointwise subtraction of the received waveform and the local clean waveform cannot be used, since (in addition to the noise) the signal amplitudes are unlikely to be equal.

This approach is termed *correlation,* and the multiply-accumulate structure requires the ability to multiply and add up signal amplitudes, as well as the ability to generate a repeating waveform. This waveform is just the same pulse waveform repeated over and over again. For the purposes of this example, we have assumed only two levels, and hence one bit per symbol interval T_s, with $K = 200$ samples of the waveform in that time. Transmitting more bits per symbol interval is certainly possible, by extending this idea.

Figure 3.53 illustrates the waveforms that occur in this structure. The pulse generator acts as a template as to what waveform is expected, while the integration or summation stage adds the pointwise product over one symbol interval. Importantly, if the noise is Gaussian, its long-term average is expected to be zero.

In formulating the solution, we need the definition of the *impulse response* of a system. This is just the output produced from a system block when the input is composed of a single pulse at the first sample instant, with all subsequent input values equal to zero. The calculation required at the multiply-summation stage is

$$y(n) = \sum_{k=0}^{K-1} x(n-k)s(n-k) \tag{3.127}$$

where the received signal $x(n)$ may be written as

$$x(n) = \alpha s(n) + g(n) \tag{3.128}$$

with α being a constant value of ± 1 according to the bit transmitted, the sequence $s(n)$ the channel *impulse response*, and $g(n)$ additive white Gaussian noise (AWGN). The output may then be simplified as follows:

$$y(n) = \sum_{k=0}^{K-1} x(n-k)s(n-k) \tag{3.129}$$

$$= \sum_{k=0}^{K-1} \overbrace{[\alpha s(n-k) + g(n-k)]}^{\text{Received signal}} s(n-k) \tag{3.130}$$

The term $\sum g(n-k)s(n-k)$ may be canceled since the noise is assumed to be uncorrelated with the impulse response, and so

$$y(n) = \alpha \sum_{k=0}^{K-1} s^2(n-k) + \underbrace{\sum_{k=0}^{K-1} g(n-k)s(n-k)}_{0} \tag{3.131}$$

Finally, if we take the very last sample at the end of each symbol, and substitute an index $n = K - 1$ (recall that there are K samples per symbol)

$$y(K-1) = \alpha \sum_{k=0}^{K-1} s^2(K-1-k)$$

$$= \alpha \sum_{k=0}^{K-1} s^2(k) \tag{3.132}$$

The last line may be deduced from symmetry – or, mathematically, we could let $m = K - 1 - k$ and change limits of the summation, so that

$$k = 0 \rightarrow m = K - 1$$
$$k = K - 1 \rightarrow m = K - 1 - k = K - 1 - (K - 1) = 0$$

Then the summation becomes

$$y(K-1) = \alpha \sum_{m=K-1}^{0} s^2(m)$$

$$= \alpha \sum_{m=0}^{K-1} s^2(m) \tag{3.133}$$

This shows that the signal will take on a peak magnitude, scaled by $\alpha = \pm 1$ at the last sampling instant $K - 1$. Returning to Figure 3.53, the decision to be made at each of the indicated sampling points is based on the magnitude of the signal $y(K - 1)$ at that point.

Although this appears to be a reasonable approach, and in fact is used in practice, it has some shortcomings. In particular, the timing is critical, and additional complexity is necessary to reset the integrator at the right moment.

To consider an alternative approach, Figure 3.54 shows the existing correlate–integrate waveforms on the left. The incoming (noisy) wave is multiplied by the channel impulse response and integrated (summed). Consider the waveforms on the right-hand side, in which the existing input waveform is shown at the top. Since the time axis is from left to right, from a "viewpoint" on the right-hand side, we may imagine that the time waveform is reversed. The incoming waveform is "seen" from this perspective. In order to match the channel impulse waveform, it is also necessary to reverse the impulse waveform with respect to the time axis. Sliding this time-reversed waveform left or right and computing the sum of the pointwise products (as was done with correlation) produces a set of output values.

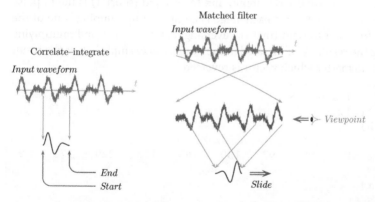

Figure 3.54 Moving from the correlate–integrate concept (left) to the matched filter (right). The correlate–integrate approach is a pointwise multiplication and summation over one symbol period. The matched filter is best thought of as reversing the time waveform according to the order we would "see" the waveform, and multiplying by the impulse response.

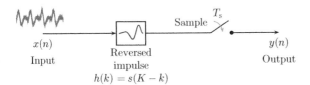

Figure 3.55 Matched filtering using a time-reversed channel impulse response. Imagine the input waveform as shown being reversed, since that is the order the filter "sees" it.

Importantly, this approach eliminates the constraint of resetting the summation output after each symbol interval. The new alternative is shown in Figure 3.55. In essence, the correlation approach uses current samples as they arrive, whereas the matched filter approach uses the current and past samples, looking back into the waveform that has "arrived" up until the current output point.

The matched filtering operation may be written mathematically as

$$y(n) = \sum_{k=0}^{K-1} h_k x(n-k) \tag{3.134}$$

This function, which is effectively a digital filtering operation, is termed *convolution*; the input signal and impulse response are said to be convolved together.

Convolution is often used for digital communications systems. Although seemingly complex, it may be explained via implementation as follows. The MATLAB function conv() is shown below, for two input sets. The first, $x(n)$, is the (longer) time series input, and the second (shorter) is the impulse response $h(n)$. The output at each stage is computed by "flipping" one of the sequences from left to right (that is, reversing its time order), and multiplying and adding the result. As illustrated in the numerical example below, the result is identical no matter which vector is reversed.

```
x = [1 2 3 4 5 6 7 8];
h = [10 11 12];
conv(x, h)
ans =
    10    31    64    97   130   163   196   229   172    96

conv(h, x)
ans =
    10    31    64    97   130   163   196   229   172    96
```

Flipping the h sequence, the first output is $10 \times 1 = 10$. The next output is $(2 \times 10) + (1 \times 11) = 31$, then $(3 \times 10) + (2 \times 11) + (1 \times 12) = 64$, and so forth.

As with the correlation approach, let the received signal be

$$x(n) = \alpha s(n) + g(n) \tag{3.135}$$

where α is a constant (according to the bit transmitted) of value ± 1, $s(n)$ is the channel impulse response, and $g(n)$ is AWGN. So the output of the matched filter is

$$
\begin{aligned}
y(n) &= \sum_{k=0}^{K-1} h_{n-k}\, x(k) \\
&= \sum_{k=0}^{K-1} h_{n-k} \overbrace{\left[\alpha s(k) + g(k) \right]}^{\text{Received signal}} \\
&= \sum_{k=0}^{K-1} \alpha h_{n-k} s(k) + \sum_{k=0}^{K-1} h_{n-k} g(k)^{\,0}
\end{aligned}
\tag{3.136}
$$

where the right-hand term cancels due to the noise $g(n)$ being uncorrelated with the channel impulse response. This leaves

$$y(n) = \alpha \sum_{k=0}^{K-1} h_{n-k} s(k) \tag{3.137}$$

We reasoned that the impulse response h_k should be the time-reversed channel impulse response $s(k)$, so mathematically over a sample interval of K samples,

$$h_k = s(K - 1 - k) \tag{3.138}$$

Substituting $n - k$ for k,

$$
\begin{aligned}
h_{n-k} &= s[K - 1 - (n - k)] \\
&= s(K - 1 - n + k)
\end{aligned}
\tag{3.139}
$$

So the output $y(n)$ is

$$y(n) = \alpha \sum_{k=0}^{K-1} s(k) s(K - 1 - n + k) \tag{3.140}$$

Using the sample at the end of one symbol period, where $n = K - 1$,

$$
\begin{aligned}
y(K - 1) &= \alpha \sum_{k=0}^{K-1} s(k)\, s[K - 1 - (K - 1) + k] \\
&= \alpha \sum_{k=0}^{K-1} s(k)\, s(k)
\end{aligned}
\tag{3.141}
$$

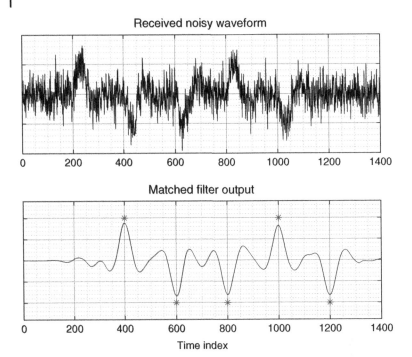

Figure 3.56 Waveforms obtained by the matched filter structure. The stars indicate the sampling point for each symbol. The output is not reset for each symbol, but rather calculated continuously using convolution.

From this, we infer that (ignoring the scaling α) the result is always positive, since we are multiplying a sample by itself. Additionally, the result takes on a maximum magnitude (scaled according to α) at the end of each bit, when sampling the value $y(K - 1)$.

For the same type of impulse response as before, the matched filter waveforms are shown in Figure 3.56. It is seen that the maximum at the end of each symbol interval may then be used to make a decision as to what the original transmitted amplitude would have been. In fact, we do not need to sample precisely at a particular instant, just around that general area.

This leads to the question of which method is better and indeed why there is a choice of two methods. The multiply-accumulate or correlation approach requires that the receiver generates a waveform similar to the impulse response, and it is necessary to restart the summation at the precise start of each symbol. Generating the channel impulse waveform may be done using either an analog circuit or a digital lookup table approach. The matched filter, however, requires the generation of the time-reversed impulse response, and

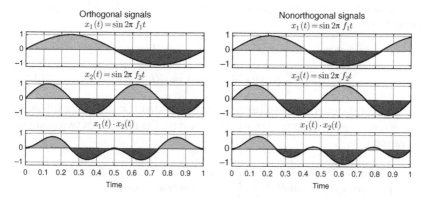

Figure 3.57 Illustrating orthogonal and nonorthogonal signals. The net area under the product of orthogonal signals is zero.

this can be problematic for analog circuits. For a digitally sampled approach, generating a waveform in reverse time simply means storing the samples and then reading them out in reverse order. This is quite simple to achieve, and thus matched filtering is preferable if digital implementation is possible. The downside is that a large number of waveform samples are required for each symbol interval, and thus proportionately faster memory and processing speed is required.

3.9.3 Orthogonal Signals

Sine and cosine signals of the same frequency may co-exist on a channel. This fact permits the capacity of a channel (in bits per second or bps) to be increased. Consider Figure 3.57 that illustrates the difference between orthogonal and nonorthogonal signals. Two input signals are shown in each case, together with the pointwise product of each. From the figure, we can see that the total area above the zero axis equals the total area below the axis. In other words, the net area of the orthogonal signal is zero. Contrast this with the nonorthogonal case, where the net area of the product is not zero.

Armed with this definition of orthogonality, we now need to know how this helps separate out the sine and cosine components. First, recall that the integral of a cosine over one period is zero. Defining the period of one cycle $\tau = (2\pi/\omega)$, the waveform for any integer multiple of the frequency ω is $x(t) = \cos k\omega t$. The net area over one cycle is $\int_{t=0}^{t=\tau} \cos k\omega t\ dt$. For any integer $k \neq 0$, this is $(1/k\omega) \sin 2k\pi$, which is always zero, irrespective of the value of k. If we were to extend this over multiple cycles, the result will still be zero.

The product of two cosine signals for *differing* integral-multiple frequencies $k\omega$ and $m\omega$ is

$$\cos k\omega t\ \cos m\omega t = \frac{1}{2}[\cos(k\omega t + m\omega t) + \cos(k\omega t - m\omega t)]$$

$$= \frac{1}{2}\{\cos[(k+m)\omega t] + \cos[(k-m)\omega t]\} \tag{3.142}$$

To find the area under the resulting product, note that if k and m are integers, $k \pm m$ is also a set of integers. Accordingly, we have the product of two cosines. Integrating this,

$$\int_{t=0}^{t=\tau} \cos k\omega t\ \cos m\omega t\ \mathrm{d}t = \frac{1}{2}\int_{t=0}^{t=\tau}[\cos(k\omega t + m\omega t)$$
$$+ \cos(k\omega t - m\omega t)]\mathrm{d}t$$
$$= \frac{1}{2}\int_{t=0}^{t=\tau}\{\cos[(k+m)\omega t]$$
$$+ \cos[(k-m)\omega t]\}\mathrm{d}t$$
$$= \frac{1}{2}\int_{t=0}^{t=\tau}\cos[(k+m)\omega t]\mathrm{d}t$$
$$+ \frac{1}{2}\int_{t=0}^{t=\tau}\cos[(k-m)\omega t]\mathrm{d}t$$
$$= \begin{cases} 0 : \text{ for all integer } k \neq m \\ \frac{\tau}{2} : \text{ for } k = m \end{cases} \tag{3.143}$$

This result shows us that the product of two sinusoids (in this case, cosines) of different frequencies related by an integral multiple is zero. In the specific case when the sinusoids are the same (mathematically, $k = m$), then the result is a constant. This means that multiple frequencies, of the same phase, can coexist and yet be separated. The separation occurs if $k = m$, and the right-hand term becomes a constant. This fact is employed in the next section, where one of the sinusoids is the received signal, and the other is the locally generated carrier wave.

Next, suppose we have two waveforms of the *same frequency* but of *differing phases*. Using the expansion from Table 3.1 with $\sin \omega t \cos \omega t$, we find that it is mathematically equivalent to $(\sin 2\omega t)/2$. Taking the integral over one cycle results in zero – in other words, sine and cosine may be separated at the receiver. Thus, if waveforms of the same frequency are $90°$ out of phase with each other, then their product will be zero and thus are orthogonal by the above definition. The special case of $90°$ phase difference is quite important and is termed phase quadrature or just *quadrature*. The usual definition is to employ cosine (as in-phase) and sine (as quadrature phase).

3.9.4 Quadrature Amplitude Modulation

If we utilize a single sine wave as a carrier, we can detect that at the receiver and hence demodulate it. If, however, we send a cosine, we can also detect it at the receiver and demodulate the signal it is carrying. As shown in the previous section, sine and cosine are able to coexist in the same channel space or frequency band. If we send a sine with amplitude A_s, and a cosine with amplitude A_c, we can determine A_s and A_c at the receiver, provided we know the phase of each (so as to keep track of which is which). Sine and cosine are said to be orthogonal in a vector sense, and in quadrature, or 90° out of phase, in a time sense.

Digital data consists of a serial stream of binary data – sequential 1s and 0s – which is typically converted into a bipolar sequence of amplitudes ± 1. We could modulate each bit in turn serially, using just two amplitude levels, or two phase values. This may be visualized using the axes as shown in Figure 3.58a, where one data point is shown, with its corresponding amplitude and phase. The combined amplitude and phase corresponds to a particular selection of sine amplitude and cosine amplitude.

Figure 3.58b shows four points on the sine–cosine or *IQ* plane, and four points will permit representation of two binary digits (00, 01, 10, 11). So the question arises, can we extend this approach? If we scaled the appropriate *I* and *Q* amplitudes carefully, we could place 8 points on the plane as illustrated in Figure 3.58c. Because the phase is changing (but not the amplitude) for each point, it makes sense to refer to this as *Quadrature Phase Shift Keying* (QPSK). Finally, Figure 3.58d shows 16 points, each with differing amplitude and/or phase. The particular combination of amplitude and phase (or, equivalently, sine and cosine) uniquely selects one point. Since there are 16 points, 4 bits could be represented at once. Of course, this approach could be extended even further, using distinct amplitude/phase combinations. This is termed *Quadrature Amplitude Modulation* (QAM) in general, since the amplitude of separate, quadrature-phase signals are employed. The points on the *IQ* plane comprise a *constellation* for a particular modulation scheme.

To develop a general approach to analyze this, consider Figure 3.58a again, where just one point on the plane is shown. The diagram shows that this combination is, in fact, a sine wave (whose amplitude is the length R of the arm from the origin to the defined point) with a phase shift φ (the angle from the cosine axis). Since the point is defined as $I \cos \omega t + Q \sin \omega t$, we can use trigonometry to rewrite this point as

$$I \cos \omega t + Q \sin \omega t = R \cos(\omega t + \varphi) \tag{3.144}$$

Applying the expansion for $\cos(x + y)$ to $R \cos(\omega t + \varphi)$ on the right-hand side, and equating in turn the corresponding $\sin \omega t$ and $\cos \omega t$ terms on the left, leads to

$$R = \sqrt{I^2 + Q^2} \tag{3.145}$$

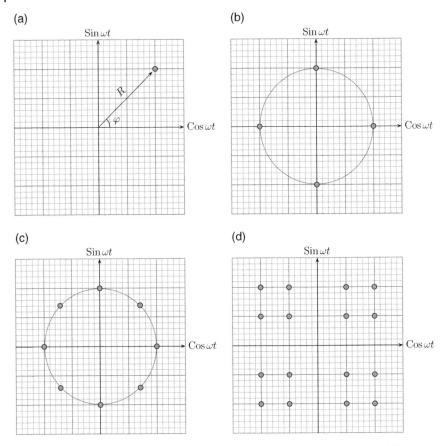

Figure 3.58 Illustrating sine and cosine signals on an *IQ* plane for quadrature modulation. (a) A single point with magnitude *R* and phase φ. (b) Four points with the same magnitude and 90° phase difference. (c) Eight points with the same magnitude but a 45° phase difference. (d) Sixteen points with differing magnitude and phase.

$$\varphi = -\arctan\left(\frac{Q}{I}\right) \tag{3.146}$$

Note that some authors prefer to emphasize that the phase angle is a delay,[5] by defining $R\cos(\omega t - \varphi)$, and as a result $\varphi = \arctan(Q/I)$.

Figure 3.59 shows a block diagram of the QAM system. The two signals to be modulated, $m_1(t)$ and $m_2(t)$, represent the amplitudes for each axis, *I* and *Q*. These are multiplied by cosine and sine carriers, respectively. The resulting signals are orthogonal, and adding them before transmission does not destroy any information. The demodulator shown in Figure 3.60 is effectively the inverse

---

5 Differing definitions of phase angle and frequency go back a long way; see, for example, van der Pol (1946).

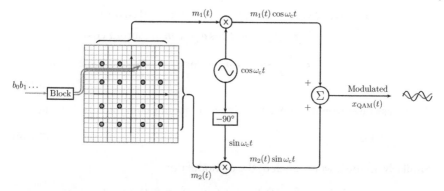

Figure 3.59 QAM modulation diagram. The input bit combination (4 bits here) selects one of 16 sine and cosine amplitude pairs within the constellation.

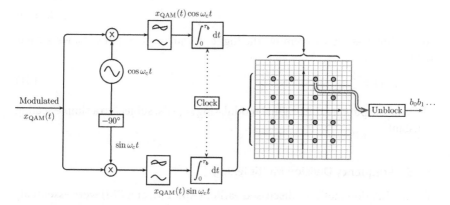

Figure 3.60 QAM demodulation. Multiplication by the sine and cosine carrier separately, followed by integration over one or more cycles, determines the amplitude and hence position in the constellation. The original bit pattern may then be looked up directly.

of the modulation, with the per symbol integrators (accumulators) added, as discussed in Section 3.9.2. Note, though, that the demodulator must know the frequency and phase of the carriers.

To see why this works, suppose the QAM system has modulation inputs $m_1(t)$ and $m_2(t)$. The output is then

$$x_{\text{QAM}}(t) = m_1(t)\cos\omega_c t + m_2(t)\sin\omega_c t \tag{3.147}$$

Demodulating this composite signal involves multiplication of the incoming modulated signal by cosine and sine waveforms, which are phase locked to the received signal. In the upper branch of the demodulator,

$$x_{\text{QAM}}(t) \cos \omega_c t = [m_1(t) \cos \omega_c t + m_2(t) \sin \omega_c t] \cos \omega_c t$$
$$= m_1(t) \cos \omega_c t \cos \omega_c t + m_2(t) \sin \omega_c t \cos \omega_c t$$
$$= \frac{m_1(t)}{2}(\cos 2\omega_c t + \cos 0) + \frac{m_2(t)}{2}(\sin 2\omega_c t + \sin 0)$$

$$(3.148)$$

A lowpass filter would remove the higher frequency components, leaving

$$y_1(t) = \frac{1}{2}m_1(t) \tag{3.149}$$

Similarly, in the lower branch of the demodulator,

$$x_{\text{QAM}}(t) \sin \omega_c t = [m_1(t) \cos \omega_c t + m_2(t) \sin \omega_c t] \sin \omega_c t$$
$$= m_1(t) \cos \omega_c t \sin \omega_c t + m_2(t) \sin \omega_c t \sin \omega_c t$$
$$= \frac{m_1(t)}{2}(\sin 2\omega_c t + \sin 0) + \frac{m_2(t)}{2}(\cos 0 - \cos 2\omega_c t)$$

$$(3.150)$$

After a lowpass filter to remove the higher frequency components, we are left with

$$y_2(t) = \frac{1}{2}m_2(t) \tag{3.151}$$

Thus the outputs are the original modulating signals, subject to a simple scaling constant.

3.9.5 Frequency Division Multiplexing

The modulation methods discussed earlier (AM, FM, and PM) were essentially concerned with modulating one signal onto a higher frequency carrier. This was perfectly reasonable for analog transmission, as there is only one signal such as voice, music, or television line scans. A problem that arises is that of sending several analog channels (for example, telephone conversations) on the same RF signal or cable – for example, a single link between major population centers. This led to the notion of *Frequency Division Multiplexing* (FDM), whereby each separate signal was given its own carrier for modulation at the transmitter, and recovered at the receiver using a local carrier.

The QAM and QPSK methods discussed in the previous sections are well suited to digital modulation, since they can encode more than one bit at a time. More recently, the combination of these two ideas – using multiple frequencies together with multiple channels – has emerged as one of the most important methods in digital transmission. Such methods are termed *Orthogonal Frequency Division Multiplexing* (OFDM), because they use multiple sine and cosine carriers.

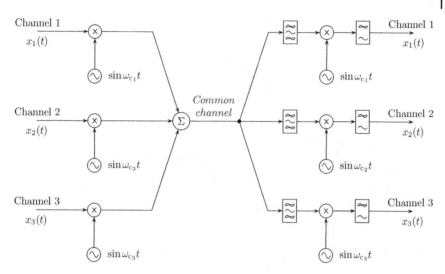

Figure 3.61 Frequency division multiplexing for multiplexing multiple channels on the one physical carrier. A separate subcarrier frequency is assigned to each channel.

OFDM is employed in many types of wireless networks, as well as for broadband over twisted pair telephone networks where the term Discrete Multitone (DMT) is also used. In this way, a data bit stream may be split up into numerous "subchannels" for parallel transmission, thus facilitating high-speed or "broadband" data transmission.

The earliest notions of what we might call FDM came with telegraphy in the 1800s (Weinstein, 2009). What we would broadly understand as FDM today is depicted in Figure 3.61. Here, we see the earlier concept of mixing, but applied using different carrier frequencies so as to move each signal source to its own separate frequency band or channel. Provided that the bandwidth of each signal does not overlap the adjacent frequency bands, individual signals may be demodulated at the receiver. The net result is that one common channel or bearer (such as microwave, coaxial, or other medium) may be used for multiple simultaneous channels. Thus, we have the notion of a subchannel.

Figure 3.62 depicts the signal-domain representation of FDM over a common channel. Each separate channel occupies a frequency band; in practice this would be a small but finite bandwidth, depending on the modulation scheme employed. For M subchannels of bandwidth B, the theoretical bandwidth requirement is something greater than the product $B \times M$, since the band edges will not be perfect, and thus a small space, termed a guard interval, is required between each subchannel.

Frequency division multiplexing

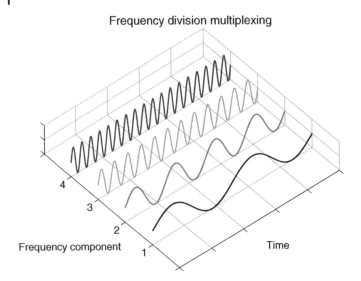

4

3

2

Frequency component 1

Time

Figure 3.62 FDM may be visualized as multiple signals evolving in time but separated in frequency.

3.9.6 Orthogonal Frequency Division Multiplexing

The basic FDM scheme may be extended to effectively double the capacity of each subchannel by using orthogonal sine and cosine carriers. As we now show, it is indeed possible to separate these at the receiver, at least in theory. Problems arise with real channels due to their dispersive nature, which usually requires relaxation of some of the channel bandwidth constraints.

This method, termed Orthogonal Frequency Division Multiplexing or OFDM, has been known to exist for some time (Weinstein and Ebert, 1971);[6] however, practical considerations with analog signal processing limited its use. Using digital or discrete-time implementations has opened up a vast field of applications for OFDM, especially in digital wireless transmission.

The types of waveform present in FDM are depicted in Figure 3.63. Here, the stream of data bits to be encoded is converted into a particular sinusoid, with a defined amplitude. The frequency defines the subcarrier, and the amplitude is defined by the particular binary value of each bit b, which of course may take on one of the two values. Extending this FDM to the case of OFDM, Figure 3.64 shows the use of simultaneous sine and cosine waveforms for each subcarrier. In the example illustrated, a pair of bits is used to define the amplitude of sine and cosine. This would yield $2^2 = 4$ points on the constellation diagram. Naturally, this may be extended to any number of amplitude levels, and adding one bit to either sine or cosine results in double the number of possible levels on that axis.

---

6 See also historical summaries in Weinstein (2009) and LaSorte et al. (2008).

Figure 3.63 Using FDM to multiplex a bit stream. The amplitude $A_{0|1}$ means that A takes on different values depending on whether the bit b is 0 or 1. Typically, these would be equal in magnitude but opposite in sign.

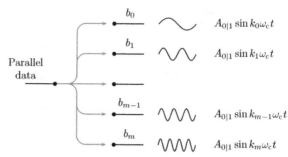

Figure 3.63 content:

Parallel data

b_0 $A_{0|1} \sin k_0 \omega_c t$

b_1 $A_{0|1} \sin k_1 \omega_c t$

b_{m-1} $A_{0|1} \sin k_{m-1} \omega_c t$

b_m $A_{0|1} \sin k_m \omega_c t$

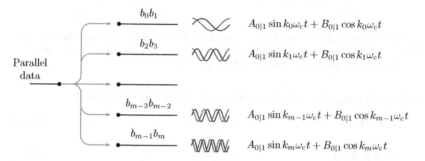

Parallel data

$b_0 b_1$ $A_{0|1} \sin k_0 \omega_c t + B_{0|1} \cos k_0 \omega_c t$

$b_2 b_3$ $A_{0|1} \sin k_1 \omega_c t + B_{0|1} \cos k_1 \omega_c t$

$b_{m-3} b_{m-2}$ $A_{0|1} \sin k_{m-1} \omega_c t + B_{0|1} \cos k_{m-1} \omega_c t$

$b_{m-1} b_m$ $A_{0|1} \sin k_m \omega_c t + B_{0|1} \cos k_m \omega_c t$

Figure 3.64 Using OFDM to multiplex a bit stream. As well as multiple subcarrier frequencies, quadrature signals are used on each subchannel.

Consider now one specific frequency. Because we have two components – sine and cosine – we can represent the particular values of amplitude R and phase φ as shown in the diagram of Figure 3.65. Furthermore, because we have shown that, mathematically, these two sine and cosine components can coexist and be separated out at the receiver, many possible combinations of R and φ may be present simultaneously. Thus in Figure 3.65, the 16 points shown may be generated by using four combinations of cosine, and four combinations of sine. This lends itself to a digital representation: the 16 possible combinations of sine and cosine amplitudes and phase can be used to represent four digital bits, since $2^4 = 16$.

It will be helpful to represent these sine plus cosine combinations as a single sine with amplitude change and phase shift. Consider the scaled sum of sine and cosine as

$$I \cos \omega t + Q \sin \omega t = R \cos(\omega t + \varphi) \tag{3.152}$$

Expanding the right-hand side

$$R \cos(\omega t + \varphi) = R \cos \omega t \cos \varphi - R \sin \omega t \sin \varphi \tag{3.153}$$

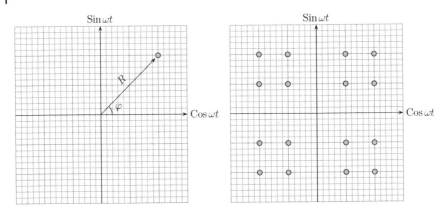

Figure 3.65 A point defined by a sine and cosine amplitude (left) is equivalent to a sine with magnitude R and phase φ. Multiple points may be represented in this way (right). The 16 points shown are then able to represent a 4-bit quantity.

Equating, in turn, the coefficients of $\cos \omega t$ and $\sin \omega t$, we find that

$$I = R\cos\varphi \tag{3.154}$$

$$Q = R\sin\varphi \tag{3.155}$$

and as a result

$$R = \sqrt{A^2 + B^2} \tag{3.156}$$

$$\varphi = -\arctan\left(\frac{Q}{I}\right) \tag{3.157}$$

The weighted sum of sine and cosine is really a sine with an amplitude change and a phase shift. In reverse, a sine with a certain amplitude and phase is the same as the sum of sine and cosine, suitably weighted in amplitude. We can go from one representation to the other via these formulas.

Finally, we turn to the task of demodulating the OFDM waveform. That is, given a received waveform apparently generated via a combination of R and φ, it is necessary to determine which particular combination of sine and cosine the received signal corresponds to. Using the amplitude of the sine and cosine component uniquely determines the point on the plane and hence the specific four bits in our current example.

The recovery of the magnitude of each component consists of multiplication and integration over one cycle, as depicted in Figure 3.66. The resulting amplitude of the component then determines one position on the plane, and the process is repeated for the cosine component. Thus, the two values fix a unique point in the constellation plane.

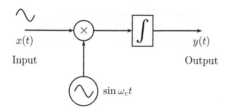

Figure 3.66 Multiplying the incoming wave by sine (or cosine) and integrating results in a scaled estimate of the amplitude of that particular component. The integration (or accumulation) is assumed to be performed over one symbol time, after which the integrator is reset to zero.

3.9.7   Implementing OFDM: The FFT

The preceding sections have shown that it is possible to encode multiple sines and cosines over a range of carrier frequencies, termed subchannels. However, this results in a very complicated system – there are a great many multiplications of waveforms required at the transmitter, as well as the receiver. Digital processing of the signals is the key to unlocking the potential of OFDM.

A significant breakthrough was the realization that the conversion from source signal to modulated signal could be performed by the *Discrete Fourier Transform* (DFT), and furthermore that the demodulation could be performed by its counterpart, the *Inverse Discrete Fourier Transform* (IDFT) (see Weinstein and Ebert (1971) and references therein). Additionally, a much faster way to implement the DFT was suggested, using the *Fast Fourier Transform* (FFT), which had recently been discovered (Cooley and Tukey, 1965). These two ideas form the basis of OFDM as it is employed today.

The DFT takes a given waveform and computes the corresponding set of sine and cosine functions, of related frequencies, which would constitute that waveform. In reverse, the IDFT takes the magnitudes of the sines and cosines and determines the corresponding time-domain waveform.

Figure 3.67 illustrates the concept of Fourier analysis. The input waveform at the top is multiplied, in turn, by sine and cosine waveforms and the result added to form weighting coefficients A and B. This is repeated for higher frequency waveforms. The resulting set of A and B coefficients specify the original waveform. This process is sometimes called *analysis*, since it analyzes the input waveform to produce a result.

In reverse, taking a set of sine and cosine waveforms at a suitable range of frequencies, and weighting them by the corresponding A and B coefficients, will reproduce the original waveform. A key issue is exactly how many waveforms, over what frequency range, is necessary to render the original waveform. This process is sometimes called *synthesis*, since it resynthesizes the original waveform using a weighted sum of prototype sine and cosine waveforms.

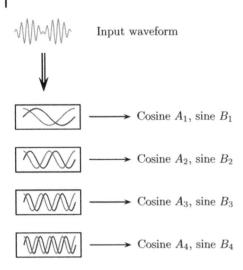

Input waveform

Figure 3.67 Fourier analysis of an input waveform determines the magnitude of each of the sine and cosine components at the various frequencies.

→ Cosine A_1, sine B_1

→ Cosine A_2, sine B_2

→ Cosine A_3, sine B_3

→ Cosine A_4, sine B_4

The DFT is most easily expressed in terms of complex exponentials, though it is important to remember that this is just a notational issue, and that these complex exponentials are essentially just sine and cosine functions that need to be kept separate for the most part (and therefore, need a method of tracking which is which).

The complex operator $j = \sqrt{-1}$ is used to separate the conventional or real parts from the j or imaginary parts. A point on the complex plain may be represented as

$$Re^{j\varphi} = R(\cos\varphi + j\sin\varphi) \tag{3.158}$$

Using this notation, the DFT, which transforms N time samples $x(n)$ into N frequency samples $X(k)$, is defined as

$$X(k) = \sum_{n=0}^{N-1} x(n)e^{-jn\omega_k}$$
$$\omega_k = \frac{2\pi k}{N}$$

Both the DFT and IDFT use the following variables:

Figure 3.68 A point on the complex plane defines the cosine magnitude (real part) and the sine magnitude (negative imaginary part).

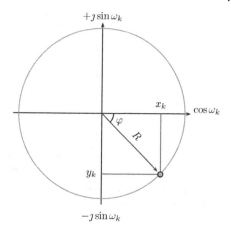

N	The total number of input samples
n	The sample index; used to reference each input sample
k	Index used to reference each output sample
$x(n)$	The value of each input sample
$X(k)$	The calculated value of output frequency point
ω_k	$\frac{2\pi k}{N}$, the frequency of the kth sinusoid

The convention adopted here is to use n, N for time indexes and data lengths, respectively, and k, K for frequency indexes and lengths. Also, x is used for time components (real values), and X for frequency components (which may be real or imaginary). The radian frequency ω_k may be thought of as the frequency of the kth sinusoid. Although the DFT produces N output samples, only $N/2$ of these are unique – the remainder are the complex conjugates of the first half. Figure 3.68 illustrates the positioning of one particular sinusoidal waveform point on the complex plane. The horizontal axis to the right indicates the relative amount of the cosine component, while the downwards direction of the vertical axis indicates the relative amount of *negative* sine. This comes about from the equation above, since the exponential defining the sine and cosine is $e^{-j\omega_k} = \cos \omega_k - j \sin \omega_k$.

To understand the application of the DFT equation, consider Figure 3.69, which illustrates how a single *cosine* waveform with samples $x(n)$ is transformed into the corresponding frequency points $X(k)$ on the complex plane. The imaginary or sine part is zero, but the cosine part is nonzero at index [7] $k = 4$. We can

---

7 Do not forget that MATLAB indexes start at 1, not 0.

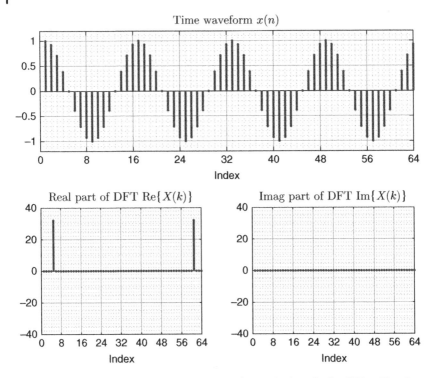

Figure 3.69 The DFT of a cosine wave corresponds to a single real value $X(k) = 32 + j0$ (here at $k = 4$) and its symmetrical counterpart at $(N - 1) - k$ (here $63 - 3 = 60$). Note that the MATLAB indexes displayed start at 1, not 0 as in the equations.

think of this in one of two ways. First, there are exactly four complete cycles in the sampled-data record. Alternatively, each cycle is comprised of 16 samples, corresponding to $2\pi/16$ radians per sample phase advance. The transformed frequency samples (64 here) map from 0 to 2π radians, and the first point is at $4/64 \times 2\pi$ radians.

Notice that there is a second point in the $X(k)$ samples, which is at the same magnitude in this case. This symmetrical point is always present, but does not provide any further useful information.

Next, consider Figure 3.70, which shows a *sine* waveform. In contrast, we have no cosine components (real values), but there are two points in the imaginary part. One has a value of $-j32$, and the other $+j32$. These two are symmetrical, and in fact complex conjugates of each other. The fact that the point at $k = 4$ is $X(k) = -j32$ informs us that it is a *positive* sine component, as per the conventions outlined in Figure 3.68.

Finally, consider the magnitude of these points. In the frequency domain, a magnitude of 32 results from a peak amplitude of one in the time domain. To

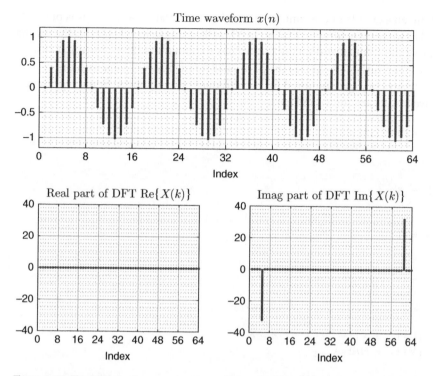

Figure 3.70 The DFT of a sine wave corresponds to a single imaginary value $X(k) = 0 - j32$ (here at $k = 4$) and its symmetrical counterpart at $(N - 1) - k$ (here $63 - 3 = 60$). Note that the MATLAB indexes displayed start at 1, not 0 as in the equations.

convert from the frequency point magnitude to the time amplitude, we must scale (divide) by $N/2 = 64/2$.

The inverse DFT works as would be expected: taking frequency points and mapping them to the corresponding time-domain waveform. As the previous examples showed, we must set the magnitude of the $X(k)$ points correctly in terms of the number of sample points, and furthermore it is necessary to ensure the correct placement on the cosine or negative sine axes. Finally, the complex-conjugate symmetry must also be obeyed.

The equation for the IDFT is

$$x(n) = \frac{1}{N} \sum_{k=0}^{N-1} X(k) e^{jn\omega_k}$$

$$\omega_k = \frac{2\pi k}{N}$$

Notice that, except for the positive sign in the exponent and the scaling $1/N$, it is virtually identical to the DFT. The behavior of the IDFT is best illustrated with an example in the context of waveform generation — and this is precisely what we require for OFDM. The MATLAB coding below illustrates the setting up of the frequency-domain $X(k)$ values to obtain the time-domain $x(n)$ samples. The variable k defines the particular frequency required, with the correct complex conjugate required for each frequency component. Some representative results are shown in Figure 3.71.

```
N = 64;
X = zeros(N, 1);

% subcarrier number, start at 1 up to N/2
k = 3;

% amplitude of this subcarrier
A = 1;

% start at component index 1 (2 in MATLAB)
% scale amplitude by N/2
% complementary component has to be complex conjugate
% -1j for sine, +1 for cosine

X(k+1) = -1j*N/2*A; % choose for sine (Q)
X(k+1) = N/2*A;     % choose for cosine (I)

X(N-k+1) = conj(X(k+1));

x = ifft(X);

% xr should be zero, but may have small imaginary
% components due to arithmetic rounding
xr = real(x);
stem(x);
```

In this example, we have used the FFT, which produces the same results as the DFT for a given set of inputs. The advantage of the FFT is that it requires considerably fewer computations. For example, a DFT of order $N = 1024$ would require approximately $N^2 \approx 10^6$ complex operations, whereas the FFT requires of the order of $N\log_2 N \approx 1000 \times 10 = 10\,000$ operations – a considerable simplification. This is especially important for real-time implementation, as would be required for a communications system. The one requirement in using the FFT is that the number of samples must be a power of two. Thus 1024 samples is acceptable (since $1024 = 2^{10}$), whereas 1000 samples is not acceptable.

Finally, we can see how the FFT is useful in both modulation and demodulation for OFDM. We simply need to specify the set of frequencies and their

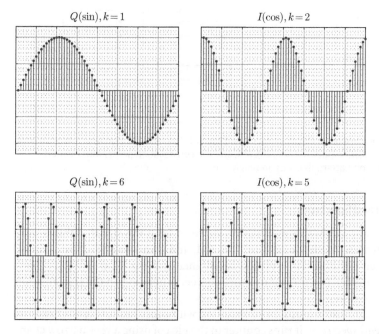

Figure 3.71 Some example *IQ* signals for OFDM, using the DFT.

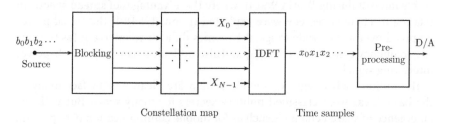

Figure 3.72 Transmission process for OFDM using the IFFT. The data is formed into blocks and used to define the constellation pattern, which is converted into the correct waveform to be transmitted using the inverse FFT.

relative sine/cosine components, as was done in the previous example. The inverse FFT (IFFT) is performed at the transmitter, to create the time-domain waveform actually transmitted from the constellation of sine/cosine points. The processing blocks are depicted in Figure 3.72.

At the receiver, the inverse operation or FFT is employed to recover the originally specified constellation points from the received time waveform. This is illustrated in Figure 3.73.

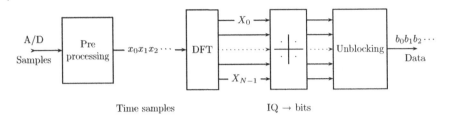

Figure 3.73 Reception of OFDM signals using the FFT. The received signal is converted back into the constellation pattern using the FFT, and the constellation points thus defined determine the bit pattern that was originally sent.

3.9.8 Spread Spectrum

Traditional approaches to modulation strive to keep the signal within a very tight bandwidth. Methods such as AM, FM, and standard PSK are examples. OFDM extends this idea to multiple, simultaneous channels, where each channel may be utilized to a greater or lesser degree, depending on the transmission channel conditions.

An approach that is useful in conjunction with other modulation schemes is that of *spread spectrum*. It runs counter to the idea of using a very narrow channel and in fact "spreads" the transmission over a wide bandwidth. The origins of spread spectrum date to the 1930s (Scholtz, 1982; Price, 1983), with considerable interest during World War II, where the advantages of spread spectrum relating to channel secrecy were exploited (Kahn, 1984). If the signal power is spread over a very wide range, it becomes difficult to intercept (as the carrier frequency continually changes) and also difficult to jam with a high-power interfering signal.

The military advantages of such a system are clear, and in fact many of the basic ideas were classified military secrets for many years. But with the emergence of shared channels such as for mobile communication (Cooper and Nettleton, 1978; Magill et al., 1994) and short-range wireless in unrestricted frequency bands, some new advantages slowly became evident.

In the case of multiuser mobile communications, there was originally only a choice between two alternatives for separating communication channels: either assign each user pair a separate frequency or use the same frequency but permit each user to access the channel only for a short duration or timeslot. The former is termed *Frequency Division Multiple Access* (FDMA); the latter, *Time Division Multiple Access* (TDMA). Each requires complete cooperation between all users accessing the radio bandwidth, and a method for assigning frequency bands or transmission timeslots. Moreover, the capacity of the local network is strictly limited once the allowable frequency channels (FDMA) or timeslots (TDMA) are used up.

A somewhat different approach is employed in a third method, *Code Division Multiple Access* (CDMA). CDMA employs the idea of *spread-spectrum* (SS), which permits multiple users to access the same channel bandwidth, with limited interference between each other's channels. As more users are added, the performance in terms of minimization of interference degrades gradually, rather than abruptly. Importantly, there is little or no configuration required in terms of frequency channel setup or timeslot allocation. This is vital in supporting low-overhead configuration of mobile communication networks, where distinct cells have overlapping geographical coverage areas.

As well as permitting multiple users to share the same channel space, spread spectrum increases the resistance to fading of the RF signal power, and interference in RF bands. The efficient use of scarce radio bandwidth is an essential characteristic of the spread spectrum approach. As demonstrated in Chapter 5, the capacity of a digital channel C (in bits per second or bps) may be related to the channel bandwidth B Hz and signal-noise ratio S/N by the channel capacity formula

$$C = B \log_2 \left(1 + \frac{S}{N} \right) \tag{3.159}$$

For a fixed signal-to-noise ratio S/N, the only way to increase the capacity C of a channel is to increase its bandwidth B. Spreading the narrow spectrum to a wider channel does this, and if we need a multi-user environment, the wider bandwidth isn't really a problem, since we would need a wider bandwidth anyway.

There are two main classes of operation of SS systems, and each has advantages and disadvantages. One is *Frequency Hopping* (FH) and is depicted in Figure 3.74. Each endpoint uses an agreed modulation and demodulation method, as usual. This is often FSK or PSK, or some higher-capacity variation. The essential difference is that instead of using a fixed channel, the center frequency "hops" in a known pattern, usually many times per second. In effect, the transmission is modulated onto a carrier frequency, which hops for a short time from one frequency to another, and stays on each channel for a short dwell time.

The hopping pattern is what gives the spread of frequencies over the transmission bandwidth. Clearly, the receiver must adjust its channel frequency to find each new transmitter frequency. This is done using a Pseudo-Noise (PN) generator, which produces at any instant one value from a set of known values. Each value is used to tune a frequency synthesizer for the channel center, and after a known time, a new value (and thus new frequency) is produced. It is important to realize that the values are not truly random, but pseudorandom. This means that the hopping pattern repeats after a certain number of hops. However, if users start at a different point in the sequence, they do not interfere with each other.

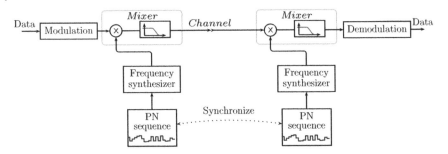

Figure 3.74 Spread spectrum frequency hopping. The center frequency for each transmission time is pseudorandom but synchronized between sender and receiver. Usually a number of bits are transmitted for each hop, making the hop rate slower than the bit rate.

The second approach to SS is termed *Direct Sequence* (DS). The signal is again modulated and demodulated, and the carrier is spread over a wide frequency range. However, whereas FH typically uses a distinct frequency for several bits of transmission, DS uses several hops for each bit. They are not distinct frequency channels as such. Each data bit to be transmitted is combined using an XOR operation with a PRBS, which has a value of 0 or 1. The period of the PRBS is very short – much shorter than each bit interval – and is termed a *chip*. Once again, there is a synchronization issue to be solved.

Figure 3.75 shows a block diagram of a *Direct-Sequence Spread-Spectrum* (DSSS) system, with representative waveforms in Figure 3.76. The input bit stream is first modulated by the high-speed chip pattern, which determines the phase angle of the carrier. This in turn is used to modulate the chip stream, and one of a number of methods such as PSK or QPSK may be employed. The end result is that, rather than distinct frequency hops as in *Frequency-Hopping Spread-Spectrum* (FHSS) as described previously, the DSSS approach blends the signal over a wide bandwidth. This has considerable advantages in mobile communications, where an unknown number of users may try to use the same bandwidth. The likelihood of interference is greatly reduced, and if interference does occur, it is only for a relatively short time. Moreover, error-checking codes may be used to recover the correct bit sequence.

The chip pattern is clearly important, and a simpler approach that has found use in practical systems is the 11-bit Barker code. The Barker code is one of a defined set of integer plus/minus sequences with certain mathematical properties (Weisstein, 2004), which are useful for receiver synchronization, as we now demonstrate. This code is defined as the pattern of positive and negative levels, and the length-11 code is

$$\mathbf{b} = [+ + + - - - + - - + -]$$

The receiver's task is to find this pattern, given the data that was transmitted. Consider that there may be an arbitrary delay from transmission to reception.

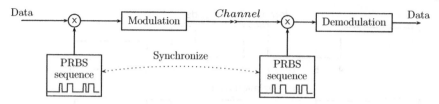

Figure 3.75 Spread spectrum direct sequence. Each bit is subdivided into several chips for transmission, using a pseudorandom binary sequence that is synchronized between sender and receiver. Thus, several chips make up each bit.

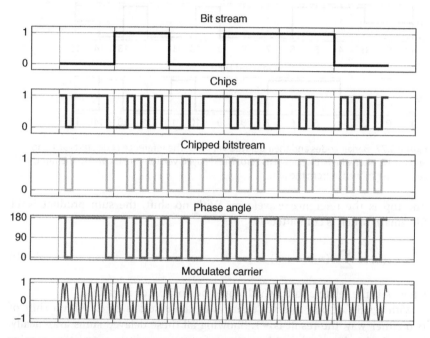

Figure 3.76 Waveforms associated with a direct-sequence spread spectrum design. The carrier itself is phase modulated according to the input bit stream and the chip stream. In this example, the bit stream is used in conjunction with the chip stream to determine the carrier phase, and only one cycle of carrier is shown per chip for clarity.

The problem for the receiver is to appropriately synchronize with the transmitter. It must search for the given pattern and must be able to find that pattern reliably.

The problem then becomes one of pattern matching, which is performed using a correlation algorithm. The received sequence is searched for the known pattern, which is mathematically where the correlation is maximized. Figure 3.77 illustrates the waveforms that have to be considered by the receiver.

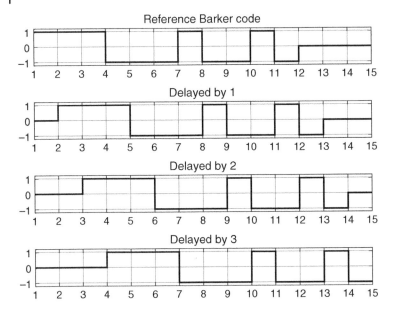

Figure 3.77 Barker codes and their delayed versions. The reference code starts at 1 and ends at 12, after which it is shown as zero. The delayed versions are moved to the right, with zero values moved in from the left.

The top is the reference waveform. With no shift, the sum product is 11 (decimal), which is computed as

$$c(k) = \sum_{n=0}^{L-k} b(n)b(n+k) \tag{3.160}$$

where k is the relative shift, which could be positive or negative.

With a shift of one interval either way, the result is zero. With a shift of two either way, the result is −1. Showing all these delayed sums graphically, Figure 3.78 indicates that this pattern of no or very little correlation continues for all possible shifts. Contrast this with the situation if we did not use an appropriate code. Effectively, the chip stream is all-1s, and the correlation of this sequence is shown in Figure 3.78 (top). A shift of one time unit (positive or negative) looks quite similar to no shift at all, and thus there is a good chance that a receiver would incorrectly interpret the received signal if noise was present.

For the 11-bit Barker code, the peak sum is 11, but the sums at other delays are either 0 or −1. This means that the worst-case similarity is 1/11, and so the relative similarity for delays other than zero is approximately 20 dB less than the similarity at a delay of zero.

The Barker code is a simple and attractive choice for the spreading function, but not the only one available. Each of the SS methods requires the generation

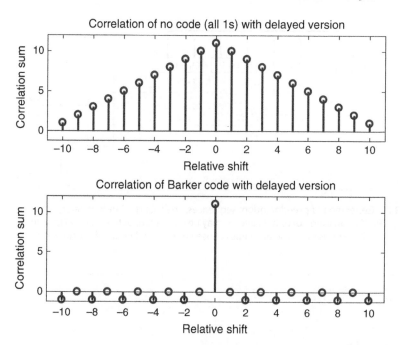

Figure 3.78 All-1s correlation (top) compared to the Barker code correlation (bottom).

of a pseudorandom sequence – one which is seemingly random but repeats after a certain time. Either a set of binary values is required, where the 1/0 value at each bit interval is random, or a set of values drawn from a predetermined range. The first is termed a PRBS or Pseudo-Random Binary Sequence, while the latter is a Pseudo-Noise or PN sequence. Both can be generated using the arrangement of Figure 3.79. Here, we have a shift register containing a number of bits – 8 in the diagram, but typically many more. The bits are shifted from each storage element to the immediate right on receipt of a clock pulse. The leftmost bit obtains its input bit by a feedback arrangement. Some (not all) of the register contents are exclusive-ORed together to form the feedback bit. This feedback bit itself may constitute a PRBS, while the shift register itself forms a PN sequence. The number of storage elements, together with the number of feedback taps, governs the time period over which the sequence repeats. The particular pattern is governed by the presence or absence of feedback taps. Finally, the initial starting point or seed determines where in the pattern space the starting point will occur.

The generation of PN and PRBS signals may be accomplished as shown below. We use the randi() (random integer) function to generate a random sample, which is drawn from the set $[0, 1]$ for a PRBS, or $[0, S - 1]$ for PN. The code below shows the PN sequence scaled to a maximum of $S - 1$.

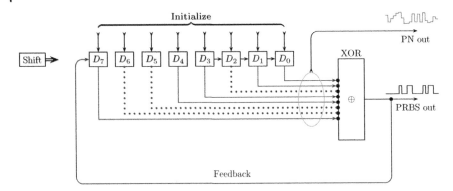

Figure 3.79 Generation of pseudorandom sequences. The Pseudo-Random Binary Sequence (PRBS), consisting of only 1s and 0s, may be generated, or the Pseudo-Noise (PN) sequence that is composed of discrete values chosen from a total range of possible values.

```
% number of samples in total
N = 4096;

% scale for 8 bit unsigned integer
S = 2^8 - 1;

% samples per bit
M = 256;

% number of bits
B = round(N/M);

% select either PRBS or PN
x = S*randi([0 1], [B, 1]); % PRBS - values 0 or 1
x = randi([0 S], [B, 1]);   % PN - values in range 0 to S-1

x = repmat(x', [M, 1]);
x = x(:);
xi = uint16(x);

plot(xi);
axis([0 N 0 S+1]);
title('Sampled Pseudo-Random Binary Sequence');
```

3.10 Chapter Summary

The following are the key elements covered in this chapter:

- The concepts of analog modulation, including AM, SSB, FM, and PM.
- Some approaches to demodulation of analog modulation.
- The notion of phase lock and the PLL and Costas loops.
- Multibit digital modulation using QAM and QPSK.
- The use of advanced modulation techniques, such as OFDM, for increased digital bit rate.
- Spread-spectrum techniques: direct-sequence (DSSS) and frequency-hopping (FHSS).

Problems

3.1 Modulation is the process of impressing a modulation signal $m(t)$ onto a carrier $x_c(t)$ in some way. Typically the amplitude, frequency, or phase of the carrier is manipulated in some way.

a) Explain how each of the modulation types shown in Figure 3.80a comes about.

b) Define each of the modulation types shown in Figure 3.80b. Explain your reasoning in each case.

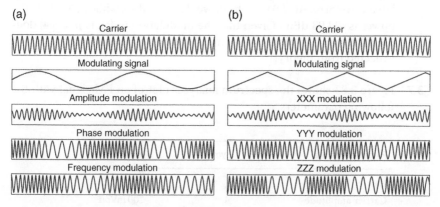

Figure 3.80 Modulation types. (a) Sine signal to be modulated. (b) Triangular ramp signal to be modulated.

3.2 Use Equations (3.28) and (3.29) to show mathematically that, given an AM waveform, A_c and A_m may be determined from the waveform graph.

3.3 Starting with the expansion for $\sin x \sin y$, show that amplitude modulation results in a carrier with amplitude A_c and two sidebands at $\omega_c \pm \omega_m$, each of amplitude $(A_m/2)$.

3.4 The sidebands in AM carry significant power, and their power depends on the modulation index.
   a) Write an equation for the power in the carrier in terms of the carrier amplitude A_c.
   b) Write an equation for the power in each sideband in terms of the sideband amplitude A_m.
   c) Using the above results, show that the total power present in an AM waveform is the carrier power multiplied by a factor of $[1 + (\mu^2/2)]$.
   d) Defining the efficiency η as the power in the sidebands divided by the total power, show that $\eta = \mu^2/(\mu^2 + 2)$. Comment on the efficiency when the modulation index is zero, and when the modulation index is unity.

3.5 Verify the parameters of the waveforms shown in Figure 3.81 using the spectrum plots only.

3.6 Figure 3.82 shows the spectrum analyzer display of an amplitude modulated waveform. At 600 kHz the power is -9.99 dBm, while at 620 kHz the power is -24.93 dBm. Given that the modulation index is $\mu = 0.4$, does the relative power difference as measured correspond to what would be expected from theory?

3.7 Figure 3.83 shows the spectrum analyzer plot of an FM signal with the following parameters:

Parameter name	Symbol	Value
Carrier frequency	f_c	600 kHz
Carrier amplitude	A_c	200 mVpp
Modulating frequency	f_m	20 kHz
Frequency deviation	Δf	80 kHz

Determine β, and from that the expected power levels of the carrier and three harmonics above the carrier.

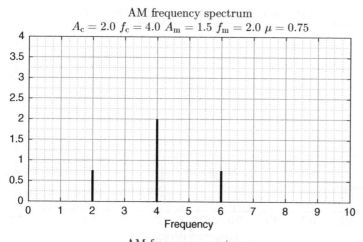

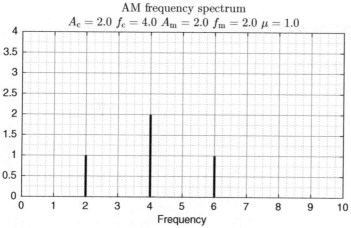

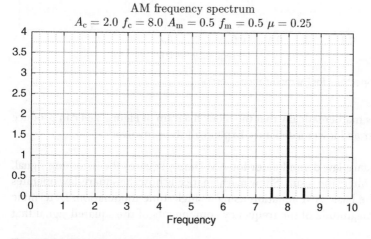

Figure 3.81 AM example spectra.

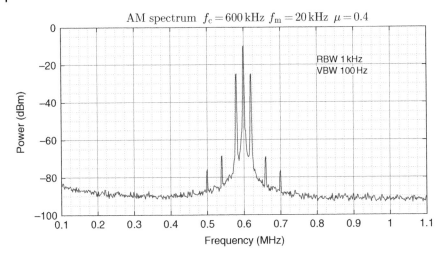

Figure 3.82 Spectrum of an AM waveform, as shown on a spectrum analyzer.

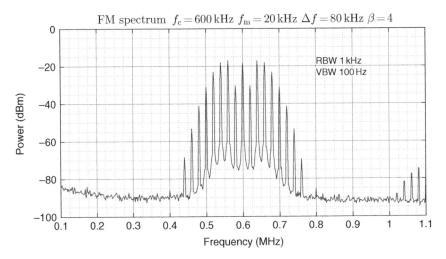

Figure 3.83 Spectrum for FM modulation question.

3.8 Show mathematically that the block diagram of Figure 3.84 can produce upper and lower sidebands. Explain all steps involved.

3.9 A square-law demodulator for AM simply squares the incoming signal, then filters the result. Show mathematically how the square-law operates on the basic AM equation using a single-tone modulation. Determine the magnitude of the frequency components of the squared signal that

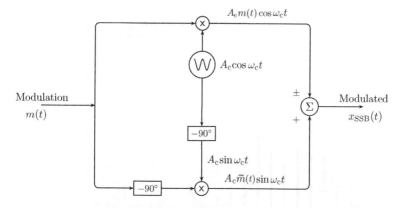

Figure 3.84 Single-sideband (SSB) generation.

would be within the range of the modulating signal's bandwidth. Which is the desired (demodulated) signal and which is unwanted distortion?

3.10 Explain how Bessel tables (Table 3.2) are used to determine the harmonics of an FM waveform.

3.11 Verify the spectral components of the waveforms shown in Figure 3.85 using the parameters given in conjunction with the table of Bessel functions (Table 3.2).

3.12 A single-tone frequency modulated signal written as

$$x_{FM}(t) = 2 \sin(2000\pi t + 2 \sin 4\pi t)$$

has corresponding frequency components

$$x_{FM}(t) = A \sum_{n=-\infty}^{n=\infty} J_n(\beta) \sin(\omega_c + n\,\omega_m)t$$

a) Work out the carrier and modulation frequencies in rad s^{-1} and Hz.
b) What is the value of β? Determine the corresponding frequency deviation.
c) Sketch the magnitude spectrum about the carrier frequency, showing all values of amplitude and frequency.
d) What would the spectrum look like if the frequencies were in MHz rather than Hz?

3.13 *IQ* demodulation may be applied to various modulation schemes, assuming that the carrier is able to be reconstructed at the receiver.

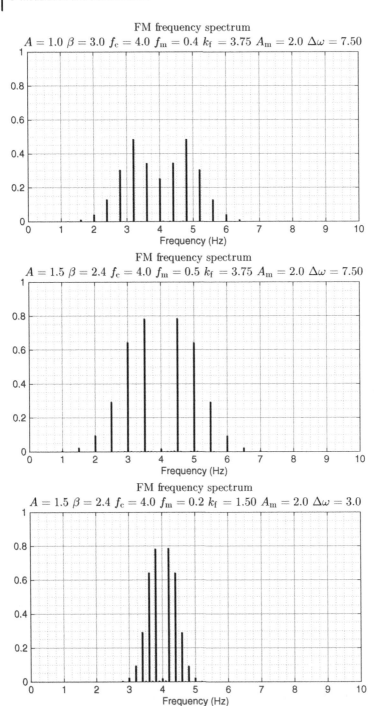

Figure 3.85 FM example spectra.

a) Section 3.8.1 showed how to demodulate a sinusoidal tone when amplitude modulated. Extend this to show how to demodulate any modulation input $m(t)$ that is amplitude modulated.

b) Section 3.8.2 showed how to demodulate a sinusoidal tone when phase modulated. Extend this to show how to demodulate any modulation input $m(t)$ that is phase modulated.

c) Section 3.8.3 showed how to demodulate a sinusoidal tone when frequency modulated. Extend this to show how to demodulate any modulation input $m(t)$ that is frequency modulated.

3.14 To prove that demodulation works for orthogonal signals, assume that the in-phase signal is $I(t)$ and the quadrature-phase signal is $Q(t)$. Then the resulting signal is the sum of these multiplied by their respective carriers, to yield a received signal

$$r(t) = Q \cos \omega t + I \sin \omega t$$

a) By multiplying the received $r(t)$ by the sine carrier $\sin \omega t$ and then integrating and averaging over one cycle time $\tau = (2\pi/\omega)$, prove that

$$\frac{2}{\tau}(Q \cos \omega t + I \sin \omega t) \sin \omega t = I$$

In this way, the I component may be recovered.

b) Similarly, show that multiplying by $\cos \omega t$, integrating and averaging,

$$\frac{2}{\tau}(Q \cos \omega t + I \sin \omega t) \cos \omega t = Q$$

which then recovers the Q component.

4

Internet Protocols and Packet Delivery Algorithms

4.1 Chapter Objectives

This chapter examines the defined standards and methods for Internet data communications, the *protocols*. This encompasses the data formats employed as well as the algorithms that operate on that data in order to facilitate data communication between physically separate devices. On completion of this chapter, the reader should:

1) Be able to define IP, TCP, and UDP, the role of protocol layers, and what functions they perform.
2) Be conversant with the key elements of each protocol, such as IP addresses and TCP sockets.
3) Be able to explain potential pitfalls in networking, such as congestion collapse, and know the approaches used to mitigate these problems.
4) Be able to explain the principles of routing and calculate the shortest path in a routing graph.

There is a vast range of detail concerning the Internet protocols that cannot be discussed in detail in one chapter – the intent here is to explain the salient points and examine some important aspects in depth. In addition to Internet standard RFC documents, a very approachable reference is Kozierok (2005). Detailed explanations of the inner workings of TCP/IP are given in Stevens (1994), while implementation details are covered in Wright and Stevens (1995a).

4.2 Introduction

Suppose we wish to connect a number of devices together for the purposes of exchanging data. What is the best way to do this? What do we mean by "best" in this context? We might want a system that uses the available infrastructure in the most efficient manner or can transmit with the highest possible data rate

Communication Systems Principles Using MATLAB®, First Edition. John W. Leis.
© 2018 John Wiley & Sons, Inc. Published 2018 by John Wiley & Sons, Inc.
Companion website: www.wiley.com/go/Leis/communications-principles-using-matlab

and the lowest possible delay. It may have to be expandable to different physical interconnection methods, from wired to optical, wireless, and satellite, each with different characteristics. For example, a wired connection may be fast and have low delay, whereas a satellite link may be fast but experience high delays. On the other hand, a wireless link may experience interference leading to frequently lost data. If we want to have many devices interconnected, how do they "discover" each other? Finally, we may want to have a system that is scalable to a large number of interconnected devices. In short, we seek the best way to set up a network for seamless connectivity of many devices.

The Internet has become a ubiquitous data communications network. This chapter considers the role of packet switching, network device discovery, data packet routing, and other aspects of what is generally known as Transmission Control Protocol/Internet Protocol (TCP/IP).

In the following, reference is made to the Internet Request For Comment (RFC) documents, which define the standards for various functions, data formats, and operational requirements; in short, the protocols for data communication. A standardization process is important, so that equipment from different vendors will interoperate without difficulties. Even issues such as byte ordering must be carefully defined. A convenient search facility for RFCs is https://tools.ietf.org/html/

Newer RFCs often update earlier ones, and likewise earlier RFCs are often updated or clarified by later RFCs. In this chapter, we endeavor to provide reference to the original RFC where appropriate. Later RFCs updating a standard are denoted by the wording "updated by." Another important role of RFCs is to remove ambiguity that may have arisen in the interpretation of a particular protocol. In addition to defining current standards, RFCs may propose new standards or clarify certain operational characteristics – or even just act as a "request for comments" on proposals and new ideas.

4.3 Useful Preliminaries

This section introduces some concepts that may be helpful in understanding the remainder of the chapter. These are: the notion of packet switching, binary or digital operations, and some data structure basics with code examples.

4.3.1 Packet Switching

One categorization of networks is into either *circuit-switched* or *packet-switched* designs. The Internet is a *packet-switched network*, so it is helpful to define just the what term means initially and why it is a useful concept.

In a circuit-switched system, a direct connection is established between devices at the start of data transfer and is maintained until the data transfer has completed – usually to the exclusion of all other data on that particular channel. On the other hand, a packet-switched system splits the data up into smaller chunks or *packets* of information. Each packet is sent separately, from source to destination, and (perhaps surprisingly) different packets may follow different physical paths. Packet switching is arguably more complex and has more overhead, since each data packet comprising a data transfer acts as a separate entity. However, it makes more efficient use of the available bandwidth for typical transmissions. Consider accessing a particular web page: In the time between one page being loaded and the user selecting a different page, is it really necessary to maintain a physical connection, to the exclusion of all other data traffic? Similarly, in digital transmission of audio conversations, a large proportion of time is spent idle, thus wasting the physical interconnection (which could be used for other data transfer if necessary). Sharing the connection bandwidth among many users drives costs down but also comes with a cost of a different sort: increased complexity.

The case for packet switching, as opposed to circuit switching, is a good one. But this is not immediately obvious and historically was not the preferred option. If we subdivide some data into smaller data "packets," consisting of perhaps hundreds or thousands of bytes each, consider what might transpire. Figure 4.1 illustrates some possibilities.

First, what is the size of each packet? Is a smaller size better or a larger size? Or is a mix of packet sizes acceptable? What about the relative delay between each data packet? Moreover, how does each packet find its way from the source to destination? What happens if the data channel is not ideal, such that one or more bits within the packet are corrupted? What if packets do not arrive at the destination with the correct data, or even not at all? These issues are all addressed by the various Internet protocols.

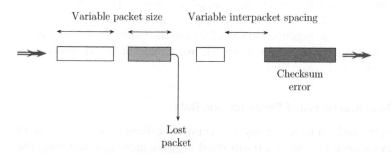

Figure 4.1 Factors affecting packet delivery: the length of each data packet, the time gap between packets, the routing of packets from one place to another, the possible loss of one or more data packets, and errors within a packet that has reached its destination.

Table 4.1 The *truth table* for standard Boolean logic operations.

Bits		Not		And	Or
A	B	$\overline{A}$	$\overline{B}$	$A \cdot B$	$A + B$
0	0	1	1	0	0
0	1	1	0	0	1
1	0	0	1	0	1
1	1	0	0	1	1

Table 4.2 Place-value representation for binary numbers.

2^7	2^6	2^5	2^4	2^3	2^2	2^1	2^0
128	64	32	16	8	4	2	1
0	1	1	0	1	0	0	1

4.3.2 Binary Operations

This chapter examines binary network addresses and utilizes binary operations for many of the routing algorithms. As a result, it is worthwhile to briefly review the important concepts regarding operations on binary numbers.

A *binary digit* or *bit* can take on values of only 0 or 1. In a similar way to defining arithmetic operations for real numbers, we can perform the fundamental logical operations as summarized in Table 4.1.

An N-bit binary number may form an unsigned integer in the range 0 to $2^N - 1$. Similar to the decimal system, where we weight each place from the right by $10^0, 10^1, 10^2, \ldots$, we weight the binary positions by $2^0, 2^1, 2^2, \ldots$ from right to left, until all places given in a number are accounted for (that is, we can add zeros to the left). For example, an 8-bit number 0110 1001 is decomposed as shown in Table 4.2. This is the equivalent of $64 + 32 + 8 + 1 = 105$ decimal. Digital logic functions are used in certain types of bit-based error checking, whereas an arithmetic approach using multibit binary numbers is used in block-based error checking.

4.3.3 Data Structures and Dereferencing Data

In developing code to handle many data communications tasks, it is useful to be able to encapsulate the data transmitted in some more compact way. The same applies to data structures for handling the various data communications subtasks, such as addressing data to the correct destination or representing data in a compressed or reduced space form.

For example, a data packet may consist of the destination address, the source address, and the error-checking status of the packet. These three items naturally belong together. In many programming languages, this grouping is represented by a *data structure*, and MATLAB provides the `struct` keyword for this purpose. Suppose we have a simple data representation problem requiring a packet status (as a character string) and packet length (as an integer). We could use

```
PacketStruct.status = 'good';
PacketStruct.ByteLength = 1024;
```

The packet structure contents may be displayed using `disp()`. If we are unsure of what data type a given variable is, we can use `class()`, and the names of the elements contained within that class are found using `fieldnames()`.

```
disp(PacketStruct)
        status: 'good'
    ByteLength: 1024

class(PacketStruct)
ans =
    struct
fieldnames(PacketStruct)
ans =
    'status'
    'ByteLength'
```

Keeping data together as in the data structure is generally considered to be a good idea. But data also requires code to operate on that data. Thus, the *object-oriented* paradigm extends the idea of a data structure so as to include functions that operate on the structure's data. The conventional procedural function then becomes a *method*, and the data definitions become the *properties* of a *class*. Each class thus defines both data and the way that data is manipulated. An *object* is a particular variable made with a certain class template. The following class definition shows a class named `PacketClass` that contains a string variable and a numeric variable.

```
% put in file PacketClass.m

classdef PacketClass           % value (by-value) class
%classdef PacketClass < handle % handle (by-reference) class
    properties
        status = 'unknown';
```

```matlab
            ByteLength = 0;
    end

    methods
        % constructor
        function PacketObject = PacketClass(InitStatus, ...
                                            InitLength)
            disp('Calling the constructor for PacketClass');
            PacketObject.status = InitStatus;
            PacketObject.ByteLength = InitLength;
        end

        % show contents
        function Show(PacketClassObject)
            fprintf(1, 'Packet status is "%s"\n', ...
                    PacketClassObject.status);
            fprintf(1, 'Packet length is "%d"\n', ...
                    PacketClassObject.ByteLength);
        end

        function [ReturnedObject] = UpdateLength ...
                                    (PacketClassObject)
            PacketClassObject.ByteLength = 100;

            % returns a new object
            ReturnedObject = PacketClassObject;
        end
    end
end
```

When a class is created, it must be initialized. That is the purpose of the method with the same name as the class. It is called the *constructor*. Other methods may be created to manipulate data within an object, according to the class definition. Below are some examples of how such a class is created (or *instantiated*) and some ways of displaying the contents of the class.

```matlab
% TestPacketClass.m

FirstPacket = PacketClass('Ready', 32);

% conventional disp() method
disp(FirstPacket);
```

```
% show method - can invoke two ways
Show(FirstPacket);
FirstPacket.Show();

% try to change the length field
FirstPacket.ByteLength = 64;

% invoke as object.method() - object passed as first
% parameter
FirstPacket.UpdateLength();

% invoke using method(object)
UpdateLength(FirstPacket);

FirstPacket.Show();

SecondPacket = FirstPacket.UpdateLength();
FirstPacket.Show();
SecondPacket.Show();
```

An interesting issue arises when we attempt to change one of the properties within a class. This is illustrated with the UpdateLength() method, which does not change the value of ByteLength in the variable NewPacket. This is because the object itself was not returned – instead, a *copy* of the object is manipulated within the method. If the desired behavior is to change the object within the method, it is necessary to return an object from the method. This type of behavior in coding is termed *pass by value*. As an alternative, the *pass-by-reference* approach does not create a copy of the data, but rather passes a reference to the object. This is variously called a *pointer* in some languages and a *handle* in MATLAB (refer code above).

Dereferencing objects via a handle will turn out to be necessary in some of the data manipulation that follows, and so another simple example is given here. Suppose we create an integer class as follows, containing just a value and one method (the constructor). It is placed in a file of the same name as the class, IntValue.m

```
% put in file IntValue.m

classdef IntValue
    properties
        TheValue = 0;
    end
```

```
    methods
        % constructor
        function [IntValueObject] =  IntValue(val)
            IntValueObject.TheValue = val;
        end
    end
end
```

We can create an `IntValue` object, make a copy of it, and attempt to change its internal property `TheValue` as shown below.

```
clear all

% by value
var1 = IntValue(7);      % assign
var2 = var1;             % copy
var2.TheValue = 999;     % overwrite

% check original and copied value
fprintf(1, 'By value: var1=%d  var2=%d\n', ...
        var1.TheValue, var2.TheValue);
```

The output is

```
By value:       var1=7      var2=999
```

If we now redefine the class as follows using handle to indicate pass by reference, we have

```
% put in file IntHandle.m

classdef IntHandle < handle
    properties
        TheValue = 0;
    end

    methods
        % constructor
        function [IntValueObject] =  IntHandle(val)
            IntValueObject.TheValue = val;
        end
    end
end
```

Calling this version with

```
clear all

% by reference
var3 = IntHandle(8);       % assign
var4 = var3;               % copy
var4.TheValue = 888;       % overwrite

% check original and copied value
fprintf(1, 'By reference: var3=%d  var4=%d\n', var3.TheValue,
  var4.TheValue);
```

results in the output

```
By reference: var3=888   var4=888
```

Notice the key difference: in the first case, the original value was not over-written, whereas in the second case, the value was changed. This is because var4 became a reference to the data, and is not the actual object data itself. This was all due to the fact that the object was declared as a handle using

```
classdef IntHandle < handle
```

Routing of Internet data packets, discussed in this chapter, is one example where data structures, classes, and handles may be used. Another is the encoding of digital data, the subject of the next chapter.

4.4 Packets, Protocol Layers, and the Protocol Stack

A logical distinction is between *local area* and *wide area* networks. Historically, a local area network or LAN consisted of devices in relatively close physical proximity: a computer and a printer or a file server, for example. There are many types of LANs, such as wired or wireless; interconnecting LANs is the primary purpose of the Internet. Figure 4.2 illustrates this problem in general terms: two networks A and B may exist separately, and devices on each can communicate within their LAN. But interconnecting them may occur through several intermediate networks, as depicted by the separate paths within the network cloud.

As with many problems, a "divide-and-conquer" approach reduces the complexity of one large problem into several smaller ones. Consider Figure 4.3 that

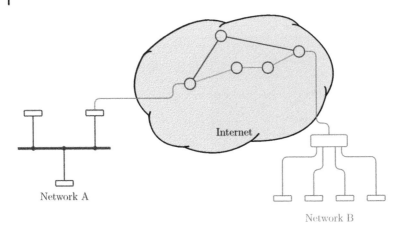

Figure 4.2 Routing from source to destination. Note the variable routing paths (defined by *hops* between nodes) and differing topologies (physical layout/interconnection) at the destination networks.

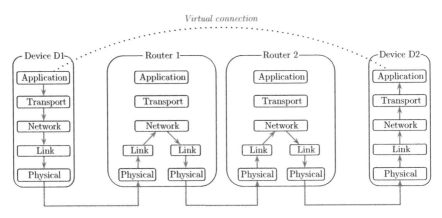

Figure 4.3 Connection between two devices, with intermediate or *forwarding hops* via forwarding devices Router 1 and Router 2.

shows two end devices D1 and D2, connected via some intermediate machines called Router 1 and Router 2, whose role is to route or direct the packet flow.

First, there is the subproblem of physical interconnection. Two or more devices, at some point, must share a physical interconnection. This is termed the *physical layer*. Aspects to be considered here are the generation of signals to represent binary data, the timing and synchronization, and access to the medium for transmission. For example, radio or wireless networks may allow only one transmitter on a specific frequency at a given time.

The next subdivision is logically the data link itself. Once we have a physical connection, we must have some way for each device to identify itself and find others and some way of determining if any errors have occurred in the physical transmission of data bits. This is termed the *link layer*, so named because it forms the data encapsulation link between separate devices.

Once we have an interconnection of devices and they can exchange packets of data, we have formed a small network. If two devices are directly connected, it is termed a point-to-point link, but if several devices are interconnected, they form a LAN. The next step is to try to interconnect several of these smaller networks together, and the *network layer* addresses these aspects. In particular, the problem of identifying which devices belong to which network must be solved. This problem is one of *routing*, or determining how each packet of data should travel from the source to destination.

In the IP suite, the network layer typically does not provide any guarantee of delivery of data packets. In fact, it may also deliver data packets to the destination out of order (just how this could occur, and what can be done about it, is discussed later in this chapter). In some applications, delivery of individual data packets is all that is needed. But in other applications, a much larger data stream must be delivered. The Internet Protocol (or IP) tries to deliver the individual data packets, but there is no guarantee of success. This is somewhat surprising to many people, but the reality is that different data services require different modes of operation. Some require guaranteed transmission of data, possibly with retransmission of portions of the data in the event of errors in transit. Others cannot wait for retransmission and require real-time delivery of data.

The next layer is the *transport layer*, which may be one of several protocols, but most commonly is termed Transmission Control Protocol (TCP) or User Datagram Protocol (UDP). TCP takes care of requesting delivery of data packets (termed *segments* in TCP) that may arrive out of sequence at the destination or may need to be retransmitted due to errors. UDP does not provide such services, but merely provides a service that sends separate data packets (termed *datagrams* in UDP) over the network.

Finally, once we have a reliable (if needed) data stream, we need to address the issue of which data service the user is requesting. For example, transmission of data packets that comprise a web page is a completely different service to transmitting voice or video, which is different again to email. Since there is a strong connection with the application the user actually sees, this is termed the *application layer*.

All this may seem a complicated way to manage things, but in fact the subdivision of responsibilities across different software and hardware subsections makes the design of networks manageable. This logical subdivision may be visualized as the *protocol stack* diagram of Figure 4.4. In this figure, each communicating device is represented by the large rounded boxes at either end. The

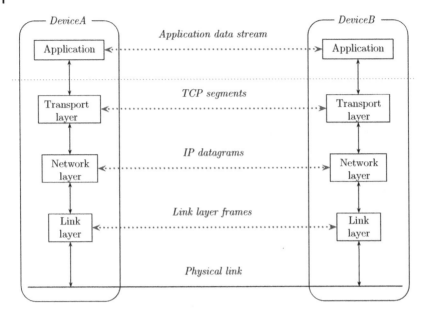

Figure 4.4 The TCP/IP protocol stack. The actual data transfer is downwards within Device A using internal memory, across the physical link, then again via memory "upwards" to the application running on Device B. Each layer performs a specific function, which allows the layers to operate independently.

data flow is conceptually across to the same layer or peer on the other device, as indicated by the dotted lines. However, there is no actual data flow in a physical sense. The data is handed down from the application on Device A, to the transport layer on Device A, to the network layer, and to the link layer. Finally, the physical layer at the bottom performs the actual, physical transmission of data bits. At the receiving Device B, the flow of data is reversed. Thus, we can focus on the role and design of one particular layer at a time, rather than the myriad interconnected issues in getting data from, say, a web server on Device A to web browser on Device B.

So, how is this actually accomplished, if the data transfer is "hypothetical" across layers of separated devices? The answer is *protocol encapsulation*, as illustrated in Figure 4.5. Starting at the top, the application layer has data that needs to be sent from Device A to Device B. This data may comprise, for example, the text or images within a web page. This data is handed off to the transport layer, here shown as TCP for reliable transmission. This layer adds its own identifying information in the header block, which is prepended to the original data (which itself remains untouched). Then, for the IP layer, an additional header is added to summarize the information required by the remote host. The link layer does likewise. As the data packets are assembled,

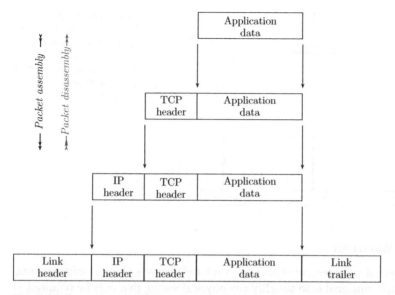

Figure 4.5 Protocol encapsulation or how layers are physically implemented. Each layer adds its own header data for communicating with its corresponding peer layer at the other end of the communications link. The diagram is not to scale, and the application data is usually much, much larger than the protocol headers.

we progress down the protocol stack. At the remote host, the disassembly is essentially the reverse of the assembly process.

So what happens at each layer? In the following, we address each of the layers in turn, with some examples. Throughout, we refer to RFCs. These are the documents that define Internet standards. Note that in considering data bit and byte ordering, the Internet by convention uses a "big-endian" order and when written has the Most Significant Bit (MSB) on the left (Refer RFC 1700, Data Notations, for details). Finally, note that RFCs use the term *octet* for a group of 8 bits, rather than a *byte*.

4.5 Local Area Networks

Starting at the local level, there may be several devices in close physical proximity that are physically interconnected. A very common example of this is wired Ethernet, and we examine this as a specific example. This may be extended to wireless LANs (WLANs), which have a surprising amount of commonality with their physically wired counterparts. Of course, there are some important differences, too.

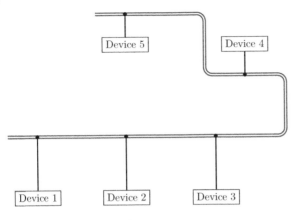

Figure 4.6 An Ethernet bus topology. Each device is connected to a common "bus," which simplifies wiring, but also means that only one device can transmit at a time.

4.5.1 Wired LANs

The physical arrangement of devices – their *topology* – is a logical starting point. Obviously, one goal is to simplify any physical wiring that may be required. If a bus topology is used as shown in Figure 4.6, then there is one common set of wires for all data. If, say, Device 1 wishes to transmit, it sends the data on the physical wire. But because there is only one set of wires, all devices hear this transmission. This means that the device being addressed must only respond when it actually hears its own address and ignore other transmissions intended for other devices. Furthermore, if one device is transmitting, then other devices must not transmit at the same time, otherwise the data will become garbled on the wire.

The bus arrangement must have some way to arbitrate access to the bus itself. Since transmission of data while another device is already transmitting would result in corrupted data on the wire, it might seem that the sender simply needs to avoid transmitting while another is doing so. But what happens if, say, Device 1 and Device 3 both happen to start transmitting data at (almost) the same time? Both will be garbled. In that case, a device can detect corruption of its data by simply listening to what was transmitted and comparing with what is received. If they do not match, then it is reasonable to assume that one (or more) other devices also tried to transmit at the same time. In the event of data corruption, the data could simply be retransmitted after a short interval. The problem is that both devices will operate on the same principles, and thus both will begin retransmission at the same time, thus resulting in infinite deadlock as this process repeats and repeats. The obvious approach is to introduce some randomness into the design, such that each device waits a random amount of time before retransmitting. If many devices were waiting to transmit, though, the chances of two or more attempting to transmit simultaneously will be non-negligible. If it transpires that a second collision occurs, a device may again wait

Figure 4.7 An Ethernet switch, which forms a star topology. This reduces the media contention problem, at the expense of wiring complexity, since there is a need for direct point-to-point wiring links to each device. The switch itself must have some intelligence, in order to route data packets to each device.

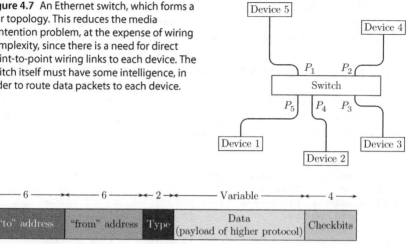

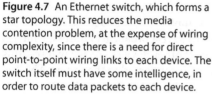

Figure 4.8 An Ethernet frame, as transmitted across a physical link. This is the lowest level of data encapsulation in the protocol stack. The numbers refer to the size in bytes of each field.

a random amount of time, but the interval over which the random wait may range is increased – the idea being that a longer timeframe will make it less likely that a second collision would occur. This process can be repeated again and again, with ever-increasing maximum wait intervals. This medium access mechanism is termed *Carrier Sense Multiple Access with Collision Detection* (CSMA/CD).

An alternative to the bus topology is the star topology as shown in Figure 4.7. In this case, an intelligent device (usually termed a *switch*) has a physical connection to each device through a port P. Most commonly, the interconnection utilizes *Unshielded Twisted Pair* (UTP), with data rates of 100 Mbps or higher. In this case, the switch has some additional processing to do. It must recognize the address of each device connected to each of its physical ports and retransmit the data frame on the physical port corresponding to the destination address. This means that the switch must have a mapping table of device address to physical port and must learn and record the device addresses as data packets are received.

The addressing problem is solved by allocating a fixed, 6-byte (48 bit) address for all Ethernet device interfaces. This number is unique and is determined by the physical hardware. These addresses are variously termed MAC or *Medium Access* addresses, the *physical address*, or *hardware address* (some literature may also refer to the NIC or *Network Interface Card*). The data packets are termed *frames* and carry both the source address and destination addresses. These are, respectively, the "from" and "to" fields shown in Figure 4.8.

The MAC addresses are typically written as bytes in hexadecimal, separated by a dash or colon – for example, 00-11-4E-56-FE-A4 or 00:11:4E:56:FE:A4. It is also useful to have a broadcast address, which is used for one device to send a message to all other devices. As we will see, this is used for discovering which devices are also connected to the same LAN. The broadcast address takes the form of all 1's or FF:FF:FF:FF:FF:FF.

The MAC or physical address may be determined using the ipconfig or ifconfig commands, depending on the operating system. An example is shown below, with the MAC address highlighted.

```
ipconfig /all
Ethernet adapter Local Area Connection:
Description        Gigabit Network Connection
Physical Address   D4-BE-D9-1C-DF-73   The physical or MAC address
DHCP Enabled       Yes                 IP address allocated from DHCP
                                       server
IPv4 Address       172.17.1.111        IP address of this device
Subnet Mask        255.255.0.0         bitmask for this subnet
Lease Obtained     5:41:30 PM          DHCP address allocation start time
Lease Expires      6:41:30 PM          DHCP address allocation end time
Default Gateway    172.17.137.254      Gateway to wider Internet
DHCP Server        172.17.137.254      DHCP Server which allocates IP
                                       addresses
DNS Servers        172.17.137.254      Domain name to IP mapping server
```

The LAN may take many other forms, of course – for example, wireless data transmission. Invariably, each physical type of transmission brings its own unique set of problems, which must be addressed by the link layer. The next higher layer, the network layer, makes few assumptions about the physical and link layers. It assumes that there is a data channel, but that the data may or may not get to the destination. But it is not concerned with, for example, the physical modulation method or CSMA/CD access arbitration.

4.5.2 Wireless LANs

Wireless Local Area Networks (WLANs) are similar in many ways to wired Ethernet LANs and share many similarities with their wired counterparts. Chief among the commonalities is the use of the 48-bit physical addressing system, which in practice means simplified management and wider deployment. However, there are clearly some important differences between wired LANs and their wireless counterparts – some obvious, and some not so obvious. One primary difference is that it is not possible for a device to simply listen to its own wireless transmission to determine if it was successful. Recall that this is the method by which shared-bus Ethernet LANs determine if another device has tried to transmit simultaneously, with the resulting data bits becoming garbled. In the case of wireless, the transmission power is many orders of magnitude higher than the received power, and thus such a check will always appear as though the data was successfully broadcast. This necessitates a different medium access procedure.

As with wired LANs, the initial and obvious strategy is not to transmit while the carrier signal indicates that another device is in the process of transmitting a frame. In place of the physical carrier, a virtual carrier sense is utilized. This entails a *Request To Send* (RTS) data frame, which includes a 15-bit duration field in the header. This duration forms the *Network Allocation Vector* (NAV) and is a field in the header that reserves the medium for a certain time period.

This effectively informs other wireless devices that the medium will be reserved while a subsequent transmission takes place. Because wireless devices may be physically separated and the wireless signal radiation decays quickly with distance, it is entirely possible that one device may receive an RTS frame while another, still within the coverage area, does not. If the co-location is such that the intended recipient receives the RTS, but another device further from the transmitter does not, then that other device will not be aware that the wireless medium is to be reserved. This is sometimes referred to as the hidden node problem. For this reason, an acknowledgment, or *Clear to Send* (CTS), is then sent by the intended recipient, rebroadcasting the NAV. In this way, all devices within the range of *both* the transmitter and receiver are informed of the network reservation time slot.

After the sender has transmitted a frame, the receiver must acknowledge it. This is another area where wired and wireless LANs differ. Wired Ethernet includes no acknowledgment process, whereas wireless transmissions are normally acknowledged, and the absence of acknowledgment implies the need for retransmission. After the transmission and acknowledgment cycle is complete, the medium is then available for other devices to begin the same request–acknowledge cycle. This is when collisions can occur. At this time, a random backoff period is employed by all wireless devices, to reduce the possibility of collisions. Clearly, any collisions of data transmission reduces the utilization of the wireless bandwidth and thus overall performance. The random backoff period is also exponentially increased (as in wired networks) so as to reduce the possibility of multiple collisions.

Since the process does not revolve around collision detection (CD) as in CSMA/CD wired protocols, it is termed *Carrier Sense Multiple Access with Collision Avoidance* (CSMA/CA). Further details are available from many sources, for example, Gast (2002), with the formal specification defined in IEEE (2012).

4.6 Device Packet Delivery: Internet Protocol

The LAN, as its name implies, was originally intended to interconnect several devices co-located within relatively close physical proximity. It is assumed that all devices are reachable by transmitting onto the external network interface. However, what if several local networks are to be interconnected? This is the role of the network layer, which is handled by the IP.

4.6.1 The Original IPv4

The data frames in IP are termed *datagrams* by convention. The layout of a datagram used in IP version 4 (usually referred to as IPv4) is shown in Figure 4.9. Since the IP must deliver datagrams from one device to another, it must have a method for addressing each device. This is done with the IP address, which (for IP version 4) is a 32-bit or 4-byte number. By convention, these are written in dotted decimal format, so, for example, 192.168.4.7 and 172.168.20.45 are valid addresses. Because the primary goal of IP is to deliver data via several interconnected networks, the addressing must be hierarchical in some manner. In broad terms, the higher-order bits (leftmost as written) refer to the network address, and the lower-order bits (rightmost) identify a specific device on that network.

The network layer is responsible for matters of addressing to the device level. IP as originally defined in RFC 760 is denoted as version 4 (abbreviated IPv4). The 32-bit address space of IPv4 is essentially exhausted for new allocations, but it is still very widely deployed. IP version 6 (IPv6) builds upon the foundations of IPv4.

The IP datagram will usually traverse several different networks (point-to-point links), and each of these may have different link layer protocols and configurations. Physical links generally have a limit on the maximum number of bits transferred at once, and this in turn determines the maximum number of bytes in a link layer frame. For example, wired Ethernet has a standard payload limit of 1500 bytes. Thus, the IP layer must subdivide the IP segments into the smallest frame size that will be encountered from the source to destination. This limit is termed the MTU or Maximum Transmission Unit. The IP standard also requires 576 bytes as the *minimum* size (512 data bytes + 64 header bytes). Thus, IP datagrams may be broken into smaller fragments at some point between the ultimate source and destination. This is covered in RFC 791 with further clarification in RFC 879.

4.6.2 Extension to IPv6

The IP version 4 (IPv4) standard has served well for a long time and continues to serve well in many situations. However, there are some shortcomings with it,

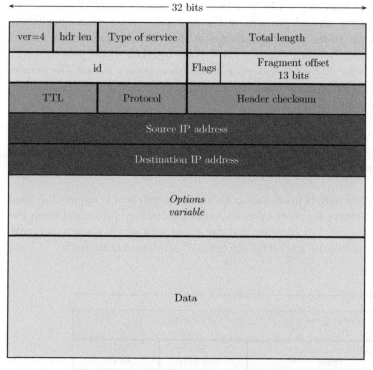

Figure 4.9 The composition of an IPv4 datagram. Note the source and destination addresses. The *Time-To-Live* is denoted by TTL and is decremented each time the datagram is forwarded on. For this reason, it is often called a *hop count*.

which have been revealed over time. Some of these shortcomings relate to the physical size of the Internet, while others relate to new types of services which were not envisaged originally.

RFC 2460 is the primary starting point for understanding the succession from IPv4. This protocol, IPv6, has been evolving for some time, but deployment is not universal. IPv4 will continue to be used for some time. One of the most obvious limitations of IPv4 is the number of addresses possible. The theoretical address limit of 2^{32} endpoints does not equate to this many addresses in practice, due to the way IPv4 addresses are allocated to organizations. Any given organization may only use a small proportion of the possible address block allocated to it. This, combined with the fact that many embedded devices are connected to the IP network, means that the available pool of IPv4 addresses is insufficient.

The use of Network Address Translation (NAT), discussed further in Section 4.6.6, has alleviated the lack of IPv4 32-bit addresses, but it is a somewhat inelegant solution in many ways. IPv6 redefines the protocol and

IP headers to have larger 128-bit address spaces, whereas NAT uses (rather ingeniously) the unused portions of the protocol header fields. Specifically, it rewrites the 16-bit port field (discussed further in Section 4.8) as packets are transferred from one side of a LAN to the interface to the outside world. This extends the 32-bit address space to 48 bits, in theory at least. The penalty associated with this is the need to copy and rewrite each and every datagram when forwarded.

The IPv6 basic datagram structure is shown in Figure 4.10. The addresses are now 16 bytes, or 128 bits. Not only is this a significant increase over IPv4 but the way the address is defined permits simpler routing (by aggregating routes) and configuration (by uniquely defining the host endpoint based on its MAC address).

The header length is fixed, and so no header length field is required. A fixed size header makes for more efficient router processing. This might seem like a small advantage, but consider that the speed with which routers can make decisions and forward packets is a potentially significant bottleneck.

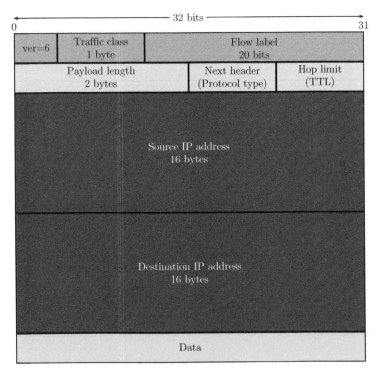

Figure 4.10 The composition of an IPv6 datagram. As well as larger address space, the simplified layout permits faster packet forwarding.

The fields named *Source IP Address* and *Destination IP Address* are each 128 bits or 16 bytes long, while the *Payload Length* specifies the length of the data after the header. Originally, RFC 1883 specified a 4-bit Priority field and 24-bit Flow Label. This is obsolete, and RFC 2460 specifies an 8-bit Traffic Class field and a 20-bit Flow Label. A "flow" in this context refers to a related and continuous stream of data, originating at a source IP address and having the same flow identifier. Since some transmissions are designed to cope with dropped packets, these flows may be partially interrupted in the event of router buffer overflow or bandwidth-related congestion. Some flows, such as real-time streaming data, may not require that every single packet makes it to the destination. These may be able to sacrifice some quality (dropped packets) as a tradeoff for maintaining continued high throughput.

Auto-configuration is another major advantage of IPv6. In IPv4, protocols such as DHCP (Section 4.7.2) and ARP (Section 4.7.1) were required to assign IP addresses and keep track of them. Once again, the benefit of hindsight shows that while these protocols work, there are more elegant solutions if designed from the ground up. A case in point is the assignment of IP addresses to a certain host. More specifically, an IP address is assigned to a specific network interface, since a given host may have several network interfaces. In the case of Ethernet, the MAC addresses may be used to form the IPv6 address, as illustrated in Figure 4.11. The 48-bit address is encoded in hardware and thus defines a unique endpoint. The upper 64 bits of the address are defined by other connected devices on the interface, and so the full 128-bit address is defined automatically. Note that forming the address in this way is not mandatory – some systems form the address randomly. Of course, it must be unique on a given local network, otherwise the correct routing will be unable to be determined. It should be noted that this algorithm may not be desirable in some situations – for example, a server that should keep the same IP address, irrespective of hardware changes.

This takes care of the lower 64 bits of the entire 128-bit address space. As described in RFC 3587 *IPv6 Global Unicast Address Format*, the upper 64 bits are used to define the higher-level part of the address and thus facilitate hierarchical routing. The rapid growth of the Internet has meant that routing

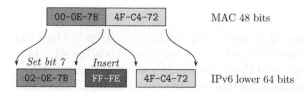

Figure 4.11 One possible method of mapping 48-bit MAC addresses into the 64-bit host portion of the 128-bit IPv6 address according to RFC 4291 Appendix A. In the case illustrated, the seventh bit from the left is set and two bytes inserted as shown.

tables have become extremely large and forwarding packets is a significant burden. The CIDR approach (discussed in Section 4.6.4) defined classless routing with variable-length bitmasks. With IPv6, the opportunity presented itself to improve routing. RFC 4291 *IP Version 6 Addressing Architecture* permits the upper 64 bits of the address to be subdivided into a 48-bit Global Routing Prefix and a 16-bit Subnet ID. The latter means that subnetting and subnet masks, so much a part of IPv4, are no longer necessary. Point-to-point addressing (unicast) is identified by the bit pattern 001 as the initial (leftmost) part of the 48-bit portion, thus leaving 45 bits for the routed part of an address.

Of course, other useful IPv4 concepts, such as local (private) addresses, multicast addresses, and the local loopback address, are defined in IPv6. IPv6 addresses, as introduced in RFC 3513 *IPv6 Addressing Architecture*, are written in a different format to IPv4. First, the addresses are written in hexadecimal, to make conversion to bit patterns easier, with a colon (:) separator. The addresses are split into 16-bit blocks, and thus each block consists of four hexadecimal digits. Suppose we select an address prefix of the form `2001:0DB8::/32` according to RFC 3849 *IPv6 Address Prefix Reserved for Documentation*. The /32 mask specifies only the first 32 bits are meaningful. Next, suppose the link-local address is, as per the previous MAC mapping example, `02-0E-7B-FF-FE-4F-C4-72`. The full address is then

`2001:0DB8:0000:0000:020E:7BFF:FE4F:C472`

This is rather cumbersome, and thus two rules are employed to simplify writing such addresses. First, leading zeros in a 16-bit block may be deleted. So `0DB8` becomes `DB8`. Next, consecutive 16-bit blocks of all zeros may be removed altogether and replaced with double colons. Finally, this leaves

`2001:db8::20e:7bff:fe4f:c472`

as the compressed address. Note that only one `::` is permitted, otherwise ambiguity may arise as to how many zeros to replace in each omitted block. Furthermore, only full 16-bit zero blocks may be replaced in this way.

Many more details of the IPv4 and IPv6 headers may be found in Kozierok (2005).

4.6.3 IP Checksum

Once a data packet is received, it needs to be checked for errors. But how can this be done, if we don't have a second copy of the data? This is the role of the *checksum*. The IP checksum field checks the header only, not the data payload. The latter is primarily the task of the transport layer (for which TCP also uses a checksum), though different link layers also include error checking of their own. Even though the end-user does not utilize the IP header directly, the checksum

45	00	00	3C	*Version, header length, total length*
75	02	00	00	*Flags, fragment info*
20	01	**C7**	**1F**	***TTL, header checksum***
AC	10	03	01	*Source address 172.16.3.1*
AC	10	03	7E	*Destination address 172.16.3.126*

Figure 4.12 An example IP header, as captured on a data link. This should be compared with the IPv4 header layout of Figure 4.9. The header checksum is C7 1F hexadecimal.

for the IP header is still necessary, because there is no point in routing a packet with a corrupt header (the addresses may well be wrong).

Figure 4.12 shows an example of the bytes in an IP header, which are used to compute the checksum. The checksum is calculated by adding the 16-bit words and adding any overflow over 16 bits back in. This is termed end-around carry. At the sender, the checksum is set to zero, then the sum of the 16-bit words is calculated with the end-around carry added in. The one's complement (that is, inverting all bit values) of this is used as the checksum in the packet header. Importantly, the same computation at the receiver including the embedded checksum that was calculated at the sender should yield a zero result if no errors occurred, which is easy to check for.

Furthermore, the procedure specified for checksum computation results in the same checksum value on either big-endian devices (which store the high-order byte in the lower memory address) or little-endian devices (which store the low-order byte in the lower memory address). This is of course important for interoperability, since the CPU of any given receiver may be built with either little- or big-endian architecture.

An example using the real-world data of Figure 4.12 will help to clarify the checksum calculation. The data packet and calculations for both big-endian and little-endian architectures are shown in Figure 4.13. On transmission, the checksum value is unknown, so it is initially set to zero. If the 16-bit data words are added in big-endian order (shown on the left), the sum is 238DE hex. Adding the overflow of 2 to form the end-around carry gives 38E0. The complement of this is C71F, which is the value placed in the header.

If the complement of this packet (including the checksum) is formed at the receiver, the sum is 2FFFD. The end-around carry is FFFD + 2 = FFFF, and the complement is 0000 – indicating that no errors have occurred. Computing the checksum in this way at the receiver simplifies matters, since it includes the received checksum as bytes in the regular calculation, and a zero result indicates that no errors have occurred with the data being checksummed. Importantly, this check implicitly includes the checksum value itself.

If the 16-bit data words are in little-endian order (shown on the right of Figure 4.13), the sum is 0E038 hex. With end-around carry it remains as 0E038,

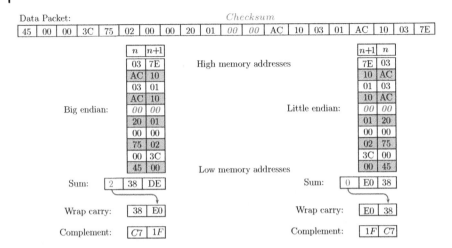

Figure 4.13 Calculating the checksum, using big-endian machine architecture (left) and little-endian architecture (right). The end result must be the correct packet data ordering, independent of the machine byte ordering.

which when complemented is 1FC7. Note that this is the same value, but byte reversed, with respect to the big-endian calculation. This is as it should be, since a big-endian machine will place its calculated value of C71F in memory in the order C71F to form the packet (high-order byte first), whereas a little-endian machine will place its calculated value of 1FC7 in memory in the order C71F to form the packet (low-order byte first). Thus, both architectures result in the same bytes for the checksum, in the same order, when placed in the packet. If the complement of this packet sum including the checksum is formed at a receiver with little-endian architecture, the result is FFFF, the end-around carry is FFFF, and the complement is 0000. Once again, this indicates that no errors have occurred while the data was in transit.

The following shows how to compute the checksum using MATLAB. First, the division by 2^{16} and subsequent discarding of the remainder effectively shifts the data right by 16 bits, leaving only the carry digits. These carry digits shifted left by 16 bits (using multiplication by 2^{16}) are then subtracted from the checksum to form the remainder, thus effectively keeping only the lowest 16 bits of the checksum. The end-around carry is then performed by adding the lowest 16 bits of the checksum to the carry bits. Finally, the complement, which is just an inversion of all 16 bits, is performed by subtracting the value from $2^{16} - 1$. Note that arithmetic operations would not be used in a real packet router. Rather, bit mask and shift operations are implemented directly using the processor's instructions, to maximize speed.

```
% set calculation order to be big or little endian as
% desired
UseBigEndian = false;

pkthex = [    '45'  '00'  '00'  '3C'  ...
              '75'  '02'  '00'  '00'  ...
              '20'  '01'  '00'  '00'  ...      % checksum 00 00
              'AC'  '10'  '03'  '01'  ...      % 172.16.3.1
              'AC'  '10'  '03'  '7E'  ];       % 172.16.3.126

Nchars = length(pkthex);
cksm = 0;

for k = 1:4:Nchars - 1

    if( UseBigEndian )
        i = [k+0 k+1 k+2 k+3];      % big-endian
    else
        i = [k+2 k+3 k+0 k+1];      % little-endian
    end

    % convert string representing hex to decimal for
    % calculations
    wordstr = pkthex(i);
    wordval = hex2dec(wordstr);

    cksm = cksm + wordval;
end

fprintf(1, 'Raw checksum %d decimal  %s hex\n', cksm,
  dec2hex(cksm));

% compute end-around carry
carry = floor(cksm/(2^16));
rem16 = cksm - carry*(2^16);
cksmea = rem16 + carry;

% complement
cksmnot = ((2^16) - 1) - cksmea;

fprintf(1, 'carry %s  cksm with carry %s  final cksm %s\n',
    dec2hex(carry), dec2hex(cksmea), dec2hex(cksmnot));
```

It should be noted that the checksum is not infallible – it cannot guarantee 100% detection of all errors. However, the error detection performance is extremely good and very trustworthy for the types of errors that typically occur in practice.

4.6.4 IP Addressing

Each device or endpoint in the IP network must have a unique network-facing address, so that packets can be routed correctly to their destination. The present situation is one of transition between IPv4 and IPv6. However, IPv4 continues to be very widely used within organizational networks, homes, and small businesses. This is due to the fact that internal addresses can be assigned within those networks independently of the "outside" network, using NAT, discussed in Section 4.6.6.

The IPv4 specification defines an address space of 32 bits, and no two devices are permitted to share a given address, since conflicts would arise. Because the protocol was to provide routing via different networks, the address space was hierarchical and subdivided into network and host (device) portions. The main classes of address, which were originally defined, are illustrated in Figure 4.14.

The MSBs define the address class. A zero in the MSB denotes a Class A network, and as shown, the upper 8 bits define the network address, with the lower 24 bits for the host or device address. This then provides 7 bits for network identification, but there are fewer than 2^7 networks possible, since certain loopback and broadcast addresses are reserved, as defined in RFC 3330 (IANA, 2002).

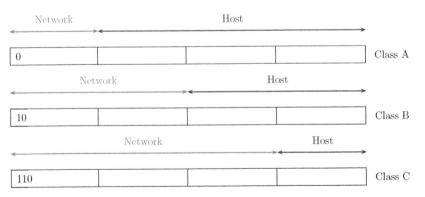

Figure 4.14 The arrangement of the original IP address classes. The leading (leftmost) bits determine the address class, then the next block of bits determines the network, and finally the rightmost bits determine the device or host within that network. This turned out to be a very inefficient way to allocate address space.

In decimal, addresses starting with a number less than 128 signifies a class A network. Similarly, a bit pattern of 10 defines a class B network, with 2^{16-2} possible networks and 2^{16} device identifiers – though note that the all 1's address is reserved for a broadcast address for all hosts, and the network address with all zero bits in the host field defines the network itself. Class B networks thus have a range of the first (decimal) digit of 128–191. Finally, Class C follows a similar pattern, with more networks of fewer devices.

In addition, certain address ranges are reserved for use within private networks, as defined in RFC 1918. These addresses are not routed from within a LAN to the outside world, but may exist within internal networks. As a result, addresses in this range do not need approval for use, and they will not conflict with other addresses, since IP packets containing these addresses are never forwarded without address translation. These addresses are employed by NAT subnetworks (internal networks). RFC 1918 specifies these address blocks and their corresponding ranges as:

10.0.0.0   –   10.255.255.255   (10/8 prefix)

172.16.0.0   –   172.31.255.255   (172.16/12 prefix)

192.168.0.0   –   192.168.255.255 (192.168/16 prefix)

The designation is given in terms of the IP address ranges and also the network range with a bitmask prefix. The designation /8 means the first 8 bits must be used as the network address, but the remaining bits may be used for a device address within that network. Similarly, 172.16/12 means that the uppermost 12 bits are used as the network address, and 192.168/16 denotes the upper 16 bits, both with the designated network prefix.

The Class A/B/C distinction was intended to provide some form of hierarchy in addressing and thus facilitate routing. But the growth of the Internet meant that this designation became unworkable. Consider a Class A network – relatively few of these networks would be permitted, and furthermore the size of each Class A network would be unwieldy. For these reasons, the notion of IP subnetworks, or simply subnets, was introduced. This divides the network space into smaller zones, as described in the Section 4.6.5.

The large size of Class A and Class B device range definitions, in particular, meant that a large proportion of the 32-bit IP address space would remain unused. The purpose of subdivision into class ranges was to facilitate routing. This is no longer used as originally envisaged, and was superseded by Classless Interdomain Routing (CIDR). This is because the routing complexity became overwhelming with a simple three-class subdivision.

It should also be noted that all devices with an IP address always include a special loopback or *localhost* address, almost always 127.0.0.1. This is intended as a software loopback, and packets sent to this address are not forwarded – RFC 1700 documents this. RFC 3330 documents a number of special-use IPv4 addresses (IANA, 2002). The configuration example shown below shows a locally assigned IP address:

```
ipconfig /all
Ethernet adapter Local Area Connection:
Description        Gigabit Network Connection
Physical Address   D4-BE-D9-1C-DF-73   The physical or MAC address
DHCP Enabled       Yes                 IP address allocated from DHCP
                                       server
IPv4 Address       172.17.1.111        IP address of this device
Subnet Mask        255.255.0.0         bitmask for this subnet
Lease Obtained     5:41:30 PM          DHCP address allocation start time
Lease Expires      6:41:30 PM          DHCP address allocation end time
Default Gateway    172.17.137.254      Gateway to wider Internet
DHCP Server        172.17.137.254      DHCP Server which allocates IP
                                       addresses
DNS Servers        172.17.137.254      Domain name to IP mapping server
```

4.6.5 Subnetworks

The original subdivision of IP addresses into Class A, B, and C for the purposes of defining networks proved unworkable. To recap, these were Class A, with 24 bits for the host field; Class B, with 16 bits for the host; and Class C, with 8 bits. This means, in effect, that there could be a small number of Class A networks, each with a very large address space. At the other extreme, there could be a large number of class C networks, each with 254 addresses available. The problem with this system is twofold. First, it makes very inefficient use of the available 32-bit address space, with a very large proportion of the available space unused. Secondly, typical LAN access methods permit only a limited number of devices per LAN, and thus the available IP address space cannot be used at all, even if there was the will to do so. The solution is to move the dividing line within the IP address between network and host (device) portions of the 32-bit address. This smaller network is termed a subnetwork, or simply a *subnet*.

Since IP addresses are ultimately binary, it makes sense to determine the position of the subnet division line using binary operators, and in fact this is how it is done in the software protocol stack. The subnet mask is a binary value, with the same number of bits as the IP address. Where the binary 1 bits occur in the subnet mask, the corresponding bits in the IP address are used to determine the subnetwork address. Where the 0 bits occur, they define which bits of the IP address denote the device address within the subnet.

Some examples serve to illustrate these concepts. First, consider Figure 4.15. In this case, an IP address of 172.16.22.34 is to be subnetted. Note that this is in the "private" address space, and we choose this so as not to conflict with any actual address, merely for the purposes of illustration. Since 172 has the binary pattern 10 at the start, it is a Class B network. We know that a Class B network

```
            Address: 172.16.22.34    10101100  00010000  00010110  00100010
   Class B network: 172.16.0.0       10101100  00010000  00000000  00000000
         /24 mask: 255.255.255.0     11111111  11111111  11111111  00000000
   Subnet address: 172.16.22.0       10101100  00010000  00010110  00000000
           Subnet 22                                     00010110
    Addrin subnet: 34                                              00100010
```

Figure 4.15 Subnet example 1. The subnet identifier is 8 bits, and the device identifier is also 8 bits.

```
            Address: 172.16.22.34    10101100  00010000  00010110  00010110
   Class B network: 172.16.0.0       10101100  00010000  00000000  00000000
         /20 mask: 255.255.240.0     11111111  11111111  11110000  00000000
   Subnet address: 172.16.16.0       10101100  00010000  00010000  00000000
           Subnet: 1                                     0001
    Addrin subnet: 1058                                      0100  00100010
```

Figure 4.16 Subnet example 2. This is the same IP address as the previous example but a larger subnetwork size as defined by the subnet mask.

defines the upper 16 bits as the network address and the lower 16 bits as the device address. Subnetting divides this very large address space into several smaller subnetworks. Suppose that the subnet mask is given as 255.255.255.0. This could also be denoted as a /24 mask, since the upper 24 bits have the value 1. The logical ANDing of the given address with the mask yields the subnet address, which ends in zeros, because ANDing with the zero bits of the subnet mask yields zero. So the subnet address (that is, the address of the subnetwork itself) is 172.16.22.0. The subnetwork number is found by remembering that this is a Class B address and thus we have 16 bits to define the device portion, which is subdivided by the subnet mask. Thus, the subnet number is 22.

The given device address is within this network and may be found by ANDing with the complement of the subnet mask. For the given mask, this corresponds to the lowest 8 bits of the IP address and is decimal 34. It should be noted that because the subnet mask division fell on an 8-bit boundary, it is a simple matter to select out the network, subnetwork, and device portions. But, this is not the case in general, as the next example will show. In fact, using the lower 8 bits as the subnet mask may be an inefficient use of the address space for a small organization, allowing as it does 254 addresses (256 less the broadcast address of all 1's in the device selection bits, less the network address with 0's in the device selection bits).

The next example, shown in Figure 4.16, uses the same IP address but a subnet mask of 255.255.240.0, which is a larger subnet. This is because there are now 12 bits for the device bits. The dividing line between the network and device is moved 4 bits to the left. In this case, the device address is not easily determined directly from the IP address, at least not by inspection. It is necessary to resort to writing out the binary values of the address and subnet mask. The figure illustrates this process.

Note that the division of addresses into subnetworks is a local one, meaning that although the subnet mask is required to perform various routing decisions, it is not required to be carried in the IP datagram itself. An example showing a subnet mask with other configuration parameters is as follows:

```
ipconfig /all
Ethernet adapter   Local Area Connection:
Description        Gigabit Network Connection
Physical Address   D4-BE-D9-1C-DF-73   The physical or MAC address
DHCP Enabled       Yes                 IP address allocated from DHCP
                                       server
IPv4 Address       172.17.1.111        IP address of this device
Subnet Mask        255.255.0.0         bitmask for this subnet
Lease Obtained     5:41:30 PM          DHCP address allocation start
                                       time
Lease Expires      6:41:30 PM          DHCP address allocation end time
Default Gateway    172.17.137.254      Gateway to wider Internet
DHCP Server        172.17.137.254      DHCP Server which allocates IP
                                       addresses
DNS Servers        172.17.137.254      Domain name to IP mapping server
```

4.6.6 Network Address Translation

The 32-bit address space of IPv4 represents a real limitation. In addition, the cost of assigning a specific IP address to each device in an organization may also be problematic. Consider an Internet Service Provider (ISP) with hundreds of thousands of customers, each with perhaps several IP-enabled devices. These would consume many IP address slots, which may be seldom used.

Consider why the IP address must be unique: if we wish to reach a particular server for information, then its address must be known and fixed. However, the vast majority of IP connected devices do not provide services and hence do not need a fixed IP address for inbound connections. Certainly, they need an IP address, but it could well be a local address. These could be allocated from the private address ranges as mentioned (10/8, 172.16/12 or 192.168/16). This would allow such devices to have network connectivity but only at a local level. If they wish to be connected to the wider Internet, their IP packets would be dropped, as forwarding of packets with private IP addresses is specifically forbidden by RFC 1918.

A very widely deployed solution to this dilemma, using the private address space yet providing Internet connectivity, *is* NAT (Srisuresh and Holdrege, 1999). While there are many variations on this concept, the operating principle is illustrated in general by Figure 4.17. It requires that the external Internet gateway operate a NAT protocol server, which translates the local (internal)

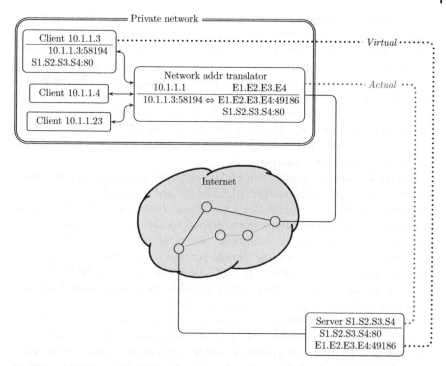

Figure 4.17 The principle of NAT using address and port translation. Port 80 is reserved for web services, but port 49186 (in this example) is allocated on a per-connection basis. The combination of 32-bit IP address and 16-bit TCP port is termed a *socket*.

requests bound for outside sites, with the reverse translation on inbound packets. In the diagram, there is only one outward-facing IP address, designated as the external address E1.E2.E3.E4. The devices internal to the private network are assigned private addresses, in this case from the 10/8 range. The NAT translator is the default gateway for internal devices, and thus all IP packets are sent to it for forwarding. Upon forwarding a packet, the NAT device performs a translation of the address and port into a distinct externally visible address.

Normally, the port number is supposed to differentiate between the end application – for example, whether the data is intended for a web browser, streaming video, or other application. The port number is a 16-bit field and is explained further in Section 4.8. The combination of IP address and application port number (16 bits) is termed a *socket* and is usually written in the form IP:Port, for example, 10.1.2.3:80 represents port number 80 on the device with IP address 10.1.2.3. If port numbers were not used, there would be no way of associating a given data packet with a particular application, once that data packet was received. Each endpoint of communication (that is, each pair of devices that are sending or receiving data) must have a socket. The socket pair thus defines a unique communication channel on the internet.

NAT, however, uses the port number in a way that was not originally intended. It uses the port number to expand the address range from the 32-bit IP address range. If a particular subnet is using NAT, only one device is connected to the outside world; the other devices must use the NAT server as their gateway to the outside world. In Figure 4.17, the gateway from the private network to the outside world has the address 10.1.1.1 on the *inside* of the private network. The NAT gateway must replace the *internal* IP address (10.x.x.x) by an external address.

This could cause ambiguity, since when response packets from the remote server S1.S2.S3.S4 destined for E1.E2.E3.E4 are received, they must be forwarded to the appropriate 10.x.x.x address. Thus, the NAT device must maintain a table of translations from internal sockets to the port number for forwarding the outgoing IP datagram. This table is again used, in reverse, to translate IP packets from the remote server (S1.S2.S3.S4:80) destined for the NAT device external socket E1.E2.E3.E4:49186. Note that 80 is the fixed server port, but 49186 is one of many possible *ephemeral* (short-lived and dynamically allocated) port numbers on the NAT translator.

In this way, the client inside the private network appears to have a connection to the server, as designated by the "virtual" connection. However, the packet connection from the server's point of view is shown as the "actual" connection. Once data reaches the NAT from either side, the translation table is used to map the IP address and port numbers. Naturally, this results in a potentially significant burden on the gateway device, as it has to look up the translation table for every single packet traversing in or out.

4.7 Network Access Configuration

So now we have two types of addresses: IP addresses, which are routable and can find their way to other destinations on the Internet, and MAC addresses, which are not routable and are confined to locally connected devices on a LAN. A natural question arises as to why this is so. Part of the answer is historical: Originally, only devices in close proximity were connected, and the idea of interconnecting such networks with IP came later. However, there are several advantages to this hierarchy of addressing. A server would normally have a fixed publicly available IP address. Other devices may expect to be able to reach this server using a given IP address. Consider, for a moment, what may occur if the server had its IP address determined by the MAC address. Since the MAC address is unique to the specific hardware, if the physical hardware were changed due to upgrades, obsolescence, or failure, then the server would have a different address. All other devices wishing to connect to it would then need to be told, somehow, of the change in physical address. The separation of

MAC addresses and IP addresses means that this problem does not occur: the new device is simply configured with the same IP address as the old one, and the services it provides can continue.

Normally (though not always), each MAC address corresponds to one IP address. Given that one device on a LAN wishes to communicate with a known IP address on the same LAN, how can it determine the MAC address that corresponds to the IP address? This needs to occur, as the network interface inspects all incoming Ethernet frames, and if the destination MAC address does not match the MAC address of the device, then the frame is discarded. Only once it is accepted, due to the MAC addresses matching, is the data packet passed to the IP layer.

4.7.1 Mapping MAC to IP: ARP

The discovery of an unknown MAC address from a known IP address occurs as follows. The protocol is known as *Address Resolution Protocol* (ARP). First, the device with a known IP address to forward a packet to (but unknown MAC address) broadcasts an Ethernet frame to all devices to which it is connected. This is done via a special broadcast address that consists of all ones (that is, 48 1 bits, or FF:FF:FF:FF:FF:FF in hexadecimal). The ARP request packet contains two pieces of information: one, the IP address that the sender wants to discover and two, the MAC address of the sender. The latter is so that a reply may be sent back. Each device on a LAN receives this message, because it is sent to the broadcast address. If the device has the same IP address as is in the ARP request, then it responds with its corresponding MAC address.

The sender then keeps this IP to MAC address in a table, called an ARP cache, to save having to repeat the process continually for each IP datagram that needs to be sent. The arp -a (or similar) command on most operating systems shows the current ARP cache. The mapping is stored for a certain time period, or cached, so as to allow for changes in the network – consider the aforementioned case of a machine being swapped out for another, where the IP address remains the same but the MAC address necessarily changes. An example is shown below:

```
arp -a
Interface: 10.1.1.3
Internet Address   Physical Address    Type
10.1.1.1           78-a0-51-1c-4f-b2   dynamic
10.1.1.255         ff-ff-ff-ff-ff-ff   static
255.255.255.255    ff-ff-ff-ff-ff-ff   static
```

The entries in the table labeled "static" are fixed and cannot be changed. In the cases above, the static addresses are also broadcast addresses, as seen by the all-one's MAC addresses. The "dynamic" address is determined via the ARP protocol as described above.

Here, we can see the IP address of the device (10.1.1.3), the IP address of a device to which it is connected on the LAN (10.1.1.1), and two broadcast addresses (10.1.1.255, which is a subnet or *directed broadcast*, and 255.255.255.255, which is termed a *limited broadcast*).

4.7.2 IP Configuration: DHCP

Another protocol commonly encountered is, *Dynamic Host Configuration Protocol* (DHCP). Consider a LAN with many devices connected to it. It may be time consuming to configure each device to its correct IP address. Even on small networks, the user expertise may be such that it is not feasible for users to set up their own configurations. Finally, in order to maintain the integrity of the network, it is imperative that no two devices use the same IP address, lest addressing conflicts occur.

In any of these scenarios, DHCP is useful. If not specifically set, the IP address of a device is initially unknown when it is powered on or joins a network. A designated server on the LAN is configured to assign IP addresses to other devices on the LAN, on request. Each device sends a DHCP request, and the DHCP server acknowledges with an address from its available pool of addresses. Normally these are also timed out, so that addresses may be reused in the event of devices leaving the LAN.

The example below shows the IP address of the DHCP server, together with the DHCP assignment time span:

```
ipconfig /all
Ethernet adapter Local Area Connection:
Description        Gigabit Network Connection
Physical Address   D4-BE-D9-1C-DF-73   The physical or MAC address
DHCP Enabled       Yes                 IP address allocated from DHCP
                                       server
IPv4 Address       172.17.1.111        IP address of this device
Subnet Mask        255.255.0.0         bitmask for this subnet
Lease Obtained     5:41:30 PM          DHCP address allocation start time
Lease Expires      6:41:30 PM          DHCP address allocation end time
Default Gateway    172.17.137.254      Gateway to wider Internet
DHCP Server        172.17.137.254      DHCP Server which allocates
                                       IP addresses
DNS Servers        172.17.137.254      Domain name to IP mapping server
```

Note that both the IP address and subnet mask, as already discussed, are assigned to this device via the DHCP server. This assignment, or lease, has an expiry time. The client must re-request allocation before the expiry of the current lease, if it wishes to continue using the same IP address.

4.7.3 Domain Name System (DNS)

IP addresses are somewhat like telephone numbers to humans: difficult to remember. For this reason, the *Domain Name System* (DNS) was developed, so as to map a hierarchical name such as example.net to a corresponding IP address. This function is performed by a Domain Name Server. When given a server name such as example.net, a device first queries the closest DNS server to find a mapping to IP address. This IP address is then used for all subsequent data packets, and the DNS mapping only needs to be queried once. For heavily loaded domains, it is not uncommon to provide multiple servers, with multiple IP addresses, in an attempt to balance the load.

Since the lookup of name-to-address mapping is performed frequently, and because it does not often change, it is cached (stored) locally. In addition to master or root DNS servers, which are said to be *authoritative*, lower-tier DNS servers exist to speed up common queries.

The DNS server address appears in a configuration listing as shown below:

```
ipconfig /all

Ethernet adapter Local Area Connection:
Description       Gigabit Network Connection
Physical Address  D4-BE-D9-1C-DF-73   The physical or MAC address
DHCP Enabled      Yes                 IP address allocated from DHCP
                                      server
IPv4 Address      172.17.1.111        IP address of this device
Subnet Mask       255.255.0.0         bitmask for this subnet
Lease Obtained    5:41:30 PM          DHCP address allocation start
                                      time
Lease Expires     6:41:30 PM          DHCP address allocation end time
Default Gateway   172.17.137.254      Gateway to wider Internet
DHCP Server       172.17.137.254      DHCP Server which allocates IP
                                      addresses
DNS Servers       172.17.137.254      Domain name to IP mapping server
```

A name server may be queried using the nameserver lookup command:

```
nslookup example.net

Server:     localrouter
Address:    192.168.0.1 the local nameserver
Non-authoritative answer:
Name:       example.net
Addresses:  2606:2800:220:1:248:1893:25c8:1946 IPv6 address
            93.184.216.34                       IPv4 address
```

In this way, the end-user only need to deal with informative names, rather than numerical IP addresses. The entire process is normally completely transparent to the user but clearly requires several data packets to be exchanged to facilitate seamless communication.

4.8 Application Packet Delivery: TCP and UDP

Now that the means for data to be sent from one device to another across a network has been established, courtesy of IP, the problems of checking for errors, packet sequencing, and delivery to specific applications may be addressed. This may be done by one of several transport protocols carried by IP, with the most prominent being TCP for byte streams and UDP for distinct data packets.

Recall that IP solves the problem of delivering data packets from one physical device to another. It does not ensure correct sequencing of the data packets and does not check the payload for errors (even though the IP header itself is checked for errors, the payload is left untouched). Importantly, IP does not deliver data to a specific application running on a device, so there needs to be some way of directing the data packets, once received, to the application for which they are intended.

In a similar way to having globally unique IP addresses, the assignment of *ports* solves the problem of knowing what application should handle a specific data flow. A port is nothing but a 16-bit number, used to identify particular applications on a device. As with IP addresses, these have to be carefully managed. The service names and transport protocol port number registry is managed by the Internet Assigned Numbers Authority (IANA,), according to the procedures in RFC 6335 (Cotton et al., 2011). Some ports are "well known," which means they are used for standard or well-known services. Probably the most common of these is port 80 for Hypertext Transfer Protocol (HTTP) transfers, used to coordinate web page delivery.

Because port numbers are 16 bits wide, there is a very large number to choose from. Lower port numbers are used to define well-known services such as hypertext web page delivery and email transfer. Higher-numbered ports may be assigned dynamically by clients. These do not run out, simply because the rate of use of ephemeral or short-lived dynamically assigned port allocations is much less than the total number available, and they may be reused after a time period. The official recommendation in RFC 6335 is for dynamically allocated ports to be allocated in the range 49152–65535 (Cotton et al., 2011).

The port field is carried in the transport protocol (TCP or UDP) header. The data frames in UDP are termed *datagrams* by convention. The layout of a UDP datagram is shown in Figure 4.18. Note the complete absence of IP addresses, since the device-to-device routing, which requires IP addresses, is assumed to be taken care of by the IP layer. If a segment reaches a device, it is assumed that it has reached the device with the correct address – that is, the IP destination address matches this device's IP address. The *source port* and *destination port* fields are then used to dispatch the datagrams to the correct application – the combination of IP address and port forming a *socket* as mentioned earlier.

UDP is a "lightweight" protocol, in that it just delivers the data from one application on one device to another. It does not check for errors in the data, nor if any data is missing, or indeed if the data arrives at the destination at all. What, then, is the use of such a data transfer protocol? Commonly, UDP services are

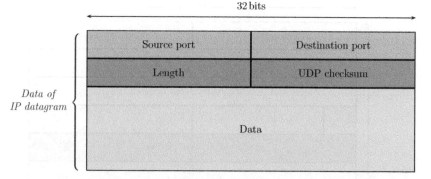

Figure 4.18 The composition of a UDP datagram. Source and destination addresses are necessary, as is the length of the datagram. The checksum checks the header, but not the contents, of the segment.

used for real-time traffic such as streamed audio and video, where retransmission of the data would be pointless, since by the time the sender was informed and the data was retransmitted, it would be too late to actually utilize that data.

The data frames in TCP are termed *segments* by convention. The layout of a TCP segment is shown in Figure 4.19. The TCP protocol addresses the problem of reliable delivery of data from one endpoint to another. Unlike UDP, it guarantees (within reason) that the data arriving at an application on a device is exactly the same, byte for byte, as the data that was sent.

Once again, IP addresses are not present in the TCP header, since that is the role of the IP layer. The source port and destination port fields perform the same role as in UDP, to form (with the IP address) a *socket* to send the data stream to the correct application. TCP provides a byte-for-byte guarantee of data delivery. It does this by arranging for the retransmission of any corrupted or out-of-order data segments. This function is almost entirely hidden from the end application, which assumes that the data is correct, unless some catastrophic error has occurred (in which case, the data transfer is terminated).

The assignment of well-known ports only solves half the problem, though. A web server, for example, may simultaneously handle many web page requests. Equally, at the other end, a web browser (client) may request many web pages (or parts of web pages) simultaneously. There needs to be a way to uniquely define the endpoints of the communication within the whole Internet, as well as within each device. Some things necessarily have to remain fixed – the IP address and the port number of the service. However, the connection for, say, the transfer of a web page, or an email message, will be relatively short-lived. The particular data transfer endpoints only need to exist while that data transfer is taking place. This is the role of the dynamic (or *ephemeral*, meaning "short-lived") ports in TCP and UDP. These are assigned dynamically on one end of the connection, for the duration of a specific transfer. This transfer may of course take

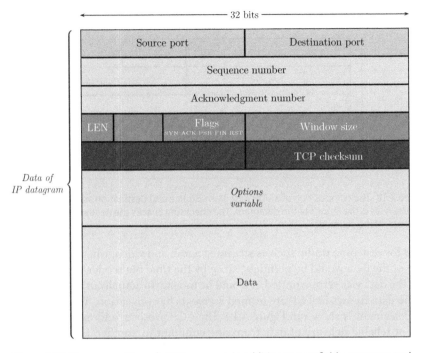

Figure 4.19 The composition of a TCP segment. In addition to port fields, sequence and acknowledgment fields are used to sequence data segments. In tandem with this, binary-valued bit-field flags are used to signal the state of the transfer, and the window size is used to maximize the data flow rate.

several data packets and may exist from milliseconds to minutes or more. The ephemeral ports are assigned for a specific data connection and thus may be reused over time without ambiguity.

The combination of IP address and port is termed a *socket*. Socket addresses are usually written in the form IP:Port – for example, 192.168.20.4:49134, where 192.168.20.4 is the IP address and 49134 is the 16-bit port number. Two sockets – one at each endpoint of the communication – define the virtual data transfer path. Figure 4.20 illustrates this concept.

In Figure 4.20, the client has address C1.C2.C3.C4 and wishes to obtain data from server S1.S2.S3.S4. If the data is to be a web page (or image from within a web page), it would send the outgoing request with a destination TCP port value of 80. In order to facilitate multiple simultaneous requests from the same client device, the port number is used to keep track of which data transfer is which. This port number is an ephemeral or short-lived port and in this case is 52196. Irrespective of the path of the individual packets through routers in the Internet, the application on the end devices uses this socket pair for that specific data transfer.

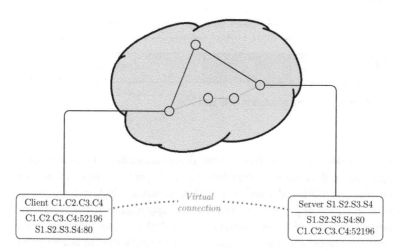

Figure 4.20 A socket pair consisting of an IP:Port combination uniquely defines the endpoints for a data transfer. Routers in the Internet use the IP address, but not the port. End devices use the port number to ensure the data reaches the correct application.

The other important field in the TCP segment header is the checksum, which is calculated using the method described in Section 4.6.3. The 16-bit checksum value is calculated using the data in the segment, and if the destination calculates a different checksum than what appears in the header, then the segment has been corrupted. The TCP protocol then requests retransmission of the data, though it does not do this directly. Rather, the stream of acknowledgments at the sender's TCP layer infers loss of data or corruption and arranges retransmission. It is important to remember that loss of a segment of data is just as bad as an error occurring, but that loss cannot normally be directly detected. Also, loss of a segment could imply that a link along the path is congested, and retransmitting the data will simply exacerbate the situation. These issues are important for network fairness, so that one device does not choke part of the network with floods of its own retransmissions. Given the complexity of these problems and their solutions, they are further explored in Section 4.9.

Finally, recall that the IP layer had a MTU. As a result, the TCP protocol has a *Maximum Segment Size* (MSS), because a TCP segment must fit into an IP datagram. Since IP requires a *minimum* MTU of 576 bytes (calculated as 512 + 64), the IP layer consumes 20 bytes for its header, and the TCP consumes a further 20 bytes for its header, resulting in a minimum MSS of 536 (RFC 879 and the more recent RFC 6691 contain detailed specifications). Note that this small size is not normally used in practice, as larger sizes result in much more efficient transfers. Typically, the MTU of Ethernet (1500 bytes) dictates the MSS used on a TCP connection as 1460 bytes.

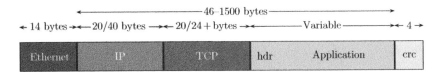

Figure 4.21 Ethernet frame encapsulation of IP and TCP.

Figure 4.21 shows each of the frame types discussed so far, and their respective encapsulation. It is worth reiterating that each protocol layer takes care of its own responsibilities (delivering a packet, creating a byte stream, using retransmission where appropriate) such that the application at the top of the stack is largely unaware of the subtleties of the underlying data network. Since the TCP protocol layer takes care of reassembling the data, the application may assume that the data is simply a "reliable byte stream." The application assumes that the data is delivered correctly and in order, which of course greatly simplifies end-user application design. This does not mean that network issues can be ignored altogether by the application, however. Data may not always arrive on a network connection when an application wants it. The application should not stall, waiting indefinitely for data that might never arrive on a possibly unreliable network. Per-socket timeouts are used extensively for this purpose in the *Application Programming Interface* (API), which gives programs access to the underlying TCP/IP services.

The application protocols such as HTTP are carried within the transport layer (TCP or UDP), often (but not always) using TCP due to its reliability guarantee. That is not to say that these protocols are fixed; they have been evolving as a result of experience with network flows and typical usage and emerging requirements such as encryption. The original HTTP 1 (Berners-Lee et al., 1996) sent and received one request at a time, which was adequate for text-only web pages, but when images became embedded, multiple requests were required. HTTP 1.1 (Fielding et al., 1999) addressed this issue by permitting multiple simultaneous requests, thus enabling more efficient use of each established connection. More recently, request prioritization and binary rather than plain-text (human readable) requests and responses are employed in HTTP 2 (Belshe and Peon, 2015). The latter also specifically addresses encryption, which was essentially a separate process in earlier versions of the protocol.

The following sections provide more detail on the TCP protocol, and as a result it will become evident why some of the abovementioned requirements are important for a secure network, which more fully utilizes the available network bandwidth.

4.9   TCP: Reliable Delivery and Network Fairness

TCP guarantees the reliable, in-order delivery of data bytes across a network from one application to another. It does this by using retransmission for corrupted or lost data packets. As well as doing this, a central goal is to maximize the data throughput for a given data stream.

Less obvious, though, is the need to share the network bandwidth with other users – it is, after all, a shared infrastructure. One application greedily transmitting as much data as possible may reduce the throughput of others. Data packet loss on a transmission path may be caused by overloaded routers, as well as congested links. In the event of data loss, transmitting more data packets may actually exacerbate the situation, with even fewer data packets successfully making it to the destination – in effect, choking the network. TCP incorporates many algorithms to incrementally adjust the packet sending rate, so as to address these issues.

The original TCP specification in RFC 793 (Postel, 1981) has had a number of enhancements and improvements deployed over several years. These are to address various issues encountered in practice. Recall that TCP uses the underlying IP services (Postel, 1991), which have no guarantee of reliability, only "best-effort" delivery of independent datagrams. The following explains the role of TCP and its salient features, using a standard client/server model. That is, a client requests data from a server, to which the server responds. This would be typical of, say, a web browser. Such an asymmetric data flow is common, though of course more symmetric data flows exist, such as in voice telephony or videoconferencing. Naturally, TCP can handle either situation, and in fact certain design aspects actually enhance the performance when bidirectional data flows are required by an application.

Consider Figure 4.22a, which shows one data packet being sent by a server to a client. It may seem intuitively obvious to acknowledge each data packet, but as the figure shows, this effectively slows down the overall data transmission process for several packets, because the sender must wait for explicit acknowledgment of successful data transfer. Suppose, then, the strategy shown in part (b) of the figure was adopted and only every second packet was acknowledged. This would considerably speed up the overall transfer in most situations. Where it would not work, though, is when an error occurs in the first data packet – both data packets would have to be retransmitted.

We could extrapolate this to have N outstanding packets before acknowledgment, rather than two. This would improve throughput in most cases, except for situations where errors or lost packets were frequent, in which case the server would unnecessarily retransmit packets by going back to packet 1 and starting the retransmission of the N packets.

Note that, by design, TCP does not include explicit acknowledgment of packets received in error. In fact, packets may not make it from one endpoint to

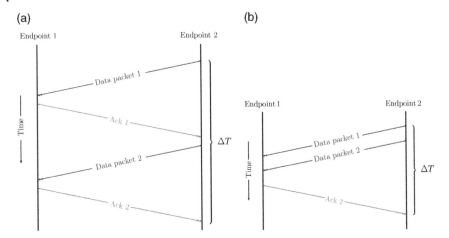

Figure 4.22 Acknowledging data packets, indicated by the *ack* lines. The sliding-window approach of acknowledging more than one packet at a time gives superior throughput, at the expense of problems in the event of errors or lost packets. (a) Acknowledging each packet as it comes. (b) Acknowledging two packets at once.

the other, in which case no acknowledgment (positive or negative) would ever be given. Thus, it is up to the sender to retransmit data in the absence of an acknowledgment. TCP uses a cumulative acknowledgment: that is, it acknowledges the data received so far, and if some is missing due to a lost packet, then the acknowledgment for the cumulative data received is repeated. This is a critical point in understanding how TCP works.

Since the network is interconnected, it makes sense to only try to transmit as many data packets as the network "pipe" can handle at any given time. This has to be adaptively estimated: The sender could be cautious and only transmit at a slow rate, suffering the penalty of low throughput. It could, on the other hand, aggressively transmit as many packets as possible, at the risk of saturating the network and having few, or none, get through. These issues are examined in more detail in subsequent sections, after discussion of the process of establishing connections.

Data in TCP is transmitted in segments, consisting of the header and actual data to be sent, if any. It is possible that no data is to be sent, in which case an empty segment comprising of only an acknowledgment of data is transmitted. Clearly, this is inefficient and should be avoided if possible. The maximum size of data that can be transmitted in TCP is called the MSS. This MSS block of data must be encapsulated with TCP and IP headers and fit within the MTU of the physical link.

4.9.1 Connection Establishment and Teardown

There are three main phases to data transfer in TCP: connection establishment, the data transfer itself, and connection teardown. The establishment and teardown is termed a *handshake*, whereby each side exchanges certain packets to verify that the other side is willing to take part in (or close down) the transfer. Although there are many states that a TCP connection goes through during its lifetime, the main ones are (i) when waiting for a connection (the listening state) and (ii) when a connection is able to transfer data (the established state). These may be demonstrated by the netstat -an command (or similar variant), as follows:

```
netstat -an
Active Connections:
Proto Local Address     Foreign Address  State
TCP   10.28.1.37:139    0.0.0.0:0        LISTENING
TCP   10.28.1.37:59769  93.184.216.34:80 ESTABLISHED
```

netstat shows the protocol (TCP, UDP, or other), the local and remote addresses as an IP:Port combination (a socket), and the state of the connection.

Figure 4.23 shows the timeline for a typical data transfer. The SYN, ACK, PSH, and FIN refer to specific bit flags in the TCP header: Synchronize, Acknowledge, Push, and Finish, respectively. The connection is established by sending a SYN request, with a sequence number in the TCP header initialized to a starting value. This is used as a counter to pace the transmission and indicates the next byte of data expected. The synchronize request (1) is followed by an acknowledgment from the server (2), and this is in turn acknowledged by the client (3). This is termed a 3-way handshake (steps 1–3). Both sides are then ready to transfer data.

The initial request for data in step (4) has the Push flag set, indicating that the data transferred so far should be sent to the application. Subsequently in (5), the FIN bit is set to indicate the end of the request. The response starts in step (6) and follows a similar sequence with the ACK bits, then PSH (7) and finally FIN (8). The final FIN is acknowledged by sending an ACK to the server (9), which ends the transfer. This indicates the typical sequence involved in sending a request from a web browser to a server, but note that the data transfer stage normally continues on for many more packets before the connection is closed down using the 4-way handshake (steps 6–9).

4.9.2 Congestion Control

Any reliable connection service clearly has to deal with the situation where the two endpoints handle data at differing speeds. This may simply be a device issue, with one end significantly faster than another, or it may be that the transaction occurring is more complex at either the receiver or the sender. For example,

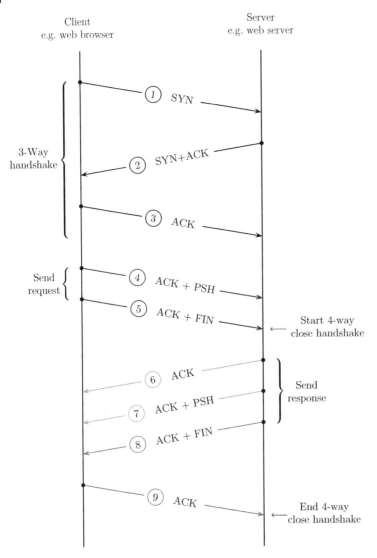

Figure 4.23 Sequence of TCP segments when setting up, sending data, and tearing down a connection. The sequence shown above is typical of an HTTP (web) request.

rendering a web page with images may be a more involved process and thus slows the client (receiver) with respect to the rate at which the server can send data. Thus, the sender of any data must be aware of how much data the receiver can reasonably cope with at any given time. This is done in TCP using the window size field (refer back to Figure 4.19), which is used by a receiver to inform the sender how much buffer space the receiver currently provides. Clearly, if that buffer space is overrun, data will be lost.

However, this is not the only way in which data may be dropped from a connection. Since the Internet consists of many interconnected devices, any device in the packet flow path may in fact not be able to keep up with the rate at which data is arriving before resending it on to the next hop in the journey. In particular, routers allocate buffer space for incoming data before forwarding, and an overloaded router may be forced to simply drop data packets. Although it may seem like increasing the router's buffer memory space is the solution to this problem, it is not. This is because no matter how fast or how much memory is provisioned, there is always the chance that a router will be overwhelmed by several devices to which it is (indirectly) connected and on whose behalf it is forwarding packets. In this case, network *congestion* is said to occur.

The issue of congestion in the Internet is a serious one, and early implementations across very low-speed links occasionally led to dramatic throughput reductions. The problems were first summarized in Jacobson (1988), and we discuss some of the consequences in terms of present-day protocol usage below. We aim to point out the main problems and reason for existence of the various TCP congestion control algorithms. A definitive guide may be found in RFC 5681 (Allman et al., 2009), and there is a considerable body of research literature on improving the performance of TCP under various conditions. More detailed explanations, apart from the RFCs themselves, may be found in a number of sources such as Hall (2000) and Kozierok (2005).

As described in the previous section, TCP relies on a sliding-window acknowledgment to ensure data reaches its destination. The timing of the acknowledgments gives useful information about the state of the network and may be used to infer how much data may be transmitted without nudging the system into overload. Consider Figure 4.24, where the data is imagined to traverse several data "pipes" of varying capacity. The area of the shaded sections represents the bandwidth-time product or how much data is in transit. Initially, a burst of data packets is sent, and these may encounter one or more bottlenecks along the way, represented by the funnels.

If the sender paced the sending of data packets such that it only introduced more data packets in response to the acknowledgments arriving back, the system would effectively become "self-clocking," and not overload the network. This, however, represents a steady-state situation, when a large amount of data is to be transferred at once. There is some time required for the sender to reach a conclusion about the state of the network. Furthermore, some applications have very small payloads and require fast response. Examples include sending one or a few keystrokes to a server or sending requests arising from clicks on a web page.

To address all of these issues, several algorithms are employed within the TCP protocol stack to maximize throughput and minimize transit delay. These

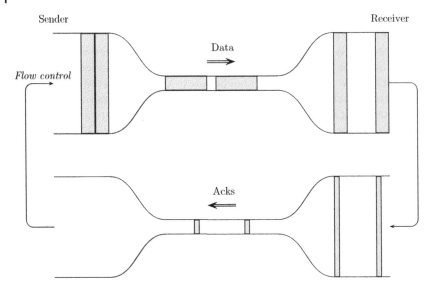

Figure 4.24 Visualizing TCP data flow as a pipe of various dimensions, corresponding to the bandwidth and delay of different sections of the network that a given exchange of data packets must traverse (*Source*: After Jacobson, 1988). The flow control acts to admit more data to the network when permissible.

algorithms have been developed over time, as understanding of the operation of networks was gained in practice. To summarize, the key requirements are:

i) To ensure that the receiver is not overrun with data.
ii) To ensure that the network is not saturated with data.
iii) To always attempt to deliver data to the end application in a timely manner.

The first of these is quite intuitive: If the sender sends more data than the receiver can process, either because the receiver is inherently slower or has more processing to do with the data (such as writing to disk), then the excess data will be lost. The second, that of network saturation, does not occur in a point-to-point link, but may well occur on interconnected networks. This is because routers at each hop have to forward data, and they have to forward data from many incoming connections. It is also possible that the capacity of the physical transmission media at one or more of the hops may be exceeded, causing data packets to bank up at an earlier router while waiting to be sent across a slower link. Finally, the timely delivery of data may need some assistance from the application itself, so as to hint whether the data should be delivered immediately (key presses, mouse clicks) or delayed for better throughput (for bulk data transfers).

To address the fast sender–slow receiver problem, it is clear that the sender should not send more than what the receiver's buffers can accommodate. This

Endpoint 1   Endpoint 2

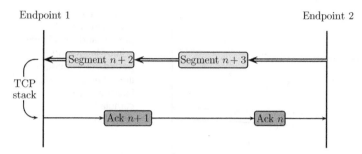

Figure 4.25 Data segments and acknowledgments on a connection. At any time, several data packets may be in-flight, with acknowledgments on their way back to the sender.

is monitored by the window field in the TCP header; the window advertises the receiver's buffer space on returning acknowledgment (ACK) packets. The sender should never try to send more than the receiver can buffer.

Addressing the problem of data lost in transit is more intricate. It is necessary for the sender to infer that data is being lost along the way, and the only way this can be done is by the acknowledgment of successfully received packets (or lack of acknowledgment, if data packets are dropped). A guiding principle is that no new packets should be introduced into the network unless other packets are exiting.

When an acknowledgment (ACK) is received, it acknowledges bytes received up to the acknowledgment number field in the TCP header. The sender maintains several variables, one of which is the *congestion window* or CWND. This is an estimate of how much data may be sent without congesting the network. This value is not fixed, but adapts to the network conditions.

Remember that there is a finite time for transmission from the source to destination, and this latency must be accounted for. The bandwidth-delay product is effectively bits per second times seconds; equivalent to bits in the "transmission pipe." Thus, in an established connection, the situation depicted in Figure 4.25 exists, with several data packets (TCP segments) in-flight, with possibly several ACKs returning, at any given time.

This would be good in a steady-state situation, but how do we determine how many packets the network pipe can handle? Assume that the connection is new and the sender wishes to fill the data pipe to the receiver in order to maximize throughput. Ideally, this would mean sending as much data as the receivers' window will allow, but there must also be an estimate of the maximum amount of data that may be sent before congestion sets in. This maximum amount must be determined reasonably quickly, or else the utilization of the network bandwidth will be poor when connections are established. This is especially important in short-lived, bursty transmissions such as web page requests.

To ramp up the connection, it is desirable to inject packets quickly at the start. Referring to Figure 4.26, one TCP segment is sent, and the receiver ensures

Endpoint 1   Endpoint 2

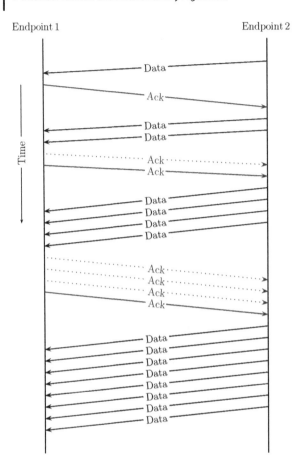

Figure 4.26 Slow-start, exponential window growth, and cumulative acknowledgments. The dotted acknowledgments are not actually sent, but inferred by a subsequent cumulative acknowledgment.

that the ACK is received. Once this happens, it is considered reasonable to send two data packets and wait for their acknowledgment. TCP uses cumulative acknowledgments and tries to delay sending a packet unless necessary (because a segment with only the ACK bit set, and no data, would be wasteful). Once the ACK for the outstanding data is received, the number of data segments sent could be increased according to the number thus far acknowledged. Then, four segments may be sent at once. Following this logic, after the next ACK, eight segments may be sent. In this way, the connection is rapidly brought up to speed. The number of segments that may be sent at each step is maintained by the TCP congestion window variable (CWND).

This rapid exponential increase is termed *slow-start* for a TCP connection. Of course, this successive doubling cannot continue forever, and if we tried to, congestion would occur when the network became overloaded. The solution to this problem is to incorporate a threshold, termed the *slow-start threshold* or

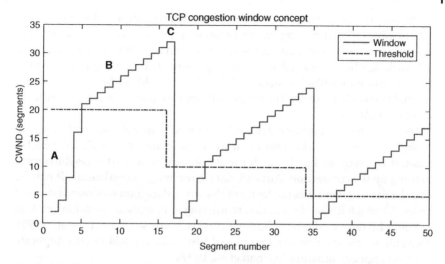

Figure 4.27 Illustrating the principle of TCP congestion avoidance. Section A is the multiplicative increase, B is the linear increase until an error occurs at C, and the threshold is halved.

SSTHRESH. The slow-start procedure continues to quickly ramp up the number of packets in transit, until the slow-start threshold is reached. At that time, the network may likely be able to handle more data but may be approaching saturation point. From then on, the exponential increase of CWND, effectively increasing per ACK, is reduced to once per round-trip time (RTT). This results in a linear increase in CWND, and thus it is less likely to saturate the network.

This exponential increase, followed by a linear increase, is illustrated in Figure 4.27. Section A is the slow-start stage with exponential growth, and B is the congestion avoidance stage with linear growth. This may continue until we reach the known receiver's limit, which is known from the window field of the returning TCP header.

Now suppose some congestion occurs in the network. This is inferred by the receiver when duplicate ACKs are received, indicating either data loss or data being received out of order. If we assume that the data is lost, then it is necessary to retransmit the lost segment. But we also need to reassess the CWND so that the congestion is alleviated. If the loss is severe, it may be that no ACKs return to the receiver at all. In that situation, the sender must rely on the *Retransmission Timeout* (RTO). Once this timeout is exceeded, we must restart the slow-start and collision avoidance procedures. This is shown at C, where CWND is reduced to a lower value, and SSTHRESH is halved. The whole process repeats, until the receiver window size is reached, or another RTO timeout occurs.

The occurrence of a timeout indicates a severe problem, and the sender must take evasive action. But suppose the situation is not as severe and that one segment was lost or received late at the receiver. If an ACK is to be sent, it cannot acknowledge the new data, due to a missing segment. The only option is to ACK the last contiguous block of data, and this ACK should not be delayed as with normal acknowledgments. The sender will see this as a duplicate acknowledgment (DUPACK).

The presence of duplicate acknowledgments is an indication of potential problems in the network. However, it would be inefficient to wait for the RTO timeout to expire and go through the subsequent slow-start procedure again to bring up the connection. After all, data is reaching its destination. It may be that the data segment was not lost, but that an IP datagram was received out of order. Although it would seem that reordering is an unusual situation and that it would be rare, evidence indicates that this is not so (Bennett et al., 1999). Two enhancements, Fast Retransmit and Fast Recovery, address the duplicate acknowledgment situation (Allman et al., 1999).

The detailed sequence is documented in RFC 2581, but, in essence, fast retransmit dictates that the receiver should wait for three DUPACKs and then retransmit the apparently lost segment. If it so happens that the receiver actually ends up receiving the same segment twice, that should not be a problem, since it keeps track of all data and its location using the sequence number. However, there must be some problem in the network for DUPACKs to occur. So, fast retransmit states that, in essence, the slow-start threshold SSTHRESH should be set to CWND/2, and CWND be increased by 3. This means that the congestion avoidance procedure will then operate when the slow-start threshold is reached.

Cumulative retransmission forms part of the original TCP reliability guarantee. The assumption is that several segments in succession will be lost due to network failures such as congestion. In the case of a single lost segment rather than a succession of lost segments, all subsequent segments must be retransmitted. This is clearly inefficient, if the failure mode resulting in single lost segments is common. Furthermore, a sender must wait one RTT to find out about the lost segment. Finally, if a sender is overly fast in retransmitting lost segments, the result may be yet more congestion.

It is clear that, to address this problem, a method of identifying which segment is lost is required. This is defined in RFC 2018, which defines Selective Acknowledgment or SACK (Mathis et al., 1996). SACK works by using the TCP options field to make the other side aware that it supports this extension. The TCP options are then used to indicate the range of data bytes missing in the event of a duplicate acknowledgment. Clearly, both sides must understand SACK. Critically, since SACK was introduced some time after standard TCP was deployed, the SACK operation must not fail with non-SACK-aware TCP stacks. In the case that one side understands selective acknowledgments but the other does not, the fallback position must always be the standard cumulative acknowledgment process.

4.9.3   TCP Timeouts

The mechanisms used in TCP for congestion avoidance are summarized in RFC 5681. The reliability guarantee of TCP comes at the price of increased complexity. There is yet more complexity involved in setting the timeout values employed for the purposes of retransmission.

In addition to the slow-start and congestion avoidance, there needs to be a mechanism whereby the sending TCP can retransmit data without receiving anything from the destination TCP (assuming the connection has already been in operation). It is necessary to have a timeout so as to retransmit data after a certain interval. The setting of this RTO is crucial; on the one hand, if set to a large value, then throughput will suffer, because on each lost data packet (or its acknowledgment), the RTO timer will have to expire before an attempt is made to resend the data. On the other hand, if the RTO value is set too low, then there is the possibility of unnecessary retransmissions, because the data (or its acknowledgment) may still be in transit.

Because network conditions may vary over time, the RTO value must be set adaptively. The timeout must be set such that it is a little greater than RTT. But the RTT may itself vary, and initially there is no reliable way to know the RTT. The solution adopted is to average the RTTs from several sent data packets and their acknowledgments. Now, it would be possible to keep a list of RTT values for all packets on a connection, but this would be impractical. This could be addressed simply by keeping a running sum of RTT values and averaging over the number of measurements, but again we run into problems. This is because RTT is not fixed, and may vary over the lifetime of the connection. Thus, we need to estimate a *smoothed RTT*, denoted as SRTT. It should be clear that we would not only want a good estimate of the SRTT but a reliable, recent estimate. If the RTT was very small to begin with, but gradually became larger, then current estimates should reflect the larger value, and vice versa. So a smoothing procedure is needed, which preferentially averages more recent measurements.

Mathematically, we can do this using a recursive equation such as the following, where $s(n)$ is the current smoothed estimate, $x(n)$ is the new measurement, and α is a scaling factor close to, but a little less than, one:

$$s(n) = \alpha \, s(n-1) + (1-\alpha) \, x(n) \tag{4.1}$$

The original TCP RFC 793 suggests a similarly smoothed RTT, which is updated as follows:

$$\text{SRTT} \leftarrow \alpha \, \text{SRTT} + (1-\alpha) \, \text{RTT} \tag{4.2}$$

where α is a smoothing factor suggested to be 0.8–0.9. A convenient choice for integer arithmetic implementation is $\alpha = 7/8$, because dividing by 2^3 is equivalent to shifting right a binary integer by 3 bits. Note that the left arrow indicates the updated quantity each time the value of SRTT is updated to the new value, using the existing value of SRTT, together with the new measurement of RTT.

The RTO needs to be larger than smoothed RTT, so using

$$RTO = \beta \, SRTT \tag{4.3}$$

with $\beta = 2$ provides a computationally efficient estimate of the RTO. However, there should be an upper and lower bound set on this important parameter estimate, so RFC 793 uses

$$RTO = \min \begin{cases} UBOUND \\ \max \begin{cases} LBOUND \\ \beta \cdot SRTT \end{cases} \end{cases} \tag{4.4}$$

where UBOUND is an absolute maximum upper bound, LBOUND is a lower bound, and β is suggested as 1.3–2.0. The definition in this way avoids pathological cases where RTO might be estimated to be significantly above what may be tolerated or when the RTT estimate is unreliable in the initial start-up phases leading to an unworkably small RTT estimate (and hence RTO).

A great many variations in the RTT calculation and the derivation of the RTO have been suggested. The problem is to find a solution that works well under all cases of normal network operation as well as unusual scenarios. In order to stop spurious retransmissions due to timeouts, Jacobson (1988) suggested that the variance of the RTT estimates should be incorporated and gave theoretical arguments to support this. Additionally, computationally efficient methods were put forward. To see the motivation, consider the formulation of the weighted moving average introduced earlier:

$$s(n) = \alpha \, s(n-1) + (1-\alpha) \, x(n) \tag{4.5}$$

If rearranged, it becomes

$$s(n) = x(n) + \alpha \, (s(n-1) - x(n)) \tag{4.6}$$

So the SRTT should reflect the current measurement, plus something proportional to the difference between the previous smoothed estimate and the new measurement. Another way to see this is to consider the statistical distribution of the RTT measurements. If they produced a bell-shaped curve, some measurements would be less than the mean and some greater. Since we want to be a little conservative and have a timeout a little longer than the mean, but encompassing most measurements, it could be based on the average plus some number of standard deviations from the mean. Since the recalculations are done per packet, they should be quite simple, and to this end Jacobson (1988) suggested an approach based on the mean deviation, which is incorporated into current standards.

RFC 6298 (Paxson et al., 2011) updates the most recent practice, which is to use the variation of the RTT estimates as follows, using the variance of the round-trip time RTTVAR. The smoothed RTT, its variance, and the timeout are initialized as

$$\text{SRTT} = \text{RTT} \tag{4.7}$$

$$\text{RTTVAR} = \text{RTT}/2 \tag{4.8}$$

$$\text{RTO} = \text{SRTT}(0) + \max \begin{cases} G \\ K \cdot \text{RTTVAR} \end{cases} \tag{4.9}$$

where $K = 4$ and G is the clock granularity. The variance is updated with a new estimate RTT' according to

$$\text{RTTVAR} \leftarrow (1 - \beta) \cdot \text{RTTVAR} + \beta \mid \text{SRTT} - \text{RTT}' \mid \tag{4.10}$$

$$\text{SRTT} \leftarrow (1 - \alpha) \cdot \text{SRTT} + \alpha \cdot \text{RTT}' \tag{4.11}$$

with $\alpha = 1/8$ and $\beta = 1/4$. The timeout is then

$$\text{RTO} = \text{SRTT} + \max \begin{cases} G \\ K \cdot \text{RTTVAR} \end{cases} \tag{4.12}$$

The timeout uses Karn's algorithm for measurements, where retransmissions are not incorporated (Karn and Partridge, 1987). In this way, a much better estimate of a "good" timeout is produced: one that smooths out variations, yet reflects more recent measurements on the link itself.

4.10 Packet Routing

The term "routing" refers to the determination of the path an IP packet is to follow through a set of interconnected networks. Routing occurs in the IP layer, but there is some interaction with the link layer in many situations. A dedicated router may perform routing at the gateway to a network, but routing may be performed by any device with more than one network interface. The routing functionality is itself distributed: That is, each device or node on a network receives an IP packet and determines whether to keep it and pass to higher layers (TCP, UDP, or other transport protocol) in its own protocol stack, or else to forward it on. There is no master controller for managing this, and each device must make its own decisions.

Simply put, a router takes an incoming IP packet from one interface and must decide which other interface to resend it out on. This is known as *forwarding*. The outgoing interface should always get the packet closer to its destination. For devices at the endpoints of a network, where there is only one upstream connection, forwarding reduces to just sending the packet to the gateway, which must in turn be able to reach the wider Internet. In that case, there is effectively no routing decision to be made.

There are two subtasks necessary to perform routing successfully. The first is to efficiently dispatch an incoming packet to the correct destination. Since this must be done for each and every packet, it must be done very quickly. Per-packet lookup of a routing table provides a rapid means for determining the next hop interface. This table contains a list of addresses and where to send packets with those addresses as a destination. Since it is not feasible to check all possible destinations, the routing table must look for the "closest" destination, rather than the exact destination – and hope that the next hop can make a more informed decision as to what is closer to the ultimate destination for a packet.

The second subtask is to actually populate the routing table itself. For the typical end-device, the routing table is relatively simple and mainly involves just determining whether an endpoint is on the local LAN or if not, to send the packet to the gateway to deal with. The routing table in this case is relatively simple to set up, and the configuration is often done automatically without user intervention. But for routers themselves, the route forwarding decision must incorporate some method of inexact matching, since knowing every single possible destination on the Internet is clearly impossible. The routing decision must also be consistent across multiple routers. Otherwise, packets may be forwarded from one router to another, only for them to be forwarded back – clearly undesirable behavior, allowing packets to circulate indefinitely (in practice, until the TTL or time-to-live counter in the IP header expires).

The initial sections below look at the first problem – that of determining the best forwarding interface for a packet, given that the routing table is available. Sections 4.10.6 and 4.10.7 then examine ways to actually build the routing table itself. First, however, we examine some realistic examples of routing tables. We use IPv4 addresses here to more conveniently illustrate the concepts using shorter addresses; the reader is referred to RFC 6177 (Narten et al., 2011) for specifics relating to IPv6 address assignment.

4.10.1   Routing Example

To place the key concepts on a firm foundation, we first briefly examine an example of how routing is typically implemented in an end-device. The device in these examples has IP address 192.168.0.131, with gateway 192.168.0.1. It is common practice – though certainly not mandatory – to configure a gateway with a low-numbered identifier such as .0.1 for the lower address portion. Some of the routes present in this device are shown below:

```
route print

Destination     Netmask           Gateway        Interface
127.0.0.1       255.255.255.255   On-link        127.0.0.1
192.168.0.255   255.255.255.255   On-link        192.168.0.131
192.168.0.0     255.255.255.0     On-link        192.168.0.131
0.0.0.0         0.0.0.0           192.168.0.1    192.168.0.131
```

The destination 127.0.0.1 is the *loopback* address: Packets sent to this address are effectively received back by the same device. Such an address is present on all IP-connected machines. The network address 192.168.0.255 is a broadcast address: This is necessary to reach all devices on the same subnetwork – recall that a broadcast address contains all 1's, and this forms 255 as the last byte of the address. The address 192.168.0.0 defines the entire subnet. This is not the broadcast address, but is effectively the wildcard address for any destination matching 192.168.0.*, where * equates to any address. Finally, the 0.0.0.0 destination refers to the default address, for any data packets for which the destination does not match the previous criteria. This gateway address is seen to be 192.168.0.1, which is the router connected to the external Internet.

We can trace the routing hops for a data packet traversing from this gateway to another. Here, the address example.net is used as a destination. Each line shows three attempts at measuring the latency (in milliseconds) to the nominated device:

```
tracert -d example.net

route to example.net [93.184.216.34]
1   11 ms    3 ms     8 ms      150.101.32.93
2   10 ms    10 ms    8 ms      150.101.34.30
3   16 ms    18 ms    27 ms     150.101.33.12
4   190 ms   204 ms   203 ms    150.101.34.42
5   221 ms   205 ms   167 ms    206.223.123.14
6   194 ms   194 ms   194 ms    108.161.249.17
7   168 ms   197 ms   164 ms    93.184.216.34
```

Note that the *latency* varies from one packet to another but is usually broadly consistent for a given destination. The variation in packet latency is termed *jitter*.

4.10.2 Mechanics of Packet Forwarding

The term *forwarding* refers to checking each incoming packet on a given physical interface, examining its destination IP address, and resending that packet on another physical interface. The outgoing physical interface should be closer, in some sense, to the ultimate destination of the packet. The act of receiving, checking, and reforwarding is termed a *hop*, with the *hop count* being the number of routing hops traversed from the source to destination. The only real changes necessary within a packet when forwarding are to decrement the TTL (time-to-live) field in the IP header (and update the checksum accordingly) and then to set up the link-layer addresses once a physical link has been chosen. The TTL field in an IP packet is decremented on each hop, so as to prevent packets circulating endlessly. Any packets with a TTL of zero must not be forwarded on (they are said to be *dropped*).

Consider the task of checking for a match to a packet's destination IP address. If the destination address matches the current device address, then the current device is in fact the final intended address, and the data packet is processed by the protocol stack. It is not forwarded on the data link. The test for exact matching can be performed using an Exclusive OR (XOR) function, since the XOR of two identical values is always zero. The routing function is then complete.

If an exact match does not occur in the initial check, then further steps are required. The destination address might happen to be physically connected on the same LAN. If that were the case, we can simply resend the IP packet with the MAC address of the destination on the local LAN, and the task is done. We can determine if the destination IP is on the same subnet as our own by comparing the IP addresses using only those bits where a binary 1 appears in the subnet mask. This is easily done by performing a logical AND of the destination IP address with our own subnet mask, then performing a logical AND of our IP address with the subnet mask. If these two results match, then we can be sure that the device is on the LAN, and we can resend it directly by setting the correct MAC address in the outgoing data frame.

This is detailed in Figure 4.28, where the calculations are performed on a device with address 192.168.128.34 and subnet mask 255.255.255.0. This mask has the effect of masking out the eight least-significant bits. The result of the mask being applied is 192.168.128.0. Using the destination address 192.168.128.12, the masking results in 192.168.128.0, which is the same subnet as the device itself. Thus, we can be sure that the destination is on the same subnet. If we now look at the second destination address 192.168.32.17 with the same mask, we see that it becomes 192.168.32.0. This is *not* on the same subnet because it differs from 192.168.128.0. In this case, the subnet mask allows us to see the results using decimal integers, but for a general subnet mask this would not be the case. Of course, the binary operations are very simple: Bitwise ANDs followed by a comparison using XOR.

Device address: 192.168.128.34	11000000	10101000	10000000	00100010
Subnet mask: 255.255.255.0	11111111	11111111	11111111	00000000
AND 192.168.128.0	11000000	10101000	10000000	00000000
Destination address 1: 192.168.128.12	11000000	10101000	10000000	00001010
Subnet mask: 255.255.255.0	11111111	11111111	11111111	00000000
AND 192.168.128.0	11000000	10101000	10000000	00000000
Destination address 2: 192.168.35.17	11000000	10101000	00010011	00010001
Subnet mask: 255.255.255.0	11111111	11111111	11111111	00000000
AND 192.168.35.0	11000000	10101000	00010011	00000000

Figure 4.28 Determining whether two addresses are on the same subnet.

If it turns out that the device is on the same subnet, then the MAC address may be found using the ARP protocol (Section 4.7.1). ARP performs a LAN broadcast asking who has a certain IP address, and the owner replies with its MAC address.

Most end-link devices are connected to the Internet via a gateway. This is the address to which IP packets are sent if the destination address is not on the directly connected LAN. The gateway device, indicating the destination address for packets not on the directly connected LAN, may be found as shown below:

```
ipconfig /all

Ethernet adapter Local Area Connection:
Description        Gigabit Network Connection
Physical Address   D4-BE-D9-1C-DF-73   The physical or MAC address
DHCP Enabled       Yes                 IP address allocated from DHCP
                                       server
IPv4 Address       172.17.1.111        IP address of this device
Subnet Mask        255.255.0.0         bitmask for this subnet
Lease Obtained     5:41:30 PM          DHCP address allocation start time
Lease Expires      6:41:30 PM          DHCP address allocation end time
Default Gateway    172.17.137.254      Gateway to wider Internet
DHCP Server        172.17.137.254      DHCP Server which allocates IP
                                       addresses
DNS Servers        172.17.137.254      Domain name to IP mapping server
```

So this takes care of link-local addressing and anything else we simply hand-off to the gateway. But what about routing between networks? The gateway device must be further connected to one or more other LANs, so as to form an internetwork. This means that there are several possible physical interfaces, and furthermore the ultimate destination may not actually be found on one of those LANs. In that case, it is necessary to send the IP packet off through one of the interfaces to another device that is "closer" in some sense to the ultimate destination. The process of choosing the closer device, and the corresponding physical interface on which to forward the packet, is the role of routing within the IP layer.

4.10.3 Routing Tasks

The decision that a router must make consists essentially of just deciding which interface to forward a packet on. However, there could be many interfaces on a router in a large organization and/or at a higher level. Furthermore, there could be a very large number of possible routes to contend with. Routing tables with hundreds or thousands of possible destinations present a considerable computational burden, even with fast and efficient lookup methods described in the following sections. There is one overall proviso: That a packet should never be forwarded back out on the same interface from which it came, otherwise a packet may bounce from one router to another and back again, circulating endlessly.

Another related problem is that of the IP address structure itself. Although the original Class A/B/C designation has some hierarchy built into it, it does not result in a very efficient use of the address space. In theory, there would be a little less than 2^{32} addresses available in a 32-bit address space. However, consider a Class A network, which allows 7 bits for the network address and 24 bits for the device address. No LAN would have anything approaching 2^{24} devices connected directly to it. Even the Class C network, with only 254 possible devices (256 less the broadcast address and network address) would be used inefficiently if, say, a small organization with a dozen connected devices used a Class C allocation. This inefficiency is compounded by the very large number of possible Class C networks.

These two problems – inefficient use of address space and router overload – gave rise to an ingenious solution known as CIDR (pronounced "cider") for Classless Interdomain Routing (Fuller and Li, 2006). This allows, in effect, the Class A/B/C distinction to be taken to the limit, while still allowing standard IP addresses to be used. CIDR performs this feat by using a network mask, in a similar (though not identical) way to subnet mask for LANs. Recall that the subnet mask subdivides an address space into smaller subnetworks.

However, the problem to be solved here is to combine several subnetworks into one, and thus the subnet idea is used, but now it represents an aggregate of several subnetworks. In this way, several subnets are combined for the purposes of routing – a router only needs to know the larger group of subnetworks, and the router at the boundary to the subnets handles routing at that level. This process has been referred to as "supernetting," as it is effectively the opposite of subnetting. In this way, routing may be simplified by distributing the routing workload. Additionally, IP address space is conserved by making more efficient use of addresses for small networks.

All of this is achieved while not having endpoints aware that it is happening – standard IP addressing appears the same to them. The routing information that must be exchanged must incorporate both a network prefix and a supernet mask, but this impacts relatively fewer routers as compared with end devices.

4.10.4 Forwarding Table Using Supernetting

The forwarding decision is simply this: Given an IP address, a device needs to work out which physical interface to retransmit a packet on. Since a router cannot possibly store every IP address it might ever want to reach, it needs to aggregate blocks of addresses, such that only one entry will suffice for a given block of devices. For example, given a packet to route to somewhere on a network that has a 192.168 prefix, then it should be unnecessary to maintain routes for all possible addresses 192.168.1.1, 192.168.1.2, and so forth.[1] That would be

---

[1] These private non-routable addresses are used for the purposes of example only in this and subsequent sections, so as not to conflict with "real" addresses. In practice, IP addresses designated as private are never forwarded on.

completely impractical. However, if the IP address of a device in closer proximity was known, but which could in turn reach the 192.168 address block, then surely it would be possible to just forward an IP packet with destination 192.168.1.2 onto that other device and let it worry about what to do then. Thus, we need not worry about the final route, only the next "hop" that gets the packet closer to its desired destination.

The router has to select an IP address that corresponds to the closest network, according to the routing table. It must check each entry in the routing table, not for complete match but for the *maximum number of bits that match*. For a given routing mask, say /18, the search proceeds to compare the bits in the packet destination with the corresponding bits in each candidate route entry. The more that match, the better.

Finally, if a closer route cannot be found, then there must always be a default route present – this determines the interface on which a data packet should be forwarded in the event that no closer match is found. This is always address zero, mask zero, or 0.0.0.0/0.

Consider Figure 4.29, which shows a router with three physical interfaces. Suppose a packet arrives on interface 0, with destination IP address 192.168.2.1. The routing table is then consulted. The table of routes gives the best interface connection for routes that the router knows about. The table must have both an IP address and a prefix for each possible route. It may seem that this is similar to subnet masking, and in many ways it is. However, the prefix is used a little differently: It specifies the aggregate route to several other subnetworks.

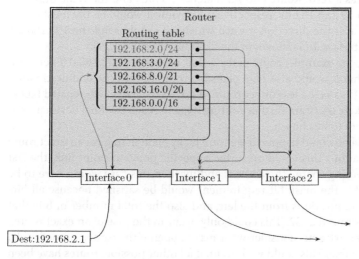

Figure 4.29 Diagram showing a route forwarding table example. There are five routing table entries and three physical link interfaces.

				Bits matching
Address:	192.168.2.1	11000000	10101000 00000010 00000001	
Route entry 1 192.168.2.0/24		11000000	10101000 00000010 00000000	(24)
Route entry 2 192.168.3.0/24		11000000	10101000 00000011 00000000	(23)
Route entry 3 192.168.8.0/21		11000000	10101000 00001000 00000000	(20)
Route entry 4 192.168.16.0/20		11000000	10101000 00010000 00000000	(19)
Route entry 5 192.168.0.0/16		11000000	10101000 00000000 00000000	(16)

Figure 4.30 Routing table showing IP addresses and netmasks, corresponding to the diagram of Figure 4.29. The network prefix bits are shaded. The routing lookup goal is to select the specific route that maximizes the number of matching bits.

Figure 4.30 shows the routing table with the component part essential for this discussion written out in binary, with the network prefix bits shaded. Note that for the purposes of exposition, we have only shown a small subset of what might be a very large forwarding table. There could be more entries before the start of those shown, with longer prefix masks, but they may point to completely different networks. The table is shown organized in decreasing order of prefix mask length (/24, /24, /21, /20, /16) since a longer prefix indicates a more specific route, with a greater chance of matching more bits.

Many route entries would be excluded very early on, simply because their prefix bits are quite different. In the example, we only show those starting with 192.168, which can be seen by inspection to match at least the upper 16 bits of the desired destination.

Considering each route in turn, we find that route entry 1 matches 24 bits from the left, which happens to match all bits in the prefix. So, this would appear to be a good candidate as the most direct route. Route entry 2 matches 23 bits from the left, since the match is broken where the route entry has a one and the destination has a zero. Route entry 3 matches 20 bits, and the remaining two routes match 19 and 16 bits, respectively. So, route 1 would be the most direct route, and that is the direction toward which we should send the 192.168.2.1 packet, on interface 1 (as shown in the diagram of Figure 4.29).

Reviewing, for example, route entry 5 that covers the 192.168.0.0 network, we can see that it would also be a plausible route toward the end destination of 192.168.2.1 – albeit a less direct route. It points to a larger aggregated block, and so it is likely there are further routing hops to reach the final destination if that route were to be chosen.

Consider what would happen if we were lucky enough to have an exact route to the destination. This could occur for a specific point-to-point link, the last one in the routing chain. The address and mask in this case would have to be 192.168.2.1/32. The mask bit requirement would be satisfied because all bits match in all 32 positions from the left, and also the total number of bits that match is maximized at 32. This could only occur in the case of an exact route.

At the other extreme, consider what may happen if the router cannot find a match in the table. This could well occur, if all other possible routes have been tried (using the address-and-mask approach) without success. For this reason, routing tables always have a *default route*, specified as 0.0.0.0/0. That is, an all-zeros address with mask of zero. This means that, by definition, the required number of bit matches when using the mask is always satisfied, because the number of bits that have to match is zero. In the case where no other routes match, the default route will always match. If even one other route partially matched, then it would have been chosen as the longest match. In this way, the default route acts as it should: a route of last resort.

The difference between matching in a routing table and subnet masking should now be apparent. In both, we mask off those bits to the right that are zero in the mask. However, in prefix routing, the bits within the prefix mask are taken into consideration, and the prefix with the largest number of matching bits from left to right as compared with the destination address is declared as the winner. The bits do not all have to match, as they do in checking a local subnet, but a contiguous match from left to right, up to the first nonmatching bit, is required.

Now consider another case: A packet arrives destined for 192.168.17.1. We know the first two bytes corresponding to 192.168 match already. The third byte (decimal 17) is 0001 0001 in binary. Comparing this with the route table given, the number of bits matching is 19 for routes 1, 2, and 3, and this is less than the mask length in each case. For route 4, 20 bits match, and the match is limited due to having used up all of the /20 mask bits. Route entry 5 matches only 16 bits (all of the mask bits). So, in this case, route 4 has the longest match length.

Finally, note one possible problem that may occur, as documented in RFC 4632 (Fuller and Li, 2006). This is illustrated in the new arrangement of Figure 4.31, where there is one 192.168.16.0/20 aggregated network, which is subdivided into smaller networks. One is 192.168.8.0/22, which is further divided into a subordinate 192.168.9.0/24.

These differ in the third byte of the IP address, and the binary value of each is shown in the diagram, with the portion of the netmask covering that byte shaded. A packet destined for the 9.0 network, at the 192.168.16.1 router, would be forwarded to 192.168.8.1, and then to 192.168.9.1 as the gateway to the .9.0 network. That is, as it should be: The higher 16.1 router has no knowledge of the contents of the lower 9.0 network, only that it may be reached via 192.168.8.1.

Next, suppose the connection breaks as indicated by the X. Then, the same packet destined for the lower network would find at the 8.1 router that 9.0 is unreachable. However, another route entry in 192.168.8.1 gives 16.0/20, which would still cover the destination, so the packet is forwarded there. As shown on in the diagram, this is a higher-level router, and in fact the result is the packet is forwarded back to where it came from. The lines indicate the routing loop thus formed. The conclusion is that a router should never send packets back to a more general (less specific) destination (in this case, 192.168.8.1 should

Destination 192.168.9.12 (9 ⇒ 0000 1001)

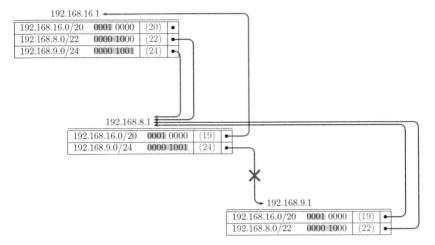

Figure 4.31 Diagram showing a routing loop caused by incorrect forwarding in an aggregated network. The incoming packet destined for the network 192.168.9.0 arrives at 192.168.16.1 for forwarding. For clarity, only the third byte from the left is shown in binary for each route table entry.

not follow the route 192.168.16.0/20), and certainly never back on the same interface from which the packet came.

4.10.5 Route Path Lookup

In the same way that a person does not want to scan the title of all books in a library in order to find just one particular book of interest, searching all possible routes in a forwarding table should not be necessary in order to find the correct forwarding address for an incoming IP packet. The packet arrival rate is typically very high, and the number of forwarding entries may be very large (in the thousands). All this will create a delay for each packet-forwarding operation and hence contribute to the total routing latency in the journey of any given IP packet.

For this reason, fast routing lookups are important. This section aims to give an overview of approaches to speeding up the routing process. It is not meant to be an exhaustive coverage of all methods, nor could it be. It deliberately avoids suggesting the "best" approach, because the criteria depends very much on the situation. Rather, the aim is to highlight the fact that there are algorithmic approaches to the routing lookup problem, and, furthermore, that an efficient algorithm is worth taking the time to understand.

The fundamental problem may be defined as follows: As stated in previous sections, the IP address blocks are aggregated into larger routing domains, so that each router then must match the destination address to the best-matching next hop. The matching is not simply looking up a single value, but rather finding the address that matches the largest number of initial prefix bits. Matching more bits indicates a closer network. So, this type of table lookup is termed *best matching prefix* or, more descriptively, *longest matching prefix* (LMP). A naïve search would consist of N-bit comparisons for an N-bit address, for each table entry – clearly an inefficient approach for anything more than a small number of entries.

Since the comparisons and other operations are binary, then a hardware architecture could be devised to perform rapid matching (Gupta, 2000). A drawback of this, apart from the evident complexity, is that the forwarding tables themselves must be updated in some way, using routing protocol messages. We have already seen a linear table of forwarding entries, and this is the simplest approach. This could be arranged as a predefined table, with space allocated for the maximum number of entries. Alternatively, a linked-list data structure could be employed, so that the number of route entries could be expanded (or reduced) as required. An extensive discussion of various data structures for IP lookup may be found in Chao (2002); finding better approaches is the subject of research (for example, (Lim et al., 2009)).

A data structure that may be employed in this situation is the *binary tree* or one of its numerous variants. Figure 4.32 shows a binary tree constructed for nodes with labels A, B, C, and D, with hexadecimal key values 4, 6, 2, and 9, respectively. The node labels A–D may store any arbitrary information and are used here for labelling. What is important in this context is the binary value of the integer key stored at each node: 0100, 0110, 0010, and 1001. The bits in this example are numbered consecutively from the left (MSB), and at each circular decision node, the choice depends on the value of the bit. A bit value of 0 (branch left) or 1 (branch right) determines the path taken at that point. Thus, we may reach node C by following the corresponding binary value to take the path from the root node $\mathcal{R}$ to the left (for 0), then left (0), right (1), and finally left (0).

The leaf nodes may hold a final value (A–D in this case) or else be null values holding no information, as represented by the square nodes. Each branch takes one of two possible paths, and the depth of the tree is the total number of branches traversed, which corresponds to the number of bits in the integer keys.

Figure 4.33 shows a slightly more complicated binary tree, this time constructed for 6-bit values rather than 4 bits. The usefulness of a binary tree in situations such as the forwarding table search is that the time it takes to reach a decision is governed by the depth of the tree, in this case the number of bits N. A full (exhaustive) search for all possible values would require 2^N comparisons. If, for example, $N = 32$, a tree search would require only 32 single-bit comparisons, as compared with 2^{32} comparisons in the exhaustive case (which,

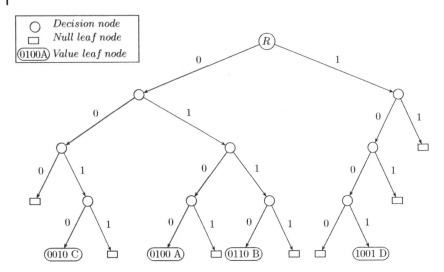

Figure 4.32 Binary tree example using a 4-bit key and node values A, B, C, and D.

clearly, is a strikingly large number). The binary tree algorithm is said to *scale* for larger N. This comes at the cost of complexity: A more complicated structure is required to maintain the information in the tree to facilitate searching. Additions to and removals from the tree are not especially difficult, so the forwarding table may be updated relatively easily for this type of tree structure.

The binary tree gives a useful direction for the development of a method for storing and processing the forwarding entries. It does not directly solve the problem, however. This is because the route-matching task requires an LMP search based on binary values, and not a precise match as we have seen thus far for the binary tree. The binary tree is generally used for searching based on some criteria in the data itself (for example, alphabetical order of characters), rather than the bits themselves. A search based on the data bits themselves is termed a *radix search*.

A generic data structure for retrieval of information is termed a "trie," as it is used for retrieval.[2] The original concept of a trie may be traced to Fredkin (1960). A specific type of trie used in IP lookups is the Patricia trie (see, for example, Sklower, 1993; Wright and Stevens, 1995b; Waldvogel et al., 1997). The Patricia algorithm was originally described in Morrison (1968), where the term Patricia was introduced (Practical Algorithm to Retrieve Information Coded In Alphanumeric). A good generic exposition of the Patricia algorithm is given in Sedgewick (1990). The specific implementation of Patricia tries for routing tables is described in Wright and Stevens (1995b).

---

2 Although "trie" comes from the inner part of "retrieval," it is pronounced variously as "try" or just "tree."

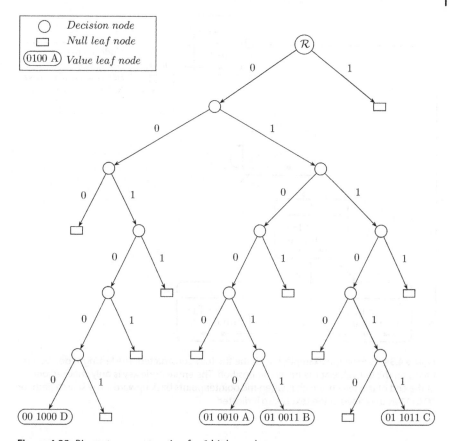

Figure 4.33 Binary tree construction for 6-bit key values.

A defining characteristic of a Patricia trie is that, when searching for a match for a given key, not all keys stored in the trie are checked. This is because only certain bit positions are checked during the search – so bit values are checked and used as a left/right branch, but the entire N-bit key is not checked until the end.

To motivate the use of the Patricia trie as applied to the forwarding table search problem (as introduced in Wright and Stevens, 1995b), consider the shortcomings of the binary tree. First, as mentioned, it provides no obvious method to incorporate the prefix-matching requirement, though this could be done with various modifications. Second, the binary tree as described contains three different types of node: decision nodes, leaf key value nodes, and terminal leaf nodes with no value. This complicates the step-by-step processing somewhat. Finally, the search space is somewhat sparse, in that not every single

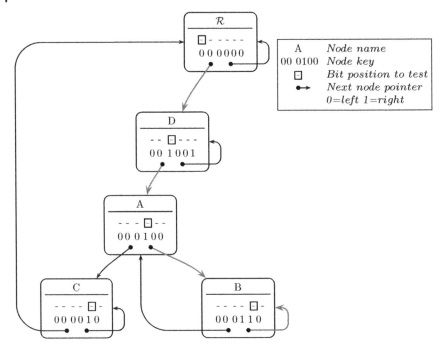

Figure 4.34 Patricia trie example 1, with the trie fully constructed. The left/right decision is based on the 0/1 value at the boxed bit position. The entire node key is only checked once, at the end of the search, which is when the pointer points back upwards. The search path for 00 0111 as described in the text is also indicated.

possible result needs to be enumerated, as would happen with a straightforward binary tree.

In the following, keep in mind that the primary problem is to find the LMP starting from the leftmost bit. Consider Figure 4.34, which shows the use of 6-bit values, with the trie populated by four entries (other than the root node, which is always present to anchor the trie). Note that each node type is identical, rather than the different decision/leaf types with a binary tree. Each node has a key name (just letters A–D in this case, along with the root node $\mathcal{R}$), a key value (which is searched per bit), and a bit position as indicated by the small box. Each node has a left/right pointer corresponding to 0/1 decision, but now we base the decision not on each successive bit, but rather on the bit position as denoted by the boxed binary digit within each node.

Each node needs to store the bit index of the first bit that differs in the same bit position from the parent node. This index of the first difference stored at each node means that subsequent lower-down nodes differ in bit positions further to the right.

Numbering the bits from the *right* as 1, 2, 3, … and starting the bit search from the left (MSB), we start at the root node, which is really just a placeholder to anchor the tree. Checking the leftmost bit (position 6) of the root node R in Figure 4.34, a value of 1 points back to itself, whereas a value of 0 points directly to node D, whereupon we check bit 4. If this bit is a 0, we go left toward node A, whereas if the bit is a 1, we go to the right node – in this case, we traverse back to the same node itself. If we have moved to node A, we must check bit 3 according to the box position. If this is a zero, we move to C, where we check bit position 2. However, if bit 3 at node A is 1, we move to node B, to check bit 2. Note that the left pointer of node C points back up to the root node; it will shortly be shown that upward pointers such as this are required so as to define a search termination criterion.

Suppose we have to search for 00 0111. It may be verified that by following the branches in a similar way to that just described, comparing bit positions for each node and traversing to the next, we eventually reach node B. The next step on from node B would be to take the right pointer, which leads back to B itself. Whenever a branch points to the same node, or a higher node, it indicates the end of the search. At that point, we know (i) that $00011x$ is the longest matching binary code (where x is either 0 or 1) and (ii) the bit position of the last matching bit is 2 (starting at 1 from the right). This is precisely the information we need for the forwarding table LMP search.

Thus, the Patricia trie constructed in this way may be used for the radix-2 LMP search. Several bits may be skipped in going from one node to the next, since each node examines not successive bits, but the next bit where the lower nodes differ. The recursion operation, where we have reached the end of the search and do not yet have a precise match but still need the LMP, is performed by checking the bit index of the current node with an ancestor node, as indicated (for example) in the link from B to A. The fact that an ancestor node is pointed to, and not a child node, is easily checked by simply checking the bit indexes, since they must *decrease* as we travel *down* the trie (Sedgewick, 1990).

Figure 4.35 shows a second example. Suppose we wish to search for 01 1000. This would lead us to node C (01 1001), and the LNP to that point would be 011. Since bit 4 is a 1, the pointer to the right is followed – which leads back to C itself. Thus, the search terminates.

It is helpful in understanding the Patricia trie algorithm to implement and experiment with code that implements a trie. A simple direct implementation in MATLAB, adapted from C code in Sedgewick (1990), is used for this purpose.

The first step is to define the data structures. The class `PatriciaTrie Node` shown below contains the data for each node – the value stored for each node (`Key`), the node name (`Name`), and the index of the bit to be tested for branching out of the node (`b`). In this simple example, the `Key` values are set

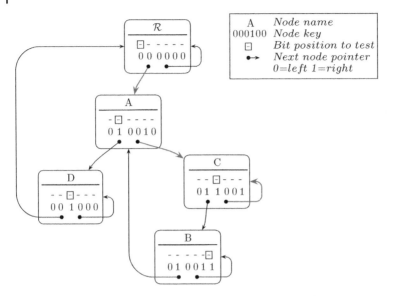

Figure 4.35 Patricia trie example 2. The search path for 01 1000 as described in the text is indicated.

to be 8-bit integers using the uint8 data type. The node names are simple strings for convenient reference.

The constructor returns a handle or pointer to the node data structure (Section 4.3.3). This is done so that a reference to the object may be passed to methods, rather than copying the entire object. This enables objects to be linked easily, simplifies the code structure, and permits more efficient code execution since references occupy a relatively small amount of memory compared with objects themselves.

The Key, Info, and b properties are set once, when each node is created, and so are defined as private class members. The Left and Right pointers, however, are changed as the trie is added to. Note that a single trie node is constructed by default to have both pointers pointing to itself, to provide a default condition for termination of searches.

```
% class for nodes within the trie
classdef PatriciaTrieNode < handle


    properties (SetAccess = private)
        Key          % the binary key for this node
        Name         % identifying name for this node
        b            % index of bit to test for this node
    end
```

```matlab
    properties
        Left            % left pointer, branch if 0
        Right           % right pointer, branch if 1
    end


    methods

      % constructor
      function node = PatriciaTrieNode(Key, Name, b)
          node.Key = uint8(Key);
          node.Name = Name;
          node.b = b;

          % initialize to self-pointers, not null
          node.Left = node;
          node.Right = node;
      end

      % display the contents of a node
      function ShowNode(TheNode)
          fprintf(1, 'PatriciaTrieNode Name="%s" Key=%d
                    b=%d\n', ...
              TheNode.Name, TheNode.Key, TheNode.b);

          if( ~isempty(TheNode.Left) )
              fprintf(1, 'Left pointer -> %s\n', TheNode.
              Left.Name);
          else
              disp('Left pointer is null');
          end

          if( ~isempty(TheNode.Right) )
              fprintf(1, 'Right pointer -> %s\n', TheNode.
              Right.Name);
          else
              disp('Right pointer is null');
          end

      end

    end % end methods

end
```

The complete trie structure is formed by grouping trie nodes and setting the left/right pointers as nodes are added. This is done with the `PatriciaTrie` class. The class always contains a root node, and in this example, 6 bits are used as the maximum bit size.

```matlab
% class for the entire trie ,
% which is comprised of individual trie nodes
classdef PatriciaTrie < handle

    properties (SetAccess = private)
        % the root node itself
        RootNode

        %maxbits = 8+1;   % for uint8 bitmask
        % should be one more than the number of bits
        % required in each node
        % as this will be stored in the root node
        maxbits = 6;      % for examples
    end

    methods

        % constructor
        function Trie =  PatriciaTrie()
            fprintf(1, 'Create Patricia Trie\n');

            b = Trie.maxbits;
            Key = 0;
            Name = 'Root';

            Trie.RootNode = PatriciaTrieNode(Key, Name, b);
        end

        %
        % insert other methods here:
        %
        % Descend()
        %
        % Find()
        %

    end   % end methods
end
```

To display the contents of the trie, a method must be added to the above class, which descends the trie and visits all branches. This is done with the Descend() method. We can descend the trie structure recursively, by starting from the present node and descending left and right from there. Each new node visited is, in effect, the start of a new trie. If not passed a trie node argument, the code assumes that the root node is the starting point.

Each new recursion is started if the bit index of the current node is greater than the bit index of the left or right node (recall that bit indexes must decrease as we go down the trie and if the bit index stays the same or increases, it indicates a back pointer).

The tests CurrNode.b > CurrNode.Left.b and CurrNode.b > CurrNode.Right.b check that the bit index is decreasing, and if so, the appropriate left/right downward branch is taken. If the bit index is not decreasing, then it indicates a back pointer to either the same node or one higher up, and as noted earlier this indicates a termination condition. The display of each node is conveniently done using the ShowNode() method for individual trie nodes.

```matlab
% descend the trie by recursion, showing each visited node's
% contents
function Descend(Trie, CurrNode)
    disp('Descend Trie');

    % Note that nargin will be 1 if called with no arguments,
    % since
    % methods are called with the first argument being the
    % object itself.
    if( nargin == 1 )
        CurrNode = Trie.RootNode;
    end

    if( isempty(CurrNode) )
        fprintf(1, 'Trie is empty\n');
        return;
    end

    % show visited node name and info
    CurrNode.ShowNode();

    % descend left
    if( CurrNode.b > CurrNode.Left.b )
        fprintf(1, 'Descend Left\n');

        Descend(Trie, CurrNode.Left);
    end
```

```
% descend right
if ( CurrNode.b > CurrNode.Right.b )
    fprintf(1, 'Descend Right\n');

    Descend(Trie, CurrNode.Right);
    end
end
```

A trie searching function must then be added as a method to the above class. The Find() method starts at the root node and traverses the trie, branching left or right according to each bit value of 0 or 1. Note that the terminating condition is when the current pointer points upward in the trie. This is stored implicitly, since the bit index of the first bit from the left to differ is stored, and so the parent's differing bit index must be greater than the child's differing bit index. This is the while(p.b > c.b) test condition. Once this loop is exited, the test for a match may be done. This is one aspect where the Patricia trie search differs from other searches: The match may not necessarily be an exact one. Some, though not necessarily all, of the bits may match; this is what provides the longest-match prefix functionality.

```
% find a given key in the trie
function Find(Trie, Key)
    p = Trie.RootNode;              % p = parent node
    c = Trie.RootNode.Left;         % c = child node
    while( p.b > c.b )
        p = c;
        if ( bitget(Key, c.b) )
            c = c.Right;
        else
            c = c.Left;
        end
    end

    if ( c.Key == Key )
        fprintf(1, 'Exact match found (value %d)\n', Key);
    else
        fprintf(1, 'Exact failed (requested %d, closest %d)\n',
                Key, c.Key);
    end
    fprintf(1, 'Name=%s  Key=%d\n', c.Name, c.Key);
end
```

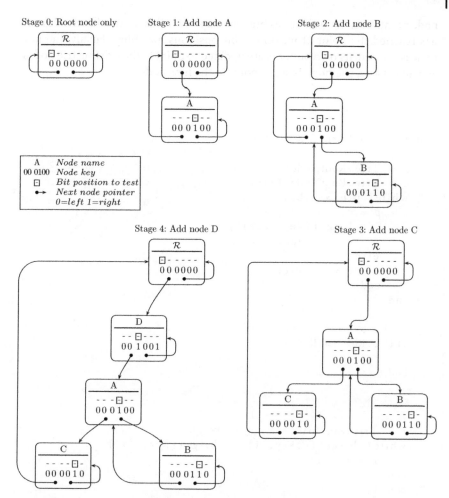

Figure 4.36 Successive steps in the construction of a Patricia trie.

The most difficult part of the trie is the initial construction. This is implemented with the Add() method below, which should be added to the Patricia trie class methods. Figure 4.36 shows graphically the stages of constructing the trie by adding nodes A, B, C, and then D.

The first stage is to descend the trie to find the closest match (it is possible, of course, that an exact match could also occur). Next, the bits of the new key are compared with the closest matching key, to find the first (leftmost) position where they differ. The trie is then descended again, using a variation for the terminating conditions. First, as before, the presence of an equal or higher bit position number in the child as compared with the parent indicates that the trie node points back to itself or has branched upward. This means that the descent has reached the point of maximum bit-positions matching. Finally, the new node is created, its own pointers are updated to point to the existing trie nodes, and an existing trie node is set to point to the new node.

```
function Add(Trie , Key, Name)
    % descend to level where there is a back−pointer
    p = Trie.RootNode;              % p = parent node
    c = Trie.RootNode.Left;         % c = child node
    while( p.b > c.b )
        p = c;
        if( bitget(Key, c.b) )
            c = c.Right;
        else
            c = c.Left;
        end
    end

    % check for a direct match
    if( c.Key == Key )
        return;
    end

    % find smallest bit index where current node and new
    % node differ
    i = Trie.maxbits;
    while( bitget(c.Key, i) == bitget(Key, i) )
        i = i − 1;
    end
    b = i;
    fprintf(1, 'smallest difference bit index %d\n', b);

    % re−descend until this level or we find the correct
    % insertion point
    p = Trie.RootNode;
    c = Trie.RootNode.Left;
    while( (p.b > c.b) && (c.b > b) )
        p = c;
        if( bitget(Key, c.b) == 1 )
            c = c.Right;
        else
            c = c.Left;
        end
    end
```

```
% create the new node
NewNode = PatriciaTrieNode(Key, Name, b);

% set pointers in new node (default is self-pointer)
if( bitget(Key, b) == 1 )
    NewNode.Left = c;          % default: NewNode.Right =
                               % NewNode;
else
    NewNode.Right = c;         % default: NewNode.Left =
                               % NewNode;
end

% set pointer to new node in parent
if( bitget(Key, p.b) == 1 )
    p.Right = NewNode;
else
    p.Left = NewNode;
end
end
```

As may be observed from the above code, the search operation is relatively short and fast, whereas adding a new node takes much more care and additional testing. This is a desirable situation for IP address lookup, since searches are done for each packet, but routing trie updates only occur when new routing messages arrive. The nature of the routing updates is discussed in the following sections.

4.10.6 Routing Tables Based on Neighbor Discovery: Distance Vector

Now that the lookup operation inherent in routing has been discussed and methods for storing and thus searching for the closest routing match have been introduced, the next question to be addressed is how the routing tables are created in the first place. Interconnected routers have some knowledge of what they are directly connected to, and so must communicate this information to other routers.

There are two main approaches to creating and maintaining the routing tables, based on related but distinct solutions to the problem. These two main approaches are usually categorized as either *Distance Vector* methods or *Link State* methods. In distance vector routing, a router uses a routing protocol to inform other routers (of which it is aware) of the networks that it can reach. The distance may simply be a *hop count* necessary to reach other networks, where a hop is defined as the traversal through a router (receiving and then retransmitting the packet). Although it may be preferable to take other factors such as bandwidth into account, the number of hops is simple to determine. The router then uses these hop counts to populate its routing table. In this context, the term "vector" refers to the particular network interface on which packets are forwarded (a direction toward the destination).

The classic distance vector protocol is the Routing Information Protocol (RIP) (Hedrick, 1988). To cope with dynamic network topology changes, the routing hop information must be periodically updated. In the original RIP, this is done by sending an update approximately every 30 s on port 520 using the UDP protocol. A route cost may be set to "infinite" if it is not updated in 180 s.

There may also be multiple routes from any given source to any given destination, and this provides some measure of redundancy, so that the entire network can be resilient in the face of outages of communication links or routers. Figure 4.37 shows an example network, connecting four networks with three routers. All that is available to aid the routing decisions is the knowledge of directly connected routers, acquired firsthand from the routers via routing protocol messages. From this, the routing path to other routers and networks, which are not directly connected but reachable via intermediate router(s), may be inferred.

This class of incremental update algorithms is categorized as a relaxation approach and is solved in general terms by the Bellman–Ford family of

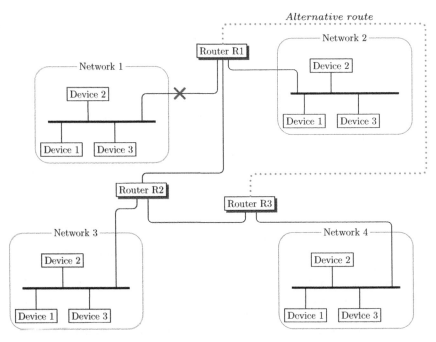

Figure 4.37 Example network routing layout. Two routes are possible from Network 1 to Network 4.

Table 4.3 The initial routing tables for routers R1 and R2.

Router R1			Router R2		
Destination	Route	Hops	Destination	Route	Hops
Network 1	—	0	Network 1	R1	1
Network 2	—	0	Network 2	R1	1
Network 3	R2	1	Network 3	—	0
Network 4	R2	2	Network 4	R3	1

The dash indicates that there is no current known route – although that may change as routing messages propagate. A "hop" is defined as a packet being forwarded by a router.

algorithms (of which the routing exchange described here is a specific case). This method is used in many other situations, such as computing the shortest path when traveling a path from one city to another via intermediate towns. For the specific problem addressed here, the solution is a distributed implementation, since each router maintains and updates its own tables based on information that it receives, rather than the algorithm running on one particular computer. In the ideal case, all routers would reach the same decisions and hence have the same routing tables, at all times. But since messages take a finite time to propagate through a network, the ideal case is not always realistic.

This distributed solution approach would appear to be satisfactory, while providing for a redundant system, but there are a number of situations where the inference of correct routing turns out to be incorrect. To illustrate, consider one alternate path from R3 to R1 in Figure 4.37 and the problems that occur if the connection from R1 to Network 1 is broken (as shown by the X in the figure).

Imagine that the link is not broken initially; the following sequence of events then takes place:

1) Router R1 advertises that it is directly connected to Network 1.
2) Router R2 then knows that it is one hop away from Network 1 (this is via R1).
3) Router R3 tells R2 that it is two hops away from Network 1 (which is via R2 then R1).

After this, the routing tables would then appear as in Table 4.3. Now suppose the link to Network 1 breaks, as shown by the X in Figure 4.37. Remember that the updates are periodic, but not synchronized. Router R1 advertises that Network 1 is unreachable. It does this by sending a hop count of infinity for Network 1. Router R2 *should* hear this and update its tables and propagate this message.

Under many circumstances this would be fine. However, suppose R2 happens to advertise its route to Network 1 *before* it has received and processed the message from R1 about the broken link. Router R2 will effectively say to R1 and R3 that it is one hop away from Network 1 – which it was – but it does not yet know about the breakage.

Then R1 hears this update and decides that a metric of one (via R2) is better than infinity, so it updates its routing table to point to R2 for destination addresses contained in Network 1, having a hop count of one more (that is, two). This may well be legitimate, since there may be multiple routes to a destination (this is one of the main strengths of the Internet, after all).

Subsequently, what may occur is that R1 advertises to R2 that it is two hops from Network 1 (through R2). This is wrong, but the router has no way of knowing that it is mistaken. R2 then thinks that it is three hops from Network 1, via R1. The entire process is summarized in Table 4.4. As shown in the table, as this continues, a convergence problem results. This is termed the "count to infinity" problem. The convergence time depends on how long this process takes to complete and depends on whether periodic updates (at timed intervals) or triggered updates (when a message is received) are used.

Although it might seem that it would be best to send router update messages at random intervals so as to reduce the network load, it turns out that this is not necessarily good. Curiously, even though some randomness is employed, it has been observed that routing messages tend to synchronize over time (Floyd and Jacobson, 1994).

Several approaches may be employed to try to rectify this problem, although each brings further issues of their own. The first is termed *split horizon*, which simply disallows a router from advertising a route to the router on a particular interface from which it heard that route. As seen in Table 4.4, this is the root cause of the problem, because the route information is propagated backward.

Table 4.4 The routing tables for R1 and R2 after the Network 1 connection breaks.

	Router R1			Router R2		
t	Destination	Route	Hops	Destination	Route	Hops
1	Network 1	—	0	Network 1	R1	1
2	Network 1	—	∞	Network 1	R1	1
3	Network 1	R2	2	Network 1	R1	1
4	Network 1	R2	2	Network 1	R1	3
5	Network 1	R2	4	Network 1	R1	3
6	Network 1	R2	4	Network 1	R1	5
⋮	Network 1	R2	⋮	Network 1	R1	⋮
⋮	Network 1	R2	∞	Network 1	R1	∞

However, split horizon alone does not solve the problem in all circumstances, as we will shortly see.

Instead of just omitting the route in the routing update as with split horizon, another possibility is simply to include the update but with a metric (hop count) of infinity. This is termed a *poison reverse*, as the route is effectively "poisoned" by being flagged as unreachable. This is a slight improvement, since the recipient of such a routing update would not use an entry with a metric of infinity. In effect, it has concrete and immediate information about the route, rather than having to infer that the route is unavailable by way of a timeout.

Rather than sending out periodic updates (which, after all, consumes bandwidth and processor time), an alternative is to only send out updates when it is necessary – that is, when a routing entry changes at one particular router. A delay is also helpful here, so as to prevent a sudden rush of updates. Since RIP updates are sent using UDP, there is a possibility that a routing message may be lost (recall that UDP does not guarantee delivery of datagrams, only a best effort). So, this means that the routes may become inconsistent in the event of the loss of an update.

Unfortunately, there is no guarantee that all these additions will solve the problem where there are multiple routes to a destination, as was illustrated by the alternate route in Figure 4.37. In the case where the R1–Network 1 link breaks, the following may transpire:

1) Router R1 advertises that Network 1 is unreachable. It does this by sending a hop count of infinity for Network 1.
2) Router R2 hears this and updates its tables and propagates this message back to R1 with an infinite metric (poison reverse).
3) Router R2 propagates a new route to R1 via R3, which in turn sends it back to R1. Since each router adds one to the metric, a count-to-infinity problem still exists.

This issue may be addressed by employing a *hold-down* interval. When a router learns that a route it was using is now unreachable, it ignores routing updates for the hold-down interval. This allows the "destination unreachable" routing message to propagate and thus prevents the reinstatement of stale routes. The combination of hold-down interval with a triggered update makes a routing loop unlikely, provided no routers distribute information during the hold interval. The problem, of course, is how to set a sensible time interval, since it must be long enough for routing messages to propagate around, yet not too long so that it interferes with normal IP packet forwarding.

Finally, we note some security issues that have arisen due to routing updates. It is possible for an external attacker to send false routing information updates and thus redirect IP traffic to another (presumably malicious) host. To combat this, newer routing protocols incorporate some authenticity measures so as to verify the source of the update messages.

4.10.7 Routing Tables Based on Network Topology: Link State

A second type of routing algorithm is the link-state approach. Again, it uses a metric (bandwidth, delay, or some other cost function); however, it is used to build a "high-level" overview of the nearby networks. This is called a topology map. Figure 4.38 shows an equivalent topology diagram for the network examined in the previous section (shown in Figure 4.37). Note that this shows the entire local network topology, which is not stored on each router in the case of distance vector protocols. In distance vector routing, each router does not attempt to build a map as such, merely a hop count for other networks.

A link-state router must not only gather the necessary link interconnection information but also determine the shortest path from a source node in the topology map to a destination. Solving this shortest path problem is not trivial; one approach is introduced in this section. The classic link state protocol is Open Shortest Path First (OSPF) (Moy, 1998).

Consider the development of a routing algorithm to find the optimal path from source $\mathcal{N}_1$ to destination $\mathcal{N}_5$ as shown in Figure 4.39. This topology was designed to have multiple point-to-point links and so highlights the handling of multiple possible routes. To simplify the discussion, each link is assumed to have the same cost in each direction, but of course this may not necessarily be

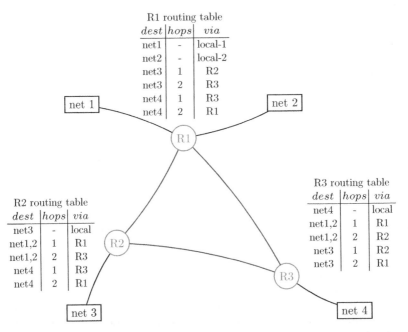

Figure 4.38 Routing topology diagram, with routing tables for each router. Rather than just a simple hop count, a metric or cost for each hop is preferable.

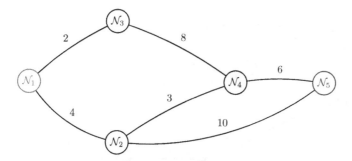

Figure 4.39 An example routing path. The goal is to find the least-cost path from $\mathcal{N}_1$ to $\mathcal{N}_5$.

the case in practice. Note that not all nodes are directly connected to all others, but that all nodes may be reached, if not directly, then indirectly via other nodes as intermediate hops. The "best" path means minimizing the sum of all the hop costs, and so each intermediate node then knows which is the best forwarding path when it receives a packet with a given destination address.

There may be more than one path from the source to destination, and it may seem that we have to enumerate all possible paths. However, a stepwise method, termed Dijkstra's algorithm, turns out to be far more efficient. Dijkstra's algorithm is described in many references (for example, Aho et al., 1987) and an interesting interview with the inventor may be found in Frana and Misa (2010). The general idea is to step through each node in turn, while maintaining the incremental path to get to that node, and then to cull those paths from the search, which are not optimal. Exactly how the paths are removed turns out to be critical.

Returning to Figure 4.39, the goal is to determine the best path from $\mathcal{N}_1$ to $\mathcal{N}_5$. From $_1$ it is evident that there are two possible paths outwards. An attractive – but flawed – approach is to take the lowest cost at each successive step. Suppose we decide to take the lowest-cost path out from $\mathcal{N}_1$, which would lead us to node $\mathcal{N}_3$. From that point, we would have no choice other than to take the path with cost 8. This leads us to node $\mathcal{N}_4$, at which point we have a choice between two outgoing paths. We might again choose the lowest-cost path, which is 3, leading to $\mathcal{N}_2$. Finally, the path with cost 10 leads to $\mathcal{N}_5$. But is this the best path? Is it the shortest, which really means the lowest cost path? In this case, the answer is an emphatic no, because clearly there exists a path via $\mathcal{N}_2$, with cost $4 + 10$. Our algorithm would have chosen a path with cost $2 + 8 + 3 + 10 = 23$ – clearly much inferior.

Furthermore, consider Figure 4.40. In that case, it is possible that we might *never* get to the end node by employing such a naïve strategy. Our objective in this case is to find the lowest cost to $\mathcal{N}_8$. But the first choice, to $\mathcal{N}_3$, would lead us down an isolated path, with no way to get back to the branch containing $\mathcal{N}_8$.

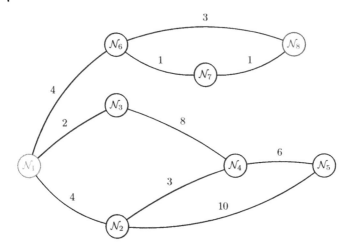

Figure 4.40 A routing example with two "islands." In trying to find the best path to $\mathcal{N}_8$, we have to avoid getting stuck in the lower branches, where there is no path to $\mathcal{N}_8$ (except back where we came from).

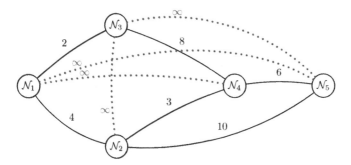

Figure 4.41 Routing path with unconnected nodes labeled with a cost of ∞.

We need two things to address this situation. First, we need to keep track of which nodes we have visited. Second, we need to find the global optimum, rather than the local optimum at each stage. It might appear that we need to enumerate every possible path, and thus visit each node several times. This is not the case if we use Dijkstra's algorithm.

To develop the concept, we first need a way of representing interconnections that do not actually exist. This is done by simply setting the path cost to ∞ or some practically large value. This is illustrated for the present topology in Figure 4.41.

Given a knowledge of the network topology and interconnection costs, the optimal route from the current node $\mathcal{N}_1$ to a destination must be determined

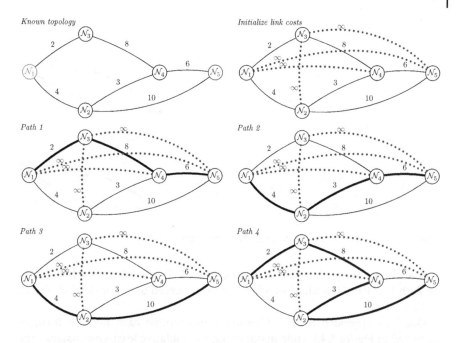

Figure 4.42 Enumerating the possible paths from the source to destination. Paths that contain an infinite cost on one or more links have not been considered, since they could not constitute a lowest-cost path.

by examination of the possible paths. Figure 4.42 shows how we could construct several possible paths through the network. We could manually determine these paths by simply tracing all possible branch paths at each node. But consider an algorithmic development of this: We need the precise steps to be able to determine not only the least-cost path but also the specific nodes that we must visit in order to traverse the identified lowest-cost path. Once this is done, the per-packet decision at each node is then to determine which interface to retransmit a packet on, given the destination address. In the figure, we have total path costs of 16 for Path 1, 13 for Path 2, 14 for Path 3, and 23 for Path 4. There are other possible paths, but they involve traversing links with an infinite cost, which clearly cannot constitute a minimal-path cost. Path 2 has the lowest cost, and so from the point of view of Node 1, a packet destined for Node 5 must be sent out on the interface that connects to Node 2.

The manual determination of routes in this way becomes problematic if we try to automate the process. Manually tracing all possible paths is not scalable – consider, say, if there were dozens, hundreds, or even thousands of nodes known in the topology. Laboriously tracing each path would take considerable time. It is clear that we should *not* simply choose the lowest-cost hop at each step. In the present example, if we did this when exiting from Node 1, then we

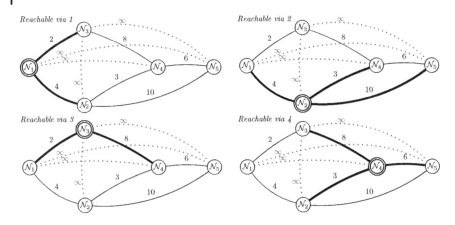

Figure 4.43 Nodes reachable in one hop from each node in turn.

would choose the path with cost 2 rather than 4. This would inevitably lead us to Path 1 or Path 4, both of which are inferior overall to the optimal choice of Path 2.

Dijkstra's approach to this dilemma is to examine each node in turn, as depicted in Figure 4.43 while maintaining a cumulative least-cost distance for each node. It is necessary to maintain not one but several possible paths, and these are updated as we visit each node in turn. It might appear that we will need a table of *all* possible routes through the network, but as we will show by example, all that is necessary is the least-cost so far for each node, as well as the predecessor node for a path that leads to this least cost.

For the example of Figure 4.43, we maintain a vector of four possible paths out from Node 1. This is obviously because we have five nodes in the network, and although the route from Node 1 to Node 1 has a cost of zero, a route to ourselves is not at all useful. Those nodes reachable via Node 1 have costs of $0, 4, 2, \infty, \infty$. These represent the cost to reach Nodes 1 to 5, respectively, and we use ∞ or some arbitrarily high value to represent an unconnected path.

Now consider those nodes reachable via Node 2. Examination of the outgoing path costs leads to the realization that we can reach Node 4 with a cost of 3, plus the cost to reach Node 2. Similarly, we can reach Node 5 with a cost of 10, plus the cost to reach the current node (Node 2). Comparing these new costs with what we already have in the list, we see that the last two yield lower costs, so must surely be better. Thus, we update our *total* costs from $0, 4, 2, \infty, \infty$ initially to $0, 4, 2, 7, 14$ after examination of Node 2.

Next, examine those nodes reachable via Node 3. We can reach Node 4 with a cost of $2 + 8 = 10$, but this is not lower than what we already have (which is 7, from the previous step), so we can discard this possibility. Finally, we perform the same set of computations and comparisons for Node 4.

Figure 4.44 Determining the new path cost at each stage of the Dijkstra algorithm, either directly or via an intermediate node. The cost via the intermediate node may be more, or it may be less.

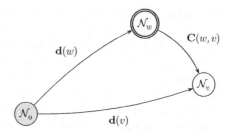

Figure 4.44 shows the computation that is performed at each stage of the above steps, traversing via each node and determining the costs. For a given intermediate node $\mathcal{N}_w$, we have the cumulative cost or distance vector $\mathbf{d}(w)$ from a predecessor node $\mathcal{N}_o$ to $\mathcal{N}_w$, and the direct cost from $\mathcal{N}_o$ to $\mathcal{N}_v$. All that is necessary is to keep a log of the lowest cost at $\mathcal{N}_v$ from the predecessor, either directly or via an intermediary. Using the cost of ∞ for unconnected nodes makes the algorithm quite general.

The decision at each stage is a restatement of the Bellman–Ford principle of optimality, as employed in distance vector routing: The final cost must be optimal if we have made the optimal decision at each node. The solution approach in link-state routing, though, is different to distance vector routing. In distance vector routing, the route costs are computed and recomputed as new information becomes available. This process is repeated until there is no change in the computed path costs (recall the problem with count-to-infinity). In link-state routing, an optimal intermediate path is determined for each node in turn. The major advantage is that each node is visited only once, not multiple times, as the topology map is built up.

To step through the current example, the cost from Node 1 to all other nodes is vector $\mathbf{d}$, and path cost in going from node i to node j is matrix $\mathbf{C}(i, j)$

$$
\mathbf{d} = \begin{bmatrix} 0 \\ 4 \\ 2 \\ \infty \\ \infty \end{bmatrix}
\qquad
\mathbf{C}(i,j) = \begin{bmatrix} 0 & 4 & 2 & \infty & \infty \\ 4 & 0 & \infty & 3 & 10 \\ 2 & \infty & 0 & 8 & \infty \\ \infty & 3 & 8 & 0 & 6 \\ \infty & 10 & \infty & 6 & 0 \end{bmatrix}
\tag{4.13}
$$

Mathematically, the decision at each step is to update the cost to reach a given node $\mathcal{N}_v$ as the smaller of the current cost, as compared with going via an intermediate node $\mathcal{N}_w$. This means the update is

$$
\mathbf{d}(v) = \min \begin{cases} \mathbf{d}(v) \\ \mathbf{d}(w) + \mathbf{C}(w, v) \end{cases}
\tag{4.14}
$$

So, we have found the lowest cost, but how does that help find the optimal path? The lowest cost must inevitably lead to the specific lowest-cost path, but as yet that has not been considered. However, when each lowest-cost cumulative path was updated, we should simply remember the node that yielded this better path. Thus, we need a vector **p** of predecessor nodes. At each stage, the via node is updated within this vector with the index of the node that we must go via (the optimal predecessor), in order to attain that lower-cost path. This gives a list of optimal predecessor nodes for each node. Since we know the last node, then we simply look up its predecessor. The predecessor of node 5 (the last node in the path) is written as **p**(5). The optimal predecessor in the case of Figure 4.43 is **p**(5) = 4; its predecessor is **p**(4) = 2, and in turn **p**(2) = 1. Thus, this gives the optimal path in reverse, and it is then a simple matter to read this path in reverse to obtain the optimal forward order. We can set up the problem in MATLAB as follows. The cost matrix is C with the cumulative distance to each node d. A flag NodeCan determines if the node is a candidate and still in contention for being in the lowest-cost path.

```
C = [ 0     4     2    inf   inf ;   ...
      4     0    inf    3    10  ;   ...
      2    inf    0     8    inf ;   ...
     inf    3     8     0     6  ;   ...
     inf   10    inf    6     0  ] ;

M = size(C, 1);

fprintf(1, 'C (cost between nodes) matrix) is \n');
for k = 1:M
    fprintf(1, ' %6d', C(k, :));
    fprintf(1, '\n');
end
fprintf(1, '\n');

d = C(1, :);    % distance vector - initial setting

fprintf(1, 'd (distance vector from origin) is initially ');
fprintf(1, ' %d ', d);
fprintf(1, '\n');

cnode = 0;      % current node we are working on
S = [1];        % set of nodes examined

NodeCan = true(M, 1);    % flag if node is still a candidate
NodeCan(1) = false;
P = ones(M, 1);          % predecessor node list
```

In the algorithm itself, we choose the node with the minimum distance so far. For this node as an intermediate, we must update the costs of all other nodes if going through this intermediate node. An update consists of checking whether the path via the intermediate would yield a lower cost, and if so updating the store of lowest cost for this node – as well as saving the index of the predecessor node that led to this lower cost.

```
for i = 1:M-1
    fprintf(1, 'Step i=%d\n', i);

    % choose a node
    dmin = inf;
    cnode = 0;
    for v = 1:M
        if( NodeCan(v) )
            if( d(v) > dmin)
                dmin = d(v);
                cnode = v;
            end
        end
    end
    fprintf(1, 'chose node cnode=%d as best dmin=%d so
far\n', cnode, dmin);

    S = [S cnode];
    NodeCan(cnode) = false;

    for v = 1:M
        if( NodeCan(v) )
            fprintf(1, 'Node %d, choice %d+%d < %d\n', ...
                v, d(cnode), C(cnode, v), d(v));
            if( d(cnode) + C(cnode, v) > d(v) )
                d(v) = d(cnode) + C(cnode, v);
                P(v) = cnode;
            end

            %pause
        end
    end
    fprintf(1, 'd (distance vector from origin) is now ');
    fprintf(1, ' %d', d);
    fprintf(1, '\n');

    fprintf(1, 'P (backtrack path) is now ');
    fprintf(1, ' %d', P);
    fprintf(1, '\n');

    fprintf(1, 'S (set of nodes we have checked) is now ');
```

```
      fprintf(1, ' %d ', S);
      fprintf(1, '\n');

      pause
end
fprintf(1, 'Path cost = %d\n', d(M));
fprintf(1, 'P is now ');
fprintf(1, ' %d ', P);
fprintf(1, '\n');
```

Finally, since each node has an optimal predecessor, it is necessary to start at the *last* node, determine its optimal predecessor, and repeat until we reach the starting node.

```
% backtrack to work out shortest path
i = M;
optpath = [i];
while( i ~= 1 )
    i = P(i);
    optpath = [i optpath];
end
fprintf(1, 'optpath is ');
fprintf(1, ' %d ', optpath);
fprintf(1, '\n');
```

The code output for the present example network topology is shown below.

```
C (cost between nodes) matrix) is
        0       4       2     Inf     Inf
        4       0     Inf       3      10
        2     Inf       0       8     Inf
      Inf       3       8       0       6
      Inf      10     Inf       6       0

d (Distance Vector from origin) is initially  0 4 2 Inf Inf

Step i=1
chose node cnode=3 as best dmin=2 so far
Node 2, choice 2+Inf < 4
Node 4, choice 2+8 < Inf
Node 5, choice 2+Inf < Inf
d (Distance Vector from origin) is now  0 4 2 10 Inf
P (Backtrack Path) is now  1 1 1 3 1
S (set of nodes we have checked) is now  1 3
```

```
Step i=2
chose node cnode=2 as best dmin=4 so far
Node 4, choice 4+3 < 10
Node 5, choice 4+10 < Inf
d (Distance Vector from origin) is now  0 4 2 7 14
P (Backtrack Path) is now  1 1 1 2 2
S (set of nodes we have checked) is now  1 3 2

Step i=3
chose node cnode=4 as best dmin=7 so far
Node 5, choice 7+6 < 14
d (Distance Vector from origin) is now  0 4 2 7 13
P (Backtrack Path) is now  1 1 1 2 4
S (set of nodes we have checked) is now  1 3 2 4

Step i=4
chose node cnode=5 as best dmin=13 so far
d (Distance Vector from origin) is now  0 4 2 7 13
P (Backtrack Path) is now  1 1 1 2 4
S (set of nodes we have checked) is now  1 3 2 4 5


Path cost = 13
P is now  1 1 1 2 4
optpath is  1 2 4 5
```

Suppose now the path costs alter as shown in Figure 4.45. The output of the optimal path search is as follows:

```
C (cost between nodes) matrix) is
       0       15       2      Inf      Inf
      15        0      Inf       3        1
       2      Inf       0        8      Inf
     Inf        3       8        0        6
     Inf        1      Inf       6        0

d (Distance Vector from origin) is initially  0 15 2 Inf Inf

Step i=1
chose node cnode=3 as best dmin=2 so far
Node 2, choice 2+Inf < 15
Node 4, choice 2+8 < Inf
Node 5, choice 2+Inf < Inf
d (Distance Vector from origin) is now  0 15 2 10 Inf
```

```
P (Backtrack Path) is now   1 1 1 3 1
S (set of nodes we have checked) is now   1 3

Step i=2
chose node cnode=4 as best dmin=10 so far
Node 2, choice 10+3 < 15
Node 5, choice 10+6 < Inf
d (Distance Vector from origin) is now   0 13 2 10 16
P (Backtrack Path) is now   1 4 1 3 4
S (set of nodes we have checked) is now   1 3 4

Step i=3
chose node cnode=2 as best dmin=13 so far
Node 5, choice 13+1 < 16
d (Distance Vector from origin) is now   0 13 2 10 14
P (Backtrack Path) is now   1 4 1 3 2
S (set of nodes we have checked) is now   1 3 4 2

Step i=4
chose node cnode=5 as best dmin=14 so far
d (Distance Vector from origin) is now   0 13 2 10 14
P (Backtrack Path) is now   1 4 1 3 2
S (set of nodes we have checked) is now   1 3 4 2 5

Path cost = 14
P is now   1 4 1 3 2
optpath is   1 3 4 2 5
```

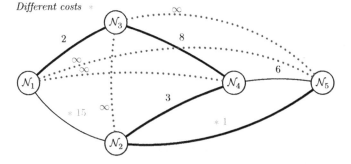

Figure 4.45 A more convoluted path results when the hop costs change as indicated. Dijkstra's algorithm still works successfully in this case.

4.11   Chapter Summary

The following are the key elements covered in this chapter:

- The concept of protocol layering and frame encapsulation.
- The Internet Protocol and IP addressing.
- The Transmission Control Protocol, the sliding-window used in TCP, optimizing throughput, and avoiding congestion.
- The method used to calculate checksums in IP and TCP.
- The key concepts involved in routing packets from the source to destination.
- The use of the Patricia trie to facilitate fast prefix searching in route tables.
- The distance vector and link-state algorithms for routing.

Problems

4.1   IP packets transported across a link have a maximum size, and packets larger than this must be broken up or fragmented. The ping command may be used to send a test packet to a destination and measure the round-trip time of an IP link. The packet size may be set with the -l (lowercase letter L) length option, and the -f option sets the *do not fragment* bit in the IP header. Packets larger than the maximum will not be sent. The following was executed on a device on a 10.0.0.0/24 subnetwork:

```
ping 10.1.1.1 -f -l 1472
    Pinging 10.1.1.1 with 1472 bytes of data:
    Reply from 10.1.1.1: bytes=1472 time=3ms TTL=64
    Reply from 10.1.1.1: bytes=1472 time=69ms TTL=64
    Reply from 10.1.1.1: bytes=1472 time=20ms TTL=64
    Reply from 10.1.1.1: bytes=1472 time=45ms TTL=64
```

And subsequently with a one-byte larger packet:

```
ping 10.1.1.1 -f -l 1473
    Pinging 10.1.1.1 with 1473 bytes of data:
    Packet needs to be fragmented but DF set.
```

The subnet is known to be a wired Ethernet with a MTU of 1500 bytes. The ping command sends an Internet Control Message Protocol (ICMP) packet with the requested number of bytes, and an 8-byte header, over IP. With reference to the packet header structures and encapsulation at each layer, explain why the 1472-byte packet was successfully transmitted, but the 1473 byte packet was not.

4.2 An IPv4 frame header contains the following bytes in hexadecimal:

45 00 00 3C

75 02 00 00

20 01 C7 1F

AC 10 03 01

AC 10 03 7E

a) What are the source and destination addresses?
b) What is the header checksum?
c) Calculate, using big-endian ordering, the checksum of the frame with the checksum field included.
d) Calculate, using big-endian ordering, the checksum of the frame with the checksum field replaced by 0000.
e) Calculate, using little-endian ordering, the checksum of the frame with the checksum field included.
f) Calculate, using little-endian ordering, the checksum of the frame with the checksum field replaced by 0000.

4.3 Given an IP address of 192.168.60.100 and a /27 subnet mask, determine:
a) The class of the IP address.
b) The subnet mask in binary.
c) The subnet address and the device address within the subnet.

4.4 Given the IP address 192.168.7.1, determine the bits that define the network and host address, respectively, using the following subnet masks:
a) 255.255.0.0
b) 255.255.255.0
c) 255.255.248.0
d) 255.255.240.0
e) 255.255.224.0
f) 255.255.192.0
g) 255.255.128.0

4.5 Given the following IP addresses and corresponding network masks, determine which bits form the prefix:
a) 192.168.0.0/16
b) 192.168.128.0/18
c) 172.18.128.0/18

4.6 The network aggregated as 192.168.8.0/22 is to be subdivided for the purposes of routing management into networks with a 24 bit prefix.

a) Write out the bits corresponding to 192.168.8.0/22 and show the masked bits.
b) What are the networks subordinate to this, with a /24 prefix? Write them out in binary and in decimal.

4.7 Write out the steps taken in traversing the Patricia trie of Figure 4.35, searching for the value 01 1000.

4.8 A Patricia trie is to be created with the following data (keys are in hexadecimal): (22, "A"), (13, "B"), (1B, "C"), and (8, "D").
a) Construct the Patricia trie, and compare with that given in the text for similar data.
b) Explain how to retrieve the value from key 22 hexadecimal.
c) Explain why an attempt to find an exact match for 09 hexadecimal would fail.

4.9 For the routing diagram shown in Figure 4.46:
a) Manually trace all possible paths, and determine their corresponding costs.
b) Use Dijkstra's algorithm to work out the optimal path cost for a route from $\mathcal{N}_1$ to $\mathcal{N}_4$.

Figure 4.46 Routing problem 1.

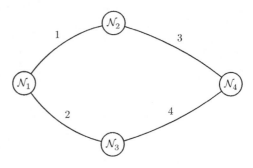

Figure 4.47 Routing problem 2.

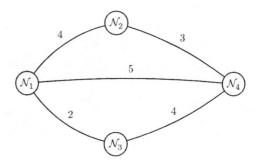

    c) Use backtracking on the predecessor matrix to work out the optimal path. Compare the path, and its cost, with the manual comparison of all possible paths.

4.10 For the routing diagram shown in Figure 4.47:
    a) Manually trace all possible paths, and determine their corresponding costs.
    b) Use Dijkstra's algorithm to work out the optimal path cost for a route from $\mathcal{N}_1$ to $\mathcal{N}_4$.
    c) Use backtracking on the predecessor matrix to work out the optimal path. Compare the path, and its cost, with the manual comparison of all possible paths.

4.11 Verify the MATLAB output given in the text for the routing topology of Figure 4.45.

5

Quantization and Coding

5.1 Chapter Objectives

On completion of this chapter, the reader should

1) Be conversant with the principles of scalar quantization and be able to explain the operation of a vector quantizer.
2) Be conversant with the principles of minimum-redundancy codeword assignment and understand the important algorithm classes for lossless source coding.
3) Be able to explain several image compression approaches, including the DCT.
4) Understand the basic approach to waveform and parametric speech encoding and be able to explain the advantages and disadvantages of each.
5) Be able to explain the key requirements for audio encoders and the building blocks that go to make up an audio encoding system.

5.2 Introduction

Quantization is the process of assigning a digital value, usually an integer, to represent one or more analog values. Since there are only a certain number of bits allocated to each of these samples, a corresponding number of discrete levels exist. Thus, we can only represent the true analog signal at its nearest approximation. Careful choice of the representation means that we can get by with as few bits as possible.

In addition to the number of representation levels, it is important to have sufficient density of sampling – either samples per second in the case of digitized audio or spatial density in the case of digitized images. The overall data rate (in bits per second or bps) for a sampled audio signal is then the number of samples per second, multiplied by the number of bits per sample. The overall rate for a sampled image is the number of samples for a given image, multiplied by the

Communication Systems Principles Using MATLAB®, First Edition. John W. Leis.
© 2018 John Wiley & Sons, Inc. Published 2018 by John Wiley & Sons, Inc.
Companion website: www.wiley.com/go/Leis/communications-principles-using-matlab

number of bits per sample. The most compact representation of digital signals (that is, the fewest bits possible for an acceptable representation) is crucial for both transmission and storage of digitized audio and images.

Because of the large amount of data required to represent sounds and images, digital sample compression is widely employed. This chapter examines the sampling process, the bit allocation process, and some of the important approaches for reducing the number of bits required.

5.3 Useful Preliminaries

This section briefly reviews some of the notions of probability, which are useful in modeling errors in communication channels, as well as difference equations, which are used extensively in signal encoding.

5.3.1 Probability Functions

Since this chapter considers quantization, or assignment of binary representations to a sample, it is useful to be able to characterize not only the possible range of values that a signal might take on but also the likelihood of any given sample taking on certain values.

If we have a continuous variable such as voltage, it may take on any value in a defined range. For example, a voltage of 1.23 or 6.4567 V may be measured. Unless precise measurement is required, we usually do not need to know the precise value, only a near-enough value. We may wish to represent the likelihood of occurrence of a certain set of ranges, for example, -2 to -1 V, -1 to 0 V, 0 to 1 V, and 1 to 2 V. Alternatively, pixels in an image may take on discrete values $0, 1, 2, \ldots, 255$ and thus already be "binned" into a set of ranges. If we measure a signal over a certain period, we can count the number of times we see the signal in each range. Graphically, this becomes a *histogram* as illustrated in Figure 5.1.

The usefulness of this type of representation is twofold. First, we can see that, in this case, there are few counts below -4 or above $+4$ and that the most likely values for our signal are in the range from -1 to $+1$. Furthermore, if we wanted to know the proportion of time in certain bins, such as the shaded ones shown with centers x_k of 0.5 and 1.0 V, we could add the corresponding counts for each bar. Since the total count is the sum of all the bars, our proportion would then be the ratio of the counts in the desired range divided by the total count.

On the other hand, if we have a continuous signal and do not wish to divide it into bins, we may use the *Probability Density Function* or PDF. An example PDF is shown in Figure 5.2, and we can see that it is broadly similar to the histogram. A key difference is that we have not binned the data into discrete intervals. Importantly, the height of the PDF does not represent a count or probability, but rather a density. We cannot, for example, find the likelihood that a value of

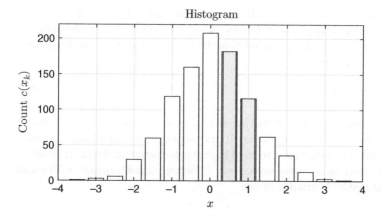

Figure 5.1 A histogram showing discrete value ranges $x_k \rightarrow x_{k+1}$ and corresponding counts. The total count of values in the bin ranges with centers 0.5 and 1.0 is just the sum of the counts of the shaded bars.

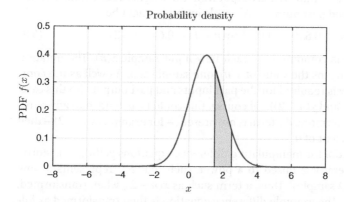

Figure 5.2 A probability density curve over a continuous sample range x. The total probability in the range from $x = 1.5$ to $x = 2.5$ is the area of the shaded region.

exactly $x = 1.234\,565$ occurred. We can, however, find the probability that the signal fell in a certain range. This is illustrated in the figure, where the area of the shaded portion corresponds to the probability of the signal being in the defined range of x values. Clearly, this will change for different ranges and for different PDFs. Also, the total area under the PDF curve must be one, since the signal must be somewhere in the range of possibilities at all times (that is, it is 100% likely the signal has some value). This is equivalent to the discrete histogram, where the sum of the individual count bars must be equal to the total number of observations.

A mathematical form of the PDF is useful for analysis, and the Gaussian distribution is by far the most commonly used. In general, it is a good approximation for many types of unwanted noise that may be encountered. Its mathematical representation is

$$f(x) = \frac{1}{\sigma\sqrt{2\pi}} \, e^{-(x-\mu)^2/2\sigma^2} \tag{5.1}$$

where $f(x)$ is the probability density over a range x, σ^2 is the variance or spread of values, and μ is the average value. If only an average value obtained from a set of samples is available, it is termed the *sample average* and usually denoted by $\bar{x}$ rather than true or *population mean* μ. Figure 5.2 is a Gaussian with mean (average) of one and variance of one.

5.3.2 Difference Equations and the *z* Transform

Many building blocks for communication systems are implemented digitally, and the difference equation describes how they operate in a step-by-step manner. At sampling instant n, an input $x(n)$ produces an output $y(n)$, and the difference equation determines each output from the weighted sum of the current and past inputs, and past outputs. A simple example might be

$$y(n) = 0x(n) + 0.8x(n-1) + 1.5y(n-1) - 0.64y(n-2) \tag{5.2}$$

Each output $y(n)$ is computed for each new input sample $x(n)$. The notation such as $x(n-1)$ means the value of $x(n)$, one sample ago. As well as the computation, memory is required for the past inputs and past outputs (in this case, $x(n-1)$, $y(n-1)$, and $y(n-2)$). Of course, this could be extended to any number of terms. Furthermore, if a term is not present – for example, $x(n-2)$ – then it implies a coefficient of 0.

The analytical tool for manipulating difference equations is the z transform. This uses the operator z raised to a power, such that z^{-D} represents a time sample delay of D samples. Thus, a term such as $x(n-2)$, when transformed, becomes $X(z)z^{-2}$. The example difference equation is then transformed as follows. The aim is to obtain a ratio of output over input, or $Y(z)/X(z)$:

$$y(n) = 0x(n) + 0.8x(n-1)$$
$$+ 1.5y(n-1) - 0.64y(n-2)$$
$$\therefore \quad Y(z)z^0 = 0X(z)z^0 + 0.8X(z)z^{-1}$$
$$+ 1.5Y(z)z^{-1} - 0.64Y(z)z^{-2}$$
$$Y(z)z^0 = X(z)(0z^0 + 0.8z^{-1})$$
$$+ Y(z)(1.5z^{-1} - 0.64z^{-2})$$
$$Y(z)(1z^0 - 1.5z^{-1} + 0.64z^{-2}) = X(z)(0z^0 + 0.8z^{-1})$$
$$\frac{Y(z)}{X(z)} = \frac{0z^0 + 0.8z^{-1}}{1z^0 - 1.5z^{-1} + 0.64z^{-2}} \tag{5.3}$$

Thus, the input–output relationship is the ratio of two polynomials in z. The coefficients of the numerator **b** and denominator **a** of the z transform expression are

$$\mathbf{b} = \begin{bmatrix} 0 & 0.8 \end{bmatrix}$$
$$\mathbf{a} = \begin{bmatrix} 1 & -1.5 & 0.64 \end{bmatrix}$$

Note that the first coefficient value in **a** is one and that this is due to the output being $1 \cdot y(n)$. Furthermore, the subsequent coefficients are negated as compared with the difference equation, where they were $+1.5$ and -0.64. This is due to the rearrangement from the difference equation (5.2) into the z domain Equation (5.3).

We can implement this difference equation as follows. The input is chosen to be the simplest type, an *impulse*, which is a single $+1$ value followed by zeros thereafter.

```
% Transfer function Y(z)/X(z) has numerator b
% and denominator a.
% Take care when transferring these to a difference
% equation:
% a(1) = 1 always, and a(2:end) is the negative of the
% difference equation coefficients.

b = [0  0.8];
a = [1  -1.5  0.64];

% impulse input
x = zeros(25, 1);
x(1) = 1;

% Compute the sequence of output samples in y,
% corresponding to each input in x.
% The coefficients in b and a define the z transform
% coefficients.
% Equivalently, these are the difference equation
% coefficients,
% but see note above.
y = filter(b, a, x);

stem(y);
```

The output is shown in Figure 5.3. Several points are worth noting here. The very first output is zero. This is because the coefficient of $x(n)$ is 0 in this example. Effectively, this delays the output by one sample. More importantly, since the output depends on past outputs (as well as the input), the impulse response extends for some time. This is called a *recursive* system.

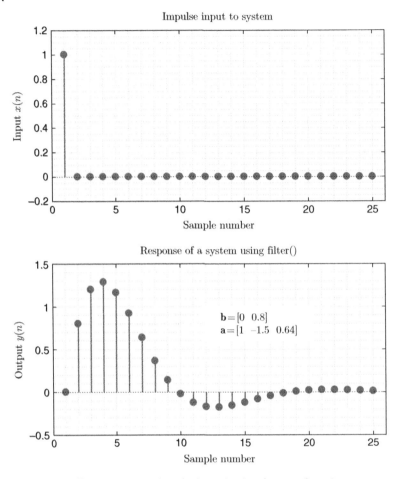

Figure 5.3 Difference equation impulse input (top) and output (lower).

A generalization of the difference equation is

$$1y(n) = [b_0 x(n) + b_1 x(n-1) + \cdots + b_N x(n-N)]$$
$$- [a_1 y(n-1) + a_2 y(n-2) + \cdots + a_M y(n-M)] \quad (5.4)$$

This form is useful, since it mirrors the way MATLAB implements difference equations using the `filter` function. Note that coefficient of $y(n)$ is one, as we have explicitly shown above, and that this leads to the first **a** coefficient being one. Also, note the negative sign for the remaining **a** coefficients, as was found in the previous numerical example.

Extrapolating the z transform to the general case, it would be

$$\frac{Y(z)}{X(z)} = \frac{b_0 z^0 + b_1 z^{-1} + \cdots + b_N z^{-N}}{1 + a_1 z^{-1} + a_2 z^{-2} + \cdots + a_M z^{-M}}$$

That is, we can enter the coefficient vectors for the `filter` function as

$$\mathbf{b} = \begin{bmatrix} b_0 & b_1 & b_2 & \cdots \end{bmatrix}$$
$$\mathbf{a} = \begin{bmatrix} 1 & a_1 & a_2 & \cdots \end{bmatrix}$$

Difference equations are implemented extensively in telecommunication systems, and their analysis is best performed in the z domain. The *zeros* of any system are the values of z, which make the numerator equal to zero. The *poles* of any system are the values of z, which make the denominator equal to zero. The stability of a system, as well as its frequency of oscillation, is determined by the poles. Fundamentally, this is because the poles determine the coefficients of the recursive or fed-back terms.

5.4 Digital Channel Capacity

A natural question that arises when we wish to transmit digital information is this: How much information can we transmit in a given time? Or, equivalently, is there a maximum rate of data transmission? The first assumption is that we transmit binary information, or bits – an abbreviation for *binary digit*, a term suggested by J.W. Tukey (Shannon, 1948). Early work on the capacity of a channel was done in relation to telegraph systems using on–off or Morse code type signaling, where the idea was that several telegraph streams could be multiplexed onto one long-distance carrier.

Hartley (1928) suggested the use of a logarithmic measure for the amount of information in a message. If we are to use a binary system, the choice of base for the logarithm is naturally 2. So if we have M levels (say, voltages), then this could encode $\log_2 M$ bits. For example, if we could transmit a voltage of $0, 1, 2, \ldots, 15$ V, there are 16 levels, and this clearly encodes 4 bits ($\log_2 16 = 4$). Of course, the unit of measurement does not have to be volts at all; it could be any chosen increment.

Nyquist is famous for his sampling theorem, which states in essence that a signal of bandwidth B needs to be sampled at a rate greater than $2B$ for perfect reconstruction (Nyquist, 1924a, b). Combining this with Hartley's observations regarding information content, we have a formula for channel capacity, which is generally known as Hartley's law:

$$C = 2\,B \log_2 M \qquad \text{bps} \tag{5.5}$$

where B is the bandwidth of the channel (Hz), M is the number of levels used for each signaling element, and bps denotes the transmission rate in bits per second.

An example serves to clarify this formula and underscore its importance. If sending a voice signal over a transmission line, a minimal bandwidth to preserve the intelligibility of the speech is about 3000 Hz. Thus, the line would have (at a minimum) a bandwidth of $B \approx 3$ kHz. If, on this same channel, we instead decide to send digital data encoded as two voltage levels (one bit per signaling interval), we have $M = 2$. Then, the total channel capacity is $C = 2 \times 3000 \times \log_2 2 = 6$ kbps. Since in practical terms the bandwidth is limited, the only way to obtain a higher information rate is to increase the number of levels M.

This formula is useful as an upper bound but neglects the presence of noise. Shannon (1948) later deduced the well-known maximum capacity formula (sometimes also known as the Shannon–Hartley law). Subject to various assumptions, this predicts the channel capacity C as

$$C = B \log_2 \left(1 + \frac{S}{N}\right) \qquad \text{bps} \tag{5.6}$$

with the channel bandwidth B as before (Hz), S is the power of the signal used to transmit the information (in Watts, W), N is the power of the noise added to the signal whilst it travels on the channel from transmitter to receiver (also in W). The term S/N is a ratio of signal power to noise power. Usually, the signal-to-noise ratio (SNR) is expressed in decibels (dB), where

$$\text{SNR} = 10 \log_{10} \left(\frac{S}{N}\right) \qquad \text{dB} \tag{5.7}$$

To illustrate this, consider again an analog circuit with the same bandwidth of $B = 3$ kHz but with a SNR of 1000. Thus $S/N = 1000$ and the SNR $= 10 \times \log_{10} 10^3 = 30$ dB. We need the logarithm to base 2 of 1000, and noting that $2^{10} = 1024$, the required logarithm is approximately 10 (precisely, $\log_2 1000 = \log_{10} 1000 / \log_{10} 2 = 3/0.69 \approx 9.96$). So we have $C = 3000 \times \log_2 (1 + 1000) \approx 3000 \times 10 = 30$ kbps.

Note the interplay between these two formulas. We have two formulas for C,

$$C = 2 B \log_2 M$$
$$C = B \log_2 \left(1 + \frac{S}{N}\right)$$

Equating these, we find that

$$M = \sqrt{1 + \frac{S}{N}}$$
$$\approx \sqrt{\frac{S}{N}} \tag{5.8}$$

The approach seems reasonable, since carrying more information in the same bandwidth implies more information at each sample and thus higher M. The tradeoff is that as M increases, the difference in successive levels becomes smaller, and thus noise asserts itself to a greater degree. In the present example, the above approach implies a requirement of about 32 levels (being $\approx \sqrt{S/N}$). Also, since the square root of the SNR yields an RMS voltage, we can say, roughly, that the errors are proportional to the voltage differences.

Of course, all this is only a theoretical development – it does not show how to encode the signal so as to maximize the rate, only what that maximum rate might be. To see the reasons behind the derivation of the Shannon–Hartley capacity bound, we may use Shannon's original paper. First, assume both the signal and noise are zero-mean Gaussians with PDF of the form

$$f(x) = \frac{1}{\sigma\sqrt{2\pi}} e^{-x^2/2\sigma^2} \tag{5.9}$$

Section 5.6.1 explains the idea of *entropy* or information content. The result needed here is that the entropy $H(X)$ of a signal, which can take on values over the range X, with given probability density f_X, is

$$H(X) = \int_{-\infty}^{\infty} f_X(x)\log_2\left(\frac{1}{f_X(x)}\right)\, dx \tag{5.10}$$

Substituting the expression for the PDF of the Gaussian distribution with zero mean Equation (5.9), the information content of the Gaussian signal is found to be

$$H(X) = \frac{1}{2}\log_2 2\pi e\sigma^2 \tag{5.11}$$

Let P be the power in the signal and σ^2 be the noise power. The entropies of the signal and noise, respectively, are

$$H_s(X) = \frac{1}{2}\log_2 2\pi e(P + \sigma^2) \tag{5.12}$$

$$H_n(X) = \frac{1}{2}\log_2 2\pi e\sigma^2 \tag{5.13}$$

These are written so as to make the dependence upon the channel amplitude, modeled as a random variable X, explicit. Shannon reasoned that both the signal and noise have information content and that the capacity in bits is the entropy or information content of the total signal (that is, signal plus noise) minus the entropy of the noise,

$$\mathcal{H}(X) = \mathcal{H}_s(X) - \mathcal{H}_n(X)$$

$$= \frac{1}{2}\log_2(P + \sigma^2) - \frac{1}{2}\log_2\sigma^2$$

$$= \frac{1}{2}\log_2\left(\frac{P + \sigma^2}{\sigma^2}\right)$$

$$= \frac{1}{2}\log_2\left(1 + \frac{P}{\sigma^2}\right)$$

$$= \frac{1}{2}\log_2(1 + \text{SNR}) \tag{5.14}$$

The capacity in bits per second is this entropy figure multiplied by $2B$ samples per second, so

$$C = B\log_2(1 + \text{SNR}) \tag{5.15}$$

This serves as a basis for our subsequent investigations. There are two problems to be addressed. First, we need to determine the best way to allocate amplitude steps, termed quantization levels, to the incoming signal (speech, audio, images, or other data to be transmitted). Secondly, it is necessary to determine how to allocate available binary digits (bits) to these quantization levels, so that we form the encoded symbol stream. The decoding is essentially the reverse: Map the symbols back into binary codes from which the original amplitude is approximated.

5.5 Quantization

The term *quantization* is usually understood to mean scalar quantization. This means taking one sample at a time and determining a binary value that can represent the analog input level (usually a voltage) at that instant. There are two important issues at play in analog to digital or A/D conversion. The first is to determine the number of levels required. Although it is always possible to employ more levels than necessary for a given signal type, a higher bit rate would be the result. The second issue is ensuring that the range of the signal is spanned. For example, speech in a typical conversation will have a large range of values from very small to very large, corresponding to the range from soft to loud sounds. Thus, there is a balance: Spanning a larger dynamic range necessarily means that for a fixed number of levels, the accuracy is reduced. On the other hand, if the maximum range is reduced and the same number of levels used to span the smaller range, then smaller amplitudes will be quantized more accurately. However, due to the reduced range, signal amplitudes at the extremes may not be able to be represented. Both of these situations lead to distortion of the reconstructed analog signal. The aim, then, is to balance these competing requirements – accuracy and dynamic range.

5.5.1 Scalar Quantization

Figure 5.4 gives a simple illustration of a quantized signal. The continuous wave-form – in this case, a sine wave – is sampled at discrete time instants. At each instant, a representational level is allocated. For N bits available to represent the level, one of 2^N levels can be specified for each sample. It is clear from the figure that there is a penalty in terms of accuracy of the value represented. The example shows a very coarse quantization, using eight levels. This could be represented as 3 bits per sample. If, for example, we used 8 bits, then there would be 256 levels. If we doubled this number of bits, we would have 65 536 levels. The number of levels is exponentially related to the number of bits, or, put another way, adding one bit to each sample doubles the number of levels.

This process is termed A/D conversion and is performed using an analog to digital converter (ADC). The reverse, or digital to analog (D/A) conversion, converts a given binary value into its corresponding analog voltage level. In the process, some accuracy is lost – the smooth sinusoid shown in Figure 5.4 is replaced by a stairstep type of approximation.

It is necessary to formalize the description of quantization and what, precisely, it does. Figure 5.5 shows quantization as a mapping from the input, along the horizontal axis, to the output on the vertical axis. The input amplitude can range anywhere along the *x* axis, and the value at which it will be reconstructed is read from the corresponding level on the *y* axis. The input levels are termed the *decision* levels, since that is the level at which a decision is made as to which

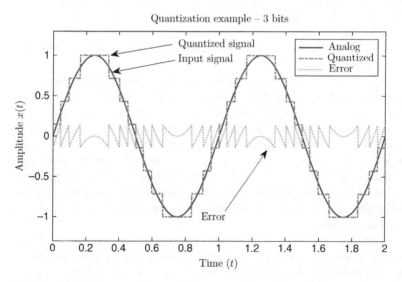

Figure 5.4 Quantizing a peak-to-peak sine wave using a 3 bit, mid-rise quantizer.

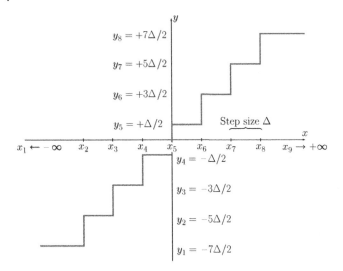

Figure 5.5 Quantization input–output mapping for a 3 bit, mid-rise quantization characteristic. *x* is the analog input, y_k represents the quantized values.

bin the analog level falls in. The output levels are termed the *reconstruction levels*, because that is the fixed level at which each sample will be reconstructed in the output waveform.

The quantizer of Figure 5.5 has a step size Δ. Any input x_k value from, say, 0 to Δ will be mapped into a corresponding reconstruction y_k value, in this case $\Delta/2$. Similarly, an input from -3Δ to -2Δ would be reconstructed as $-5\Delta/2 = -2.5\Delta$. This seems perfectly reasonable, but what happens to input values above 3Δ or below -3Δ? They are clamped to their respective maximum and minimum values of $\pm 3.5\Delta$.

It is worth noting that the characteristic shown is not the only possibility. The type illustrated may be characterized as a zero-offset, uniform-step, mid-rise quantizer. The step size may be nonuniform, if lower amplitude values are to be represented more accurately than higher amplitudes. The characteristic may not be centered on zero, but on some other positive level. Finally, the rise in the middle of the characteristic may be replaced with a "tread," giving the advantage that a value of zero is quantized exactly to zero. However, that then implies that there is an odd number of steps, so that one level may be unused (since the number of steps is 2^N), or alternatively that there are more steps on one side of the axis than the other.

Now to analyze the quantization errors. The purpose is to understand how errors arise, since such errors give rise to distortion in the sampled waveform. By analyzing the error, we can quantify how much error will occur for a given signal, and also for a "typical" signal, by using statistical approaches. It should be apparent that the average place where the signal spends a lot of time is the area which should be quantized most accurately, so as to reduce the average overall distortion.

For L levels in the quantization characteristic, denote the kth decision level as x_k and the kth reconstruction level as y_k. The error for an analog input x mapping to reconstruction y_k is thus $(x - y_k)$, and we square this because the error could be positive or negative.

The probability density of the signal $f(x)$ tells us which values of x are more (or less) likely. Thus, the error is weighted by this value and summed over the range from x_k to x_{k+1}, because this is the range of values for which y_k is output.

Finally, this situation occurs for each of the L quantization steps, so the final error figure of merit is (Jayant and Noll, 1990)

$$e_q^2 = \sum_{k=1}^{L} \int_{x_k}^{x_{k+1}} (x - y_k)^2 f(x) \, dx \tag{5.16}$$

where L is the number of output levels of the quantizer, $f(x)$ is the PDF of the input variable, x_k is the kth decision level, and y_k is the corresponding reconstruction level.

To explain the usefulness of this equation in simple terms, suppose we have a one-bit (two-level) mid-rise quantizer. This reconstructs the signal according to the following:

$$x \text{ values between } -\infty \text{ and } 0 \rightarrow y \text{ output is } -\frac{\Delta}{2}$$

$$x \text{ values between } 0 \text{ and } +\infty \rightarrow y \text{ output is } +\frac{\Delta}{2}$$

Suppose the input signal has a probability density $f(x)$ that is uniform – that is, all values are equally likely. Of course, we would have to know this in advance or at least make an assumption that this distribution would adequately represent the input. A uniform PDF for input x means that the signal is equally likely between two values. Assume that the mean is zero and the maximum amplitude of the values is $\pm A$. This PDF would be written mathematically as

$$f(x) = \begin{cases} 0 & : \ x < -\frac{A}{2} \\ \frac{1}{A} & : \ -\frac{A}{2} \le x \le +\frac{A}{2} \\ 0 & : \ x > +\frac{A}{2} \end{cases} \tag{5.17}$$

This shows that the probability density in the range from $-A/2$ to $+A/2$ is equal to $1/A$. This has to be the case, since the area under a PDF has to equal unity (that is, the signal has to be somewhere in the known range), and the rectangle so formed has an area of one.

So for the assumed input PDF, we need to work out what quantizer parameters would minimize the quantization error. First, we need a *cost function*. This requires defining a quantity that equates to cost and relating that back to the design parameters. In this case, the cost could be defined as the average quantization error, and we can define the step size of the quantizer characteristic to minimize this cost. So the question may be phrased as "is there an optimal quantizer step size which minimizes the average error of the quantized signal?"

Equation (5.16) shows that the error is squared and weighted by the probability density. In this particular case, the quantizing error is

$$e_q^2 = \sum_{k=1}^{2} \int_{x_k}^{x_{k+1}} (x - y_k)^2 \cdot \frac{1}{A} \, dx \qquad (5.18)$$

The reconstruction levels y_k are plus and minus half a step size, so using those and writing out the summation over the two levels for this example, the squared error becomes

$$e_q^2 = \frac{1}{A} \int_{-\infty}^{0} \left(x - \frac{-\Delta}{2} \right)^2 dx + \frac{1}{A} \int_{0}^{+\infty} \left(x - \frac{+\Delta}{2} \right)^2 dx$$

$$= \frac{1}{A} \int_{-\infty}^{0} \left(x^2 + x\Delta + \frac{\Delta^2}{4} \right) dx + \frac{1}{A} \int_{0}^{+\infty} \left(x^2 - x\Delta + \frac{\Delta^2}{4} \right) dx$$

The integrals now need to be evaluated. Since the distribution is uniform, the infinite range of the integrals can be put in terms of the maximum signal amplitudes,

$$e_q^2 = \frac{1}{A} \left(\frac{x^3}{3} + \frac{x^2 \Delta}{2} + \frac{x\Delta^2}{4} \right) \Bigg|_{x=-\frac{A}{2}}^{0}$$

$$+ \frac{1}{A} \left(\frac{x^3}{3} - \frac{x^2 \Delta}{2} + \frac{x\Delta^2}{4} \right) \Bigg|_{0}^{x=+\frac{A}{2}}$$

$$= \frac{1}{A} \left[(0) - \left(\frac{-A^3}{24} + \frac{A^2 \Delta}{8} + \frac{-A\Delta^2}{8} \right) \right]$$

$$+ \frac{1}{A} \left[\left(\frac{-A^3}{24} - \frac{A^2 \Delta}{8} + \frac{A\Delta^2}{8} \right) - (0) \right]$$

Simplifying algebraically,

$$e_q^2 = \frac{1}{A} \left(\frac{A^3}{12} - \frac{A^2 \Delta}{4} + \frac{A\Delta^2}{4} \right)$$

$$e_q^2 = \frac{A^2}{12} - \frac{A\Delta}{4} + \frac{\Delta^2}{4} \qquad (5.19)$$

So finally we have an expression for the average squared error, which must be minimized. The only parameter that may be varied is the step size Δ. We can determine this optimal value using calculus, since this equation is, in effect, a quadratic equation in Δ. So letting the optimal step size be Δ^*, we can find the derivative

$$\frac{de_q^2}{d\Delta} = -\frac{A}{4} + \frac{\Delta^*}{2} \tag{5.20}$$

As always with this type of problem, we set the derivative to zero to find the turning point. Setting $\frac{de_q^2}{d\Delta} = 0$ gives

$$\frac{A}{4} = \frac{\Delta^*}{2}$$

$$\therefore \Delta^* = \frac{A}{2} \tag{5.21}$$

This provides an expression for the optimal step size in terms of the signal amplitude parameter A. The result – that the best step size is half the amplitude – appears to be what we would expect intuitively. This may be rewritten in terms of the signal power or variance. First, we need an expression for the variance in terms of the PDF. In general, for any uniform PDF with limits $\pm A/2$, the variance is

$$\sigma^2 \triangleq \int_{-\infty}^{+\infty} x^2 f(x) \, dx$$

$$\therefore \sigma^2 = \int_{-\frac{A}{2}}^{+\frac{A}{2}} x^2 \left(\frac{1}{A}\right) \, dx \tag{5.22}$$

This results in

$$\sigma^2 = \frac{A^2}{12} \tag{5.23}$$

Turning this around to give the amplitude in terms of the variance of signal x,

$$\sigma_x^2 = \frac{A^2}{12}$$

$$\therefore A = 2\sigma_x\sqrt{3} \tag{5.24}$$

Substituting into the equation for the optimal step size,

$$\Delta^* = \frac{A}{2}$$

$$= \frac{2\sigma_x\sqrt{3}}{2}$$

$$= \sigma_x\sqrt{3} \tag{5.25}$$

This result tells us that the quantizer step size should be $\sqrt{3} \approx 1.73$ times the standard deviation of the signal in order to minimize the error. Doing so will minimize the quantizing noise – or, equivalently, maximize the SNR.

So a general conclusion we could reach is that the optimal step size is proportional to the standard deviation for a given PDF. But how does this change when we change the number of levels, which of course changes the number of bits required to quantize each sample?

The quantizing SNR is the signal power divided by the noise power. Expressed in decibels it is

$$\text{SNR} = 10 \log_{10} \left(\frac{\sum_n x^2(n)}{\sum_n e^2(n)} \right) \tag{5.26}$$

where the error is the difference between the value $x(n)$ and the quantized value $\hat{x}(n)$

$$e(n) = x(n) - \hat{x}(n) \tag{5.27}$$

For a uniform N-bit quantizer, the step size Δ is the ratio of the peak amplitude to the number of steps:

$$\Delta = \frac{2 \, x_{\text{max}}}{2^N} \tag{5.28}$$

If the quantizing noise has a uniform distribution, then the variance is

$$\sigma^2 \triangleq \int_{-\infty}^{+\infty} x^2 f(x) \, dx$$

$$\therefore \sigma_e^2 = \int_{-\Delta/2}^{+\Delta/2} x^2 \left(\frac{1}{\Delta} \right) \, dx$$

$$\rightarrow \sigma_e^2 = \frac{\Delta^2}{12} \tag{5.29}$$

Substituting the step size Δ into this,

$$\sigma_e^2 = \frac{\Delta^2}{12}$$

$$= \frac{4x_{\text{max}}^2}{12 \cdot 2^{2N}}$$

$$= \frac{x_{\text{max}}^2}{3 \cdot 2^{2N}} \tag{5.30}$$

We can write the SNR as

$$\text{SNR} = \frac{\sigma_x^2}{\sigma_e^2}$$

$$= \frac{\sigma_x^2}{\left(\frac{x_{\text{max}}^2}{3 \cdot 2^{2N}} \right)}$$

$$= 3 \left(\frac{\sigma_x^2}{x_{\text{max}}^2} \right) \cdot 2^{2N} \tag{5.31}$$

The usual way is to express this in decibels (dB) by taking the logarithm and multiplying by 10. Using standard logarithm rules, this expression becomes

$$\text{SNR} = 10\log_{10}3 + 10\log_{10}2^{2N} + 10\log_{10}\left(\frac{\sigma_x^2}{x_{\text{max}}^2}\right)$$

$$\approx 4.77 + 6.02N + 20\log_{10}\left(\frac{\sigma_x}{x_{\text{max}}}\right) \qquad \text{dB}$$

It is necessary to make an assumption regarding the maximum value in relation to the standard deviation. If this is four standard deviations, $x_{\text{max}} = 4\sigma_x$, and

$$\text{SNR} \approx 6N - 7.3 \qquad \text{dB} \tag{5.32}$$

That is, we can say that the SNR is proportional to the number of bits N

$$\text{SNR} \propto N \tag{5.33}$$

with a proportionality constant of 6. The essential conclusion is that adding one bit to a quantizer improves the SNR by about 6 dB. Note that, because of the various assumptions made in this derivation, this result is really only valid for a large number of quantizer steps and also where zero (or only a few) samples are clipped because they are outside the range $\pm 4\sigma_x$.

5.5.2 Companding

The above discussion on quantization assumes that samples are quantized *linearly*. That is to say, each step is of equal size. What would happen if the step sizes were not equal? Furthermore, why have the step size fixed over time? We now turn to the issue of step size selection and investigate changing the step size dynamically. Although it is more common to have equal step sizes for all the levels of a quantizer, the technique of compressing and expanding the levels has been extensively used for some types of audio and also finds application in more advanced encoding schemes. The idea of companding is credited to Eugene Peterson, who observed that "weak sounds required more delicate treatment than strong" (Bennett, 1984, p. 99; Crypto Museum, 2016).

If we keep a fixed step size, it implies that all amplitudes contribute equally to the noise when quantized. However, since quantization is only an approximation, we may relax the accuracy requirement. For high amplitudes, larger step sizes may suffice. This sacrifice of some accuracy at larger amplitudes permits smaller step sizes for smaller amplitudes. Depending on the distribution of the signal, the average error may in fact decrease, if smaller amplitudes are more likely. In addition, there may appear to be less noise (in audio, for example), since perception may mask the extra noise at higher amplitudes.

This gives rise to the notion of "companding," which means compressing and expanding the range. Figure 5.6 shows this as an input–output mapping. On the horizontal axis we see the input amplitude, with the output on the vertical axis.

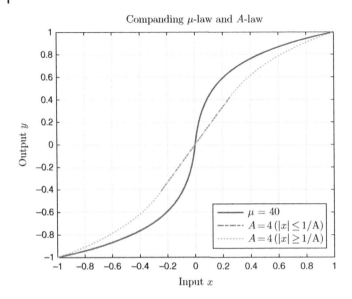

Figure 5.6 Representative comparison of μ-law and A-law companding. The μ and A values have been chosen to highlight the fact that A-law is a piecewise characteristic.

For low levels of input amplitude x, the change in output δy for a small change in input δx is fairly large. However, for high input amplitudes, the change δy is relatively small for a given input change δx. This achieves the goal of having larger step sizes for larger amplitudes.

When written as a set of equations, the process implies an analog mapping, however, it is more conveniently implemented digitally using a lookup table (LUT). In practice, we may quantize the source using a 13-bit (*A*-law) or 14-bit (μ-law) set of levels and map this into an 8-bit equivalent. This permits the use of fewer bits per sample but retains approximately the same level of perceptual distortion overall.

Note that the process as described is the encoding or compression stage. The expansion stage is the inverse of this operation. These operations are most easily performed by a precomputed LUT containing all the input–output pairs for compression and expansion.

Figure 5.6 shows a comparison of the characteristics discussed. A linear response would be represented as a 45° line, with output equalling input. The μ-law characteristic (often written as "mu-law") is defined by the mapping equation

$$c_\mu(x) = \text{sign}(x) \, \frac{\ln(1 + \mu|x|)}{\ln(1 + \mu)} \qquad 0 \le x \le 1 \tag{5.34}$$

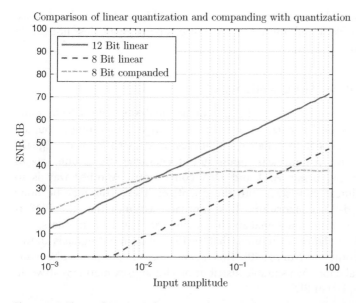

Comparison of linear quantization and companding with quantization

Figure 5.7 The performance of a companding quantizer as compared with linear quantization. The tradeoff inherent in companding is evident: better performance at low signal levels, with inferior performance at higher signal levels.

This is a "saturating" characteristic – small values of the input signal are treated differently to larger values. Also shown in the figure is the A-law characteristic. It is similar in concept to μ law but has a slightly different definition:

$$c_A(x) = \begin{cases} \text{sign}(x)\, \frac{A|x|}{1+\ln A} & 0 \le |x| \le \frac{1}{A} \\ \text{sign}(x)\, \frac{1+\ln(A|x|)}{1+\ln A} & \frac{1}{A} \le |x| \le 1 \end{cases} \tag{5.35}$$

Note that A-law breaks the range into distinct regions – linear for small amplitudes and nonlinear for larger amplitudes. The values of μ and A shown in the figure are chosen to illustrate the differences in the characteristics. The inverse step for converting a μ-law compressed amplitude signal back is

$$x = \text{sign}(c_\mu(x))\, \frac{1}{\mu}[(1+\mu)^{|c_\mu(x)|} - 1] \tag{5.36}$$

Figure 5.7 shows a comparison of 8-bit linear, 12-bit linear, and 8-bit companded signals in terms of SNR. For very low input amplitudes, μ-law companding using 8 bits even manages to perform better than linear 12-bit quantization (since the SNR is evidently superior). As the amplitude of the input increases, the linear quantizers show increasing performance with higher SNR, with the companded signal reaching a plateau for higher amplitudes. This is reasonable, since we cannot have improvement in one area without a penalty in another. However, that penalty comes at higher amplitudes that are statistically less likely and, moreover, the noise is perceptually less noticeable when the amplitude is large.

5.5.3 Unequal Step Size Quantization

As discussed in Section 5.5.2, the use of unequal steps does have some advantages. Specifically, a larger step size for larger amplitudes, coupled with a smaller step size for smaller amplitudes, works well for voice signals. Companding using a nonlinear input–output characteristic is one way to achieve this, using linear A/D and D/A conversions. Another way is to build the nonlinear step sizes into the quantizer itself. This may be done using the Lloyd–Max algorithm. This approach assumes that the PDF is nonuniform – that is, some signal amplitudes are more likely than others.

The Lloyd–Max algorithm assumes that the PDF of the input signal is known analytically (Jayant and Noll, 1990). This means an assumption must be made based on typical data. Two suitable distributions for approximating real-world data are the Laplacian PDF

$$f(x) = \frac{1}{\sigma\sqrt{2}}\, e^{-\sqrt{2}|x-\bar{x}|/\sigma} \tag{5.37}$$

and the Gaussian PDF

$$f(x) = \frac{1}{\sigma\sqrt{2\pi}} e^{-(x-\bar{x})^2/2\sigma^2} \tag{5.38}$$

To develop the Lloyd–Max algorithm, we first use the same expression for quantization error as before:

$$e_q^2 = \sum_{k=1}^{L} \int_{x_k}^{x_{k+1}} (x - y_k)^2\, f(x)\, \mathrm{d}x \tag{5.39}$$

Now, however, both the decision (input) levels x_k and the reconstruction (output) levels y_k may vary. Thus, to solve for the minimum distortion, we need to vary these and find the solution where

$$\frac{\partial e_q^2}{\partial x_k} = 0 \qquad k = 2, 3, \dots, L \tag{5.40}$$

$$\frac{\partial e_q^2}{\partial y_k} = 0 \qquad k = 1, 2, \dots, L \tag{5.41}$$

This cannot easily be solved analytically for anything but simple PDFs, and an iterative numerical solution is required. To do that, we note from the equations that:

i) The decision levels x_k are halfway between the neighboring reconstruction levels.
ii) Each reconstruction level y_k is the centroid of the PDF over the appropriate interval.

Mathematically, this becomes the Lloyd–Max algorithm (Jayant and Noll, 1990), which requires iterative solution of the following:

$$x_k^* = \frac{1}{2}(y_k^* + y_{k-1}^*) \qquad k = 2, 3, \ldots, L$$

$$x_1^* = -\infty$$

$$x_{L+1}^* = +\infty$$

$$y_k^* = \frac{\int_{x_k^*}^{x_{k+1}^*} x f(x)\, \mathrm{d}x}{\int_{x_k^*}^{x_{k+1}^*} f(x)\, \mathrm{d}x} \qquad k = 1, 2, \ldots, L \tag{5.42}$$

This set of equations is then iterated until convergence. This is illustrated in Figures 5.8 and 5.9 for $L = 4$ and $L = 8$ levels, respectively. Note that the iteration does not require any explicit calculation of the derivatives, just evaluation of the PDFs and incremental adjustment of the decision–reconstruction levels.

5.5.4 Adaptive Scalar Quantization

The quantizer discussed in the preceding sections are fixed at the time of their design. When the encoder is operating, it always uses the same decision points and reconstruction levels. However, speech and audio signals are not stationary – they change over time. This leads to the notion of changing the

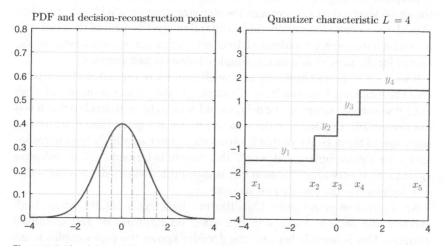

Figure 5.8 Lloyd–Max PDF-optimized quantizer with $L = 4$.

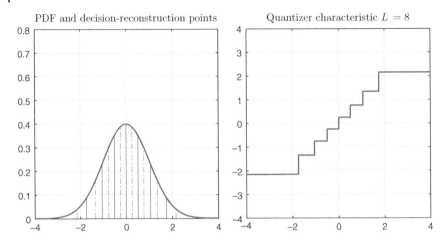

Figure 5.9 Lloyd–Max PDF-optimized quantizer with $L = 8$.

step size as the signal is being encoded. For a small-amplitude signal, the step size could be fixed but small. For a large-amplitude signal the step size could be made larger but again fixed. The decision as to whether to use a small or large step size is based on the relative magnitude of the input signal. This could be calculated based on the instantaneous value of a sample or the energy (variance) calculated over a block of samples.

This scheme requires a little attention to a few details. First, some signals such as speech tend to go rapidly from low amplitude to high amplitude. Thus, in adapting the step size, it must be changed fairly rapidly too. Likewise, when the signal level reduces on average, the step size must be quickly lowered. This should not be done too rapidly, and so attention must be paid to the block size over which the energy is estimated. In effect, such a scheme must estimate the energy for the present and future samples, based on past samples.

If we were to permit buffering of a small amount of the signal, the variance could be estimated based on "future" samples, in the sense that the decoder has not yet seen those samples. The delay would need to be kept small so that it was not perceptible. Importantly, the step size to use at a given time must be sent somehow to the decoder, so that it knows the correct amplitude level to reconstruct for a given binary codeword. This system is termed AQF or Adaptive Quantization Forwards. The step size is transmitted via a side channel – that is, interleaved in the encoded bit stream.

An alternative is Adaptive Quantization Backwards (AQB), where both encoder and decoder use a step size based on estimates of past quantized samples. This is feasible because the decoder knows the past samples it has reconstructed. This approach does not require explicit transmission of the step

Figure 5.10 The layout of a VQ codebook.
Each of the *K* codevectors has a dimension of *L*.

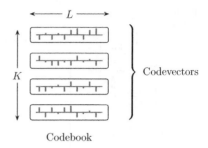

Codebook

size, however, the step size adaptation may take longer and be less accurate, because it is based on past samples.

5.5.5 Vector Quantization

All the previous discussion on quantization has assumed that one sample at a time is quantized. However, what if not one, but several samples in a block, were quantized at the same time? This technique is termed *Vector Quantization* (VQ), and by way of differentiation, conventional one-at-a-time quantization is sometimes termed scalar quantization. Figure 5.10 shows the layout of a VQ codebook, which contains the representative codevectors used to match a vector of source samples.

The term VQ might imply that a one-dimensional vector is quantized, and this is often the case. However, when quantizing blocks of, say, an image, this approach might be better termed "matrix quantization." However, the term VQ is normally used for both $N \times 1$ and $N \times M$ blocks.

The key advantage of quantizing several samples at once is that it makes it possible to exploit the correlation between samples. The basic idea is as follows. First, both the encoder and decoder have an identical *codebook* of "typical" vectors drawn from a source statistically similar to the data that is to be quantized in practice. The codebook is comprised of a large number of codevectors, each equal to the size of the data block (vector) to be encoded. The encoder then buffers the incoming samples into blocks of size equal to the vector dimension and performs a search of the codebook for the closest match. The matching criteria has to be some form of mathematical similarity; usually the codevector with the minimum mean-square error is chosen. The index of the closest match is then sent to the decoder. The decoder then simply uses this index to look up the corresponding codevector. The individual samples are then read from this codevector to reconstruct the sample stream. This is illustrated in Figure 5.11.

Several problems are apparent with this approach. The most obvious is that of how to generate the codebook in the first place. We return to that problem shortly. First, however, consider the data reduction possible with VQ. Suppose

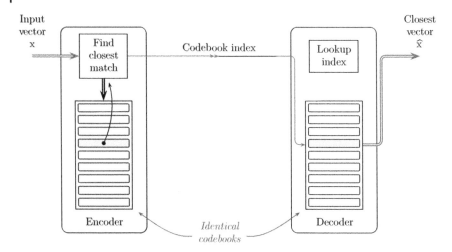

Figure 5.11 Encoding and decoding with a vector quantizer. The encoder must perform a search for the best matching vector, whereas the decoder simply looks up the vector corresponding to the index transmitted over the communications channel.

the problem is to encode a block of 8 samples, each to 8-bit precision. Using conventional scalar quantization, this would require

$$8 \; \frac{\text{Samples}}{\text{Vector}} \times 8 \; \frac{\text{Bits}}{\text{Sample}} = 64 \; \frac{\text{Bits}}{\text{Vector}} \tag{5.43}$$

Suppose, for this simple example, the codebook contained 1024 entries. That is, 1024 vectors of dimension 8. The number of bits required to index this codebook would be $\log_{10} 1024 = 10$. The average bit rate using VQ is then

$$\frac{10 \; \text{Bits/Vector}}{8 \; \text{Samples/Vector}} = 1.25 \; \frac{\text{Bits}}{\text{Sample}} \tag{5.44}$$

In other words, the number of bits per vector has been reduced from 64 (using scalar quantization) to 10 (using VQ); equivalently, the number of bits per sample has decreased from 8 to 1.25.

From this simple example, it should be clear that increasing the size of the codevectors (not the number of codevectors) would decrease the average rate, which is desirable. However, the average distortion will increase significantly, since the codebook has to approximate a larger possible number of sample vector combinations with fewer available codevectors. Each codevector effectively has to represent a wider variety of source patterns.

On the other hand, increasing the number of codevectors (that is, more entries in the codebook) means that there are more representative codevectors to choose from; hence the average distortion should reduce. This, however, means many more vectors to search – consider that adding just one bit to the vector index would double the size of the codebook.

As a result, the computational complexity may become a significant barrier. Consider what must happen when the encoder searches the codebook. If the source vector is **x** and the codebook vector is c_k, the error (or distortion) computation for a vector of dimension L is

$$d_k = \sum_{j=0}^{L-1} (x_j - y_{kj})^2 \qquad (5.45)$$

In the present example, there are 1024 entries in the codebook, each of dimension 8. The comparison of each vector requires eight subtractions and square operations, as well as additions, to find the mean-square error. If this code is implemented in a sequential loop, there is additional overhead for indexing the correct values, as well as the looping overhead for iterating over the block. Thus an approximation to the number of calculations is around 25 operations per vector, times 1000 vectors in the codebook, for 25 000 operations. This is required for each vector (block) to be encoded, leading to a not inconsiderable number of calculations. Increasing the codebook size, and/or the codevector dimension, would increase this figure. Furthermore, increasing the codebook size results in an exponential increase in complexity, since each extra bit added to the vector index doubles the size of the codebook.

The approach outlined above is termed an *exhaustive search*, since every possible codevector in the codebook is checked for each encoded block. Various ways have been proposed to reduce this search complexity, usually involving some tradeoff in terms of larger memory requirements.

Encoding using an exhaustive search may be more formally defined as follows:

1) Given a codebook of size K, with vector dimension L.
2) For each source vector **x** to be encoded:
   a) Set the distortion d_{min} to be a very large number.
   b) For each codevector (index k), complete the following steps:
      i) For each candidate c_k in the codebook **C**, calculate the distortion using the chosen metric, typically the squared error $d_k = ||\mathbf{x} - \mathbf{c}_k||^2$.
      ii) If the distortion d is less than the minimum distortion so far (d_{min}), set $d_k \rightarrow d_{min}$ and save the index $k \rightarrow k^*$. The best reproduction vector so far $\hat{\mathbf{x}}$ is saved as this codevector: $\mathbf{c}_k \rightarrow \hat{\mathbf{x}}$.
      iii) Repeat for all K codevectors in the codebook.

The decoder, however, merely has to use the transmitted index k^* to look up the corresponding codevector. Clearly, the complexity of the encoder is much greater than the decoder. Such an arrangement is sometimes termed an *asymmetric coder*.

This method is satisfactory if the codebook is a reasonable representation of the data samples to be encoded. But how is the codebook, comprising all the

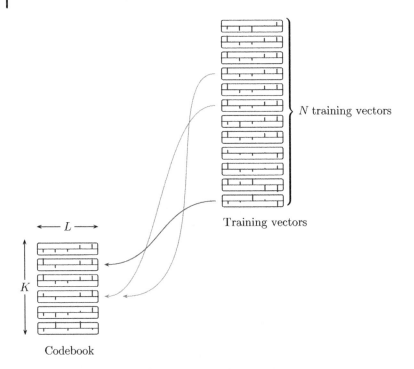

Figure 5.12 The VQ training process. More than one training vector may map into any given codebook vector.

codevectors, determined? Rather than the analytical solutions as outlined for scalar quantizers, an iterative solution using representative source vectors is usually applied. This training algorithm essentially tries to map typical source vectors into clusters, assuming that those will represent the source, in practice, with sufficient accuracy. Figure 5.12 illustrates the mapping process. In addition to the vector dimension and codebook size, it is necessary to select sufficient training vectors, which are representative of the typical source characteristics.

There are many possibilities for the training algorithm. The k-means or LBG training algorithm is essentially as follows:

1) A codebook **C** of size K, with vector dimension L, is to be determined.
2) Begin with N feature vectors $\mathbf{x}_k$: $k = 0, 1, \ldots, N - 1$.
3) Create an initial codebook **C** by arbitrarily selecting K feature vectors.
4) For each vector in the training set:
   a) Perform a search over the current codebook to find the closest approximation.
   b) Save this vector $\hat{\mathbf{x}}_k$ at index (codeword) k. Increment the count of mapped vectors for index k.

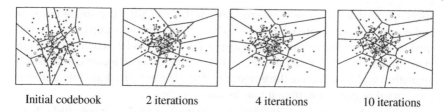

| Initial codebook | 2 iterations | 4 iterations | 10 iterations |

Figure 5.13 VQ training iterations and convergence. The small dots are training data, while the circles are the centroids.

5) For each codevector in the codebook:
   a) Compute the centroid of all training vectors that mapped into that code-vector using the sum and count above.
   b) Replace the codevector with this centroid.
6) Encode each training vector using the new codebook, and compute the aver-age distortion. If small enough, quit. Otherwise repeat from (4).

The centroid (or average) calculation step may be visualized using the Voronoi diagram of Figure 5.13, which depicts a VQ in two dimensions. Each input vector is represented by a point in space, and the two components of each vector are the horizontal and vertical coordinates. Each codevector, represented by a circle, is the best match for any input vector falling in the region indicated. The boundary of each region changes as the training process continues through each iteration. Such a Voronoi diagram may be easily produced using the `voronoi()` function in MATLAB.

5.6 Source Coding

Data on a digital channel is composed, at the lowest level, of a stream of bits. The content, however, may comprise text, images, sound, or other types of infor-mation. Thus, it is necessary to map between the content as presented and the encoded bitstream – termed *source coding*. The encoder performs this map-ping, and the decoder performs the reverse, although the steps involved are not necessarily the exact inverses of each other.

Different types of source material have differing requirements. The most fun-damental is the requirement for precise reconstruction – that is, whether or not the decoded bitstream at the receiver is exactly equal to that presented to the encoder. Although it might seem that exact bit-for-bit matching is neces-sary, that is not always so. If, for example, data files are to be reconstructed, then bit-exact transmission is clearly mandatory. But some channels are subject to error, and it may be difficult to guarantee that exact bit correctness always occurs.

Furthermore, it is possible to compress some types of source data a great deal more if we are willing to sacrifice the bit-exactness requirement. Images, speech, and video are the most common in this category. If a single pixel, or audio sample, is corrupted, it is likely to have little or no effect on the perceived quality of the sound or the picture displayed. In fact, relatively high levels of error may well be tolerated in different scenarios. If a particular compression algorithm is able to reduce the amount of data significantly, with only minimal perceptual degradation, then it may be a better choice.

Bit-exact compression algorithms are termed *lossless*, whereas *lossy* algorithms may sacrifice some exactness of reconstruction for higher compression. Both techniques are employed widely in practice and quite often are used to complement each other in a given telecommunications application.

In different applications it may be necessary to compress text or media such as audio or video. The data emitted by an encoder is comprised of *symbols*, which represent the data in some transformed way. The set of possible symbols is usually termed an *alphabet*. This is by analogy with an alphabet of letters, but it has a more generic meaning. For example, if an image only consisted of 64 colors the alphabet is comprised of each of those 64 colors and would require $\log_2 64 = 6$ bits to uniquely specify each symbol. A symbol may go further though – for example, one symbol may represent a group of several pixels in a certain order.

5.6.1 Lossless Codes

This section introduces lossless codes; that is, encoding methods that can reproduce the source data stream exactly and without error. These types of codes are sometimes termed *entropy codes*.

5.6.1.1 Entropy and Codewords

In encoding a data source, it is necessary to determine the possible range of values to be encoded. As will be shown, the probability of each individual value is important. In theoretical terms, the information rate, or *entropy* of the source, governs the minimum number of bits needed to encode data emanating from that source. Entropy answers the question "how many bits are necessary to encode this data?" If we had to encode, for example, only uppercase letters of the English alphabet, then 26 binary patterns would be required. This implies the need for a minimum of 5 bits to cover 32 possibilities (since $2^5 = 32$, which is greater than 26). Turning this around, the minimum requirement would be $\log_2 26$ bits. The implicit assumption in that calculation is that each of those patterns is equally likely. This assumption often won't hold in practice, but that is a good thing. As it turns out, unequal probably of occurrence can be used to our advantage.

This is where the concept of entropy is useful. For source symbols s_i drawn from an alphabet S of possible symbols, each with probability of occurrence $\Pr(s_i)$, the entropy is defined as

$$\mathcal{H} = - \sum_{s_i \in S} \Pr(s_i) \log_2\{\Pr(s_i)\} \qquad \text{Bits/symbol} \qquad (5.46)$$

Entropy is a theoretical lower bound – it tells us the lowest number of bits required for that particular source, taking each symbol at a time, given its probability. Importantly, we can see that if the probabilities are unequal, we may have a chance to reduce the entropy or average number of bits per symbol. So the question is this: How do we assign the bits to each symbol in order to make the most efficient use of the available bits?

For a given source, the probabilities may be fixed, or they may only vary a little over time. If this is the case, how do we assign binary codewords to each source symbol, so as to minimize the output bitrate? It would seem logical to assign shorter codewords to more likely symbols and any remaining codewords, which will invariably be longer, to the less likely symbols. Mathematically, we want to minimize the average codeword length over a block of encoded symbols. If we could do this, knowing the codeword length for each source symbol, the average codeword length would be a weighted average, computed as

$$L_{\text{av}} = \sum_{s_i \in S} \Pr(s_i)\, L(s_i) \qquad \text{Bits/symbol} \qquad (5.47)$$

This is nice in theory, but how do we work out what the optimal codeword assignments should be? To give a concrete example, suppose we have four symbols: A, B, C, and D. If we assign codewords such that symbol A is 0110, symbol B is 101, C is 011, and D is 10, then we would have a *variable-wordlength code* (VWLC). In light of the earlier discussion, this would make sense only if D were more likely than A (for example), since if it is more likely and has a shorter codeword length, then fewer bits may be required overall.

In terms of Equation (5.47), the overall average would then be less. Note that it is possible to end up with a fractional bit rate, since the *average* number of bits per symbol is the target quantity. For example, given that the symbols A,B,C, and D have bit lengths of 4, 3, 3, and 2, respectively, suppose the probabilities of each was 25%. The average codeword length, according to the above formula, would then be $0.25 \times 4 + 0.25 \times 3 + 0.25 \times 3 + 0.25 \times 2 = 3.0$. But suppose the probabilities were actually 50, 20, 20, and 10%. In that case the calculation is

```
% the symbol lengths in bits
s = [4 3 3 2];

% the probability of each symbol
pr = [0.5 0.2 0.2 0.1];

% check - sum of probabilities should be one
sum(pr)
ans =
    1.0000
```

```
% the average codeword length
sum(pr.*s)
ans =
    3.4000
```

The average is 3.4 bits/symbol. This is not an integer, even though each individual symbol was assigned an integral number of bits.

For the given codeword assignment, there is potential for confusion in the decoding. Suppose we sent the binary string 1010110. The decoder could decode that as 101 followed by 0110, which would equate to the source symbols BA. Alternatively, it could be decoded as 10 101 10, which is DBD. To get around this ambiguity, we could employ a special bit pattern to be the separator, or "comma." Suppose 110 is the separator. Then the comma-coded block ABCD becomes A,B,C,D. This would be 0110 110 101 110 011 110 10. The problem with this approach is obvious: It has increased the average number of bits in the message block. Furthermore, we still have to be very careful that the comma pattern does not appear at the start of any codeword assignment or even within it, since this would confuse the decoder. Both of these problems have an elegant solution in a method called the Huffman code (Huffman, 1952). It solves both the decoding ambiguity problem and assigning an integer minimal length code for each symbol.

5.6.1.2 The Huffman Code

Equation (5.46) makes it clear that the entropy of a source depends on the symbol probabilities. It provides a lower theoretical bound, but does not tell us how to assign codewords in an optimal fashion. Some method is needed to assign bits to each symbol and thus form the codewords. In the ideal case, L_{av} would equal the entropy $\mathcal{H}$. But this is only in theory, and in practice the average codeword length approaches the entropy but quite never reaches it. Essentially, it gives us something to work toward – a theoretical lower bound.

Huffman codes assign an integer number of bits to form the codewords for symbols in an optimal manner. In addition to reducing the average codeword length, the codes are uniquely decodable – there is no confusion as to where one codeword ends and the next starts. Importantly, no special separator bit pattern is required.

The Huffman method works as follows. Suppose we have the set of source symbols, with known probability, as illustrated in Figure 5.14. The leaf nodes have the defined symbol values (A, B, C, D, and E in this example). We have annotated the interior nodes according to what is combined at the stage before – for example, DE is the combination of D and E. This is not part of the algorithm's requirements, but it makes following the code development somewhat easier.

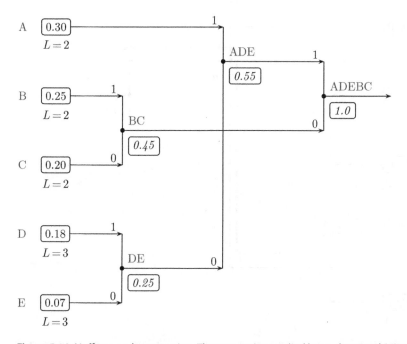

Figure 5.14 Huffman code generation. The convention applied here when combining two nodes is to assign a 1 bit to the higher probability leaf. When the probabilities at each step are combined in the way shown, the resulting average length is 2.25 bits/symbol.

First, the symbols and their corresponding probabilities are drawn in a column on the left. These are called leaf nodes. Then we proceed to combine the symbols, two at a time, to create intermediate nodes. For example, D and E are combined, with a combined probability of $0.18 + 0.07 = 0.25$. We adopt the convention of a binary 1 for the higher probability and a 0 for the lower probability (this is of course arbitrary, and any convention can be used as long as it is consistently applied). The same happens for B and C. Then, we recombine those intermediate nodes with leaf nodes or other intermediate nodes in the same fashion – nodes with summed probabilities 0.25 (labeled DE) and 0.30 (A) are combined to give 0.55 (labeled ADE), followed by combining that result with 0.45 (BC). Finally, we end up with a single node (the "root node") on the right, whose probability should clearly equal 1.0, since we have added all the source probabilities to reach that stage (although indirectly, by adding pairs in succession).

Using this (sideways) tree structure to encode a source symbol is illustrated in Figure 5.15. Suppose we wish to encode symbol D. Starting at the leaf node, we follow the path toward the root node at the far right. Along the way, we record the branches taken. So at the first intermediate node, we have a 1, since

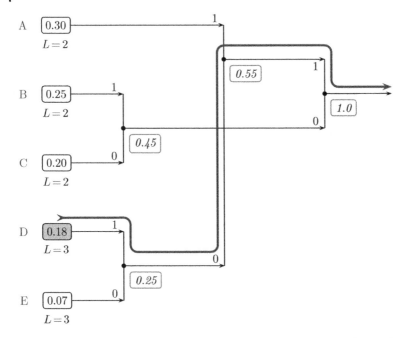

Figure 5.15 Huffman code encoding. Starting at the leaf node corresponding to the symbol to be encoded, the node joins are followed until the root node is reached. The branch from which the path entered at each join determines the bit value and is recorded.

we are coming from the upper branch that had the higher probability at the code design stage (0.18 compared with 0.07). Following the path shown, we next encounter a 0 at the junction (0.25 is lower than 0.30). Finally, at the root node, the bit value is 1 (0.55 is larger than 0.45).

Decoding at the receiver is the inverse of encoding. The same tree is illustrated in Figure 5.16, but now we traverse from right to left. This makes sense, since the decoder starts with no knowledge of the symbol expected next – it only consumes individual bits as the bitstream arrives. Starting from the root node, if we receive a 1, it means we must take the upper branch. Next, we take the lower branch shown on the diagram (0) and then the upper branch according to the bit value of 1. Finally, we arrive at the leaf, which shows symbol D.

The decoding process is, therefore, the reverse of the encoding. In encoding, we start at the leaf, assign 1/0 according to the branch on which we come to encounter the next node, and recursively repeat until we reach the root node. The decoder takes branches, starting from the root, according to whether a 1 or 0 is received. Clearly, the bitstream for a symbol has to be reversed before being sent to the decoder.

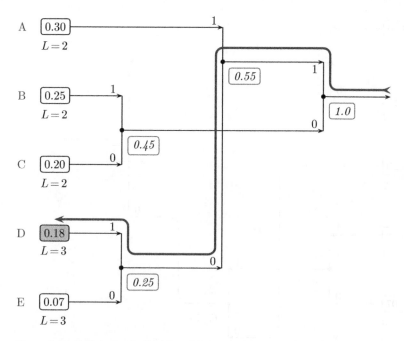

Figure 5.16 Huffman code decoding. Starting at the root node, each successive bit received determines which branch to take at each node, until a leaf node is reached. This corresponds to the symbol to be decoded.

In constructing the Huffman code tree, the 1/0 assignment for selecting the branch is arbitrary. We have used 1 for a higher probability and assigned the name "up" and conversely 0 for lower probability ("down"). As long as the encoder and decoder use the same convention, there is no problem. However, there are many possibilities for joining symbols at each node. Figure 5.17 illustrates a different set of combinations, leading to another Huffman tree. Careful examination will reveal that one difference is when the combined node DE with probability 0.25 is combined. In the case of Figure 5.14, nodes B (probability 0.25) and C (probability 0.20) are combined, whereas in Figure 5.17, already combined node DE (probability 0.25) is combined with C. This differing tree structure will result in different codeword assignments. As it happens, the average codeword length is still the same – but it should be clear that many trees are possible due to the multitude of ways in which nodes may be combined.

However, Figure 5.18 represents a different case, resulting in a shorter average codeword length. How can we construct the tree to ensure that the outcome is always a set of codes with the lowest average codeword length? A simple

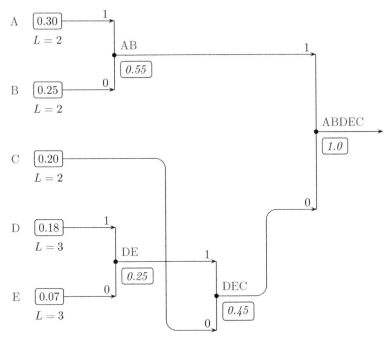

Figure 5.17 Huffman code generation, using an alternative grouping. Notice that at each stage, the two lowest probabilities are combined into a new interior node.

rule is to *only combine the pairs having the lowest probability at each itera-tion*. This is sometimes called the *sibling property* (Gallager, 1978). Comparing Figures 5.14 and 5.18, we can apply the average codeword length equation to obtain the average codeword lengths as follows.

```
% the probabilities
p = [0.30 0.25 0.20 0.18 0.07];

% codeword length, with and without sibling property
% maintained
nsib = [2 2 2 3 3];
nnosib = [1 2 3 4 4];

% average codeword length - sibling nodes combined
sum(p.*nsib)
ans =
      2.2500
% average codeword length - non-sibling nodes combined
sum(p.*nnosib)
      2.4000
```

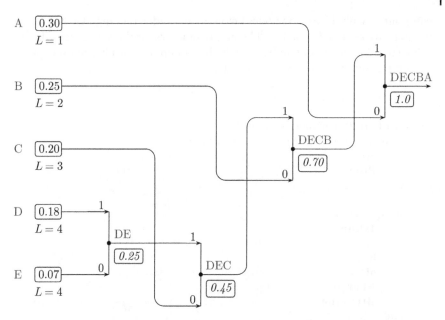

A [0.30]
$L = 1$

B [0.25]
$L = 2$

C [0.20]
$L = 3$

D [0.18]
$L = 4$

E [0.07]
$L = 4$

DE
[0.25]

DEC
[0.45]

DECB
[0.70]

DECBA
[1.0]

Figure 5.18 Huffman code generation, when nodes of the lowest probability at each stage (sibling nodes) are *not* joined in order. The average codeword length is 2.40 bits/symbol.

Thus, in the tree with the sibling property maintained, the average length for an arbitrarily long message will approach 2.25 bits per symbol. However, for the other tree, which was constructed without following the sibling convention, the average codeword length is a little longer, at 2.40 bits per symbol. The difference of 0.15 bit per symbol may be small, but over a large encoding block, this small difference may become significant. Moreover, in a typical coding scenario, the number of source symbols would be significantly larger (in the hundreds) than the five-symbol alphabet considered in this simple example.

Note that there is still some possible ambiguity, as in the two Huffman codes constructed with the same average probability. In that case, there were two nodes with equal probabilities of 0.25, and the order of combination of them is arbitrary. Naturally, both encoder and decoder must follow the same decision logic when confronted with such a situation.

Construction of a Huffman code is quite tedious for anything other than a very small example. The following illustrates how the tree construction, encoding, and decoding may be performed in MATLAB. The code uses the handle operator to create a reference (or pointer) to a node (data structures and referencing in MATLAB are explained further in Section 4.3.3). Each node has a probability and symbol associated with it (the internal nodes are assigned a compound symbol from the nodes below, to aid understanding). Up and down pointers are assigned, which will be empty pointers for the leaf nodes. The bit value stored in each node indicates whether an up or down branch was taken from the parent node to reach this node.

```matlab
% Huffman node class. Returns a handle (pointer) to the data
% structure
classdef HuffmanNode < handle
    properties (SetAccess = private)
        Sym             % symbol for this node
        Prob            % probability of this symbol
    end

    properties
        IsRoot          % the root node
        IsLeaf          % leaf or branch node
        hUp             % up pointer, branch if 1
        hDown           % down pointer, branch if 0
        hParent         % parent node
        BitValue        % 1=up, higher prob, 0=down, lower prob
    end

    methods
        function hNode = HuffmanNode(IsRoot, IsLeaf, Sym,
        Prob, hUp, hDown)
            hNode.IsRoot = IsRoot;
            hNode.IsLeaf = IsLeaf;
            hNode.Sym = Sym;
            hNode.Prob = Prob;

            if( nargin == 4 )
                hUp = HuffmanNode.empty;     % up & down
                %pointers not given,
                hDown = HuffmanNode.empty; % so initialize
                %to empty
            end

            hNode.hUp = hUp;
            hNode.hDown = hDown;
            hNode.hParent = HuffmanNode.empty;
            hNode.BitValue = '-';
        end
    end
end
```

The Huffman tree itself is composed of nodes; the root node is saved as it is the point where searches begin. The nodes may be leaf nodes (with a symbol and probability) or nodes where two branches are joined.

```
classdef HuffmanTree < handle
    properties (SetAccess = private)
        hRootNode        % the root node itself
        hAllNodes        % list of all the nodes
    end

    methods
        % constructor
        function hTree = HuffmanTree(Syms, Probs)
            hRootNode = hTree.CreateTree(Syms, Probs);
        end

        % add tree methods here

    end % end methods
end
```

Descending the tree (for decoding) from the root node consists of reading each new bit in turn and taking the appropriate branch to the upward or downward node. The search terminates at a leaf node.

```
function [DecSym] = Descend(hTree, BitStr)
    BitsGiven = length(BitStr);
    hCurrNode = hTree.hRootNode;

    BitsUsed = 0;
    while( ~hCurrNode.IsLeaf )
        BitsUsed = BitsUsed + 1;
        if( BitsUsed > length(BitStr) )
            % decoded symbol
            DecSym = [];

            fprintf(1, 'Descend(): not enough bits\n');
            return;
        end

        CurrBit = BitStr(BitsUsed);
        fprintf(1, 'Currbit %c\n', CurrBit);
        if( CurrBit == '1' )
            % up
            hCurrNode = hCurrNode.hUp;
        else
            hCurrNode = hCurrNode.hDown;
```

```
            end
        disp(hCurrNode);
    end
    fprintf(1, 'At Leaf, symbol "%s" bits used=%d (given %d)\n', ...
        hCurrNode.Sym, BitsUsed, BitsGiven);

    if( BitsUsed == BitsGiven )
        DecSym = hCurrNode.Sym;
    else
        DecSym = [];
        fprintf(1, 'Error: bits used %d, bits given %d\n', ...
            BitsUsed, BitsGiven);
    end
end
```

Ascending the tree (for encoding) starts at the leaf node for the given symbol. Each parent pointer is taken in succession, and the bit at each stage is accumulated for the resulting bit string. When descending the tree, this bit string must be reversed.

```
function [BitStr] = Ascend(hTree, Sym)
    % find starting node in list of leaf nodes
    NumNodes = length(hTree.hAllNodes);
    fprintf(1, 'Searching %d leaf nodes\n', NumNodes);
    StartLeaf = 0;
    for n = 1:NumNodes
        if( hTree.hAllNodes(n).IsLeaf )
            if( hTree.hAllNodes(n).Sym == Sym )
                StartLeaf = n;
                break;
            end
        end
    end

    if( StartLeaf == 0 )
        fprintf(1, 'Ascend() error: cannot find symbol in
                   % nodes\n');
        BitStr = [];                % signals error
        return;
    end

    % starting leaf node for the given symbol
    hCurrNode = hTree.hAllNodes(StartLeaf);
    BitStr = [];        % accumulated bit string from leaf to
                        % root
    while( ~isempty(hCurrNode) )
        disp(hCurrNode);
```

```
            if ( ~hCurrNode.IsRoot )
                % accumulate bits along the way
                BitStr = [BitStr  hCurrNode.BitValue];
            end

            hCurrNode = hCurrNode.hParent;
        end
end
```

Actually creating the Huffman tree is the most complicated step. First, the given symbols are allocated to leaf nodes. Then, all nodes that are candidates for pairing are determined and sorted in increasing order of probability. Note that nodes in this search may be leaf nodes or internal nodes and that once a node has been combined, it is flagged as not being available anymore. The comments show where each pair could simply be selected in the order they turn up, but this would not guarantee correct sibling ordering. Combining according to the sibling order requires sorting according to the probability of each node.

Once the next pair of nodes to be combined is determined, a new parent node is created. The upward pointer from each child node is set, as well as the up/down pointers to each child from the parent. The process repeats until all nodes have been combined, at which point the root node is reached.

```
function hRootNode = CreateTree(hTree, Syms, Probs)
    % create leaf nodes
    Paired = []; % zeros(N, 1);
    N = length(Syms);
    for n = 1:N
        IsRoot = false;
        IsLeaf = true;
        hNewNode = HuffmanNode(IsRoot, IsLeaf, Syms(n),
        Probs(n));
        hTree.hAllNodes = [hTree.hAllNodes hNewNode];
        Paired = [Paired false];
    end

    % combine nodes in pairs. There are N-1 pairing nodes
    for n = 1:N-1

        PairList = [];
        for k = 1:length(hTree.hAllNodes)
            if ( ~Paired(k) )
                % candidate for pairing
                PairList = [PairList k];
            end
        end

        % Just select the first two.
        % Does not preserve the sibling ordering, and so
```

```matlab
% is not guaranteed to generate the shortest possible
% codewords.
iup = PairList(1);
idown = PairList(2);

% A better method: sort in order of increasing
% probabilities, and then select the two lowest
% for combining.
ProbVals = [];
for p = 1:length(PairList)
    i = PairList(p);
    ProbVals = [ProbVals hTree.hAllNodes(i).Prob];
end

% sort all node probabilities in ascending order
% return the index of the ordering
[SortedVals, SortIdx] = sort(ProbVals);

% From the pair list, need to select those two with
% the lowest probability. This is the first two
% of the sort index list.
iup = PairList(SortIdx(2));
idown = PairList(SortIdx(1));

% flag as having been combined
Paired(iup) = true;
Paired(idown) = true;

fprintf(1, 'selected up  : %d, sym:"%s" prob:%f\n', ...
    iup, hTree.hAllNodes(iup).Sym, hTree.hAllNodes(iup).Prob);
fprintf(1, 'selected down: %d, sym:"%s" prob:%f\n', ...
    idown, hTree.hAllNodes(idown).Sym, hTree.hAllNodes(idown).Prob);

% create parent node with sum of probabilities
ProbSum = hTree.hAllNodes(iup).Prob + hTree.hAllNodes
(idown).Prob;

% create a fake node name by combining the child
% node names
NodeSym = [hTree.hAllNodes(iup).Sym  hTree.hAllNodes
(idown).Sym];

IsRoot = false;
IsLeaf = false;

hNewNode = HuffmanNode(IsRoot, IsLeaf, NodeSym, ProbSum, ...
                       hTree.hAllNodes(iup), ...
                       hTree.hAllNodes(idown));

% child nodes point to parent
hTree.hAllNodes(iup).hParent = hNewNode;
hTree.hAllNodes(iup).BitValue = '1';

% parent node up & down point to children
hTree.hAllNodes(idown).hParent = hNewNode;
hTree.hAllNodes(idown).BitValue = '0';
```

```
        fprintf(1, 'created new node with prob %f\n', ProbSum);
        fprintf(1, 'Parent is: \n');
        disp(hNewNode);

        % save new parent in list of all nodes; not paired
        % yet
        hTree.hAllNodes = [hTree.hAllNodes  hNewNode];
        Paired = [Paired false];

        if( n == N-1 )
            % this occurs only when the very last pair is
            % combined, which by definition is the root node
            hNewNode.IsRoot = true;
            hRootNode = hNewNode;
            fprintf(1, 'Root node saved\n');
            hRootNode.disp();
        end
    end

    % save root node in tree structure itself
    hTree.hRootNode = hRootNode;
end
```

To test the Huffman tree code, we can assign a test as outlined in the figures given. The symbols and their corresponding probabilities are first defined, then the tree created:

```
Syms = ['A' 'B' 'C' 'D' 'E'];
Probs = [0.3 0.25 0.20 0.18 0.07];

hTree = HuffmanTree(Syms, Probs);
```

Encoding a letter is performed from leaf to root using `Tree.Ascend()`, while decoding a bitstring is performed using `Tree.Descend()` as follows:

```
Sym = 'A';
BitStr = hTree.Ascend(Sym);
fprintf(1, 'Symbol "%s" encoded as bit string "%s"\n',
        Sym, BitStr) ;

% reverse order
TransBitStr = fliplr(BitStr);

BitStr = '101';
DecSym = hTree.Descend(BitStr);
fprintf (1, 'Bit string "%s" decodes to symbol "%s"\n',
        BitStr, DecSym ) ;
```

5.6.1.3 Adapting the Probability Table

The preceding method of building a Huffman tree works well, and all that is necessary is for the encoder and decoder to store an identical Huffman tree. However, storing this tree presupposes that the symbol probabilities are fixed and known in advance. Depending on the particular situation, this may not be the case. Furthermore, the probabilities may change over time, as a data stream is being encoded. For example, part of a document may contain text at first, followed by images.

The probability table is clearly the key to efficient encoding. If we have an accurate set of statistics, we can make efficient codes. The Huffman method as described uses a fixed or static table – it is set at the time of design. But what if we changed the probabilities as encoding progresses? We do, after all, have the incoming source symbols, and the encoder could maintain a frequency table. The decoder could also maintain such a table, since it receives and decodes the same set of symbols as the source.

The Huffman tree could be constructed for each new compression requirement. This may be suitable for compressing a data file, where one pass over the data to compute the frequency table is done, followed by encoding using that table. The table itself needs to be prepended to the data file (or transmitted), so that the decoder can create its own identical Huffman coding table. However, this is not ideal for telecommunications, as the source data may not all be available at the start. As a result, an adaptive Huffman encoder is required, as proposed in Gallager (1978). This maintains an adaptive count of source symbols as encoding progresses, which is ideal for a telecommunications system where the symbol probabilities are not known beforehand and, furthermore, may change as encoding progresses. The entire tree does not need to be reconstructed, only adapted with some nodes interchanged if their relative probabilities change sufficiently.

Of course, the decoder must be kept synchronized at all times, otherwise incorrect decoding may occur. As each symbol is received, it could be encoded and the table updated. The decoder would receive the bitstream, decode the output, and update the table. Note that the order is critical here: The encoder must not update its own table until *after* it has encoded the symbol, because otherwise the decoder's Huffman tree will potentially be different from the encoder's, and thus the transmission will lose synchronization. In effect, the encoding tree will necessarily be one symbol out of date, so as to match the decoder. Note that it would not be necessary to completely reconstruct the Huffman tree on each symbol. Looking at the Huffman trees as presented, it is clear that the update only needs to occur when one symbol probability increases to the point where it exceeds that of its sibling.

5.6.2 Block-based Lossless Encoders

The previous types of encoders work well, for encoding single symbols individually. But what if there is dependency between adjacent symbols? As a simple example, suppose we are to compress an image on a screen, broken down into colors. If we take the image one row at a time and try to compress it, we might find that there are long runs of the same color of pixel. This may give us an idea: Instead of encoding one bit pattern for a particular color, we could encode a block or run of the same color as the pair of symbols (pixel color, runlength). That way, we could exploit redundancy that extends beyond the occurrence of individual symbols.

Note that the color itself may in fact not be a single byte: It may possibly be a triplet representing red, green, and blue components. Thus, the "symbol" we are encoding might actually be a set of color values. The run length value that we are encoding could be one byte or more (or perhaps less). Clearly the limitation would be the size of this field: If we had, say, a 4-bit field for the run length, then the run value would be limited to 16 pixels at a time. There may also be an efficiency tradeoff: For individual pixels that are of a different color to their predecessors, it would still be necessary to encode a length of one. Thus, in the worst case, the runlength output might actually produce data expansion.

This approach could be made more generic, and extended to text, by considering not individual letters but groups of letters. For the purpose of explanation, it is probably easier to think of these as "words." So in compressing the text here, some words occur more frequently than others.

When compressing data, it is clearly necessary for the decoder to keep in synchronization with the encoder. This seems obvious, but careful design of the algorithm is necessary to ensure that this is always so. Furthermore, errors in the transmission can lead to catastrophic run-on effects. For example, if a run length of 45 was inadvertently received as 5 due to a single-bit error, an image would become distorted.

Several block-based lossless encoders will be introduced below. In what follows, the term "string" is taken to mean a block of symbols. This might be an English word, but that is not necessarily the case. In fact, to a compression algorithm, the notion of "word" is entirely arbitrary. Often the term "phrase" is used as an alternative. The encoders emit particular values, such as the (run, length) pair in the example above, and the term "token" is often used for individual entities or a set of values. The context usually makes that clear.

5.6.2.1 Sliding-Window Lossless Encoders

A well-known approach to encoding runs or blocks of symbols is commonly known as the Lempel–Ziv (or simply LZ) algorithms. In fact, this is not one algorithm, but a family of related algorithms with several variations.

Consider the encoding of text as you see here. At some point of time, the decoder has access to the most recently decoded text, and the encoder can see

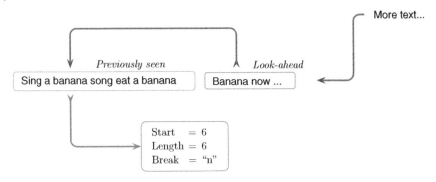

Figure 5.19 Lempel–Ziv window-style compression. To encode "banana now" we need the index 6, which is the starting offset in the previously encoded window (ignoring spaces for clarity). The length also happens to be 6. The next byte is "n" (again, ignoring the whitespace).

a block of text that has not yet been encoded. Given a block of data in the past, it is arguably likely that future data will contain the same, or very similar, runs of symbols. The simplistic limit of this is the runlength encoder, but that would not work very well for, say, English text.

The recency effect in data streams may be exploited in the following way. Both the encoder and decoder are required to maintain a block of recently decoded text, as shown in Figure 5.19. The encoder looks at the next block of text it has in front of it to be encoded. With any luck, a phrase in the future block will have appeared in the past. Thus, in order to communicate the next block of symbols to the decoder, it is first necessary to search the previously encoded text for the same pattern. The index of the pattern is then sent, together with the length of the pattern. Ideally, the longer the pattern the better, because this encodes more symbols at once. The algorithm proceeds by finding a match, and then searches for a longer match, and a longer match, and so on. At some point, the pattern matching will cease. The decoder needs to be told of the last symbol that broke the match.

Thus, it is necessary to transmit the triplet (index, length, symbol), where the last symbol is the code for the nonmatching symbol encountered. The decoder's job is then to copy *length* bytes from the previously seen text, starting at the given *index* and then to append the nonmatching symbol. At this point, the encoder and decoder can slide the window of previously encoded text along and repeat the procedure.

As design parameters, it is clearly necessary to determine the length of the previously seen buffer, for this determines the number of bits in the index parameter. It is also necessary to decide on the number of symbols in the lookahead buffer at the encoder, as this will determine the number of bits required for the length parameter. In general, these are not the same, because a substantial previous data window is required to maximize the chances of finding a matching pattern. For example, if the previously encoded block was 4 kbyte and the lookahead is 16 bytes, the index would require 12 bits and the length 4 bits. The requirement for each encoded symbol emitted would then be $12 + 4 + 8 = 24$ bits. If the average pattern length was three and the pattern was always seen in the encoding buffer, the encoding would about break even as compared with straightforward transmission of the raw byte stream. However, if the average pattern found was longer, a net reduction in the output data would result.

This family of encoders is generally termed an LZ77 algorithm (Ziv and Lempel, 1977), although there are many variants. A postprocessing stage could also be added, to utilize (for example) Huffman encoding on the (index, length, symbol) tokens encoded or one of the components (for example, the symbol only).

As well as tuning the optimal length of the buffers, computational issues may present a challenge. Each new symbol read in requires a search in the encoder's buffer, and this may take time. Various data structures, such as tree-based partitioning, may be used to speed the encoding search. The decoder, by comparison, has a simple index and copy requirement. This type of arrangement is sometimes referred to as *asymmetric*, since the complexity of the encoder is much greater than the decoder.

5.6.2.2 Dictionary-based Lossless Encoders

A related approach, which uses a table of patterns, is the LZ78 family of algorithms (Ziv and Lempel, 1978). In this case, a table (usually referred to as a *dictionary*) is employed, rather than a sliding window. This is illustrated in Figure 5.20. It is important to understand that the term "dictionary" does not refer to English words, but it is conventional to refer to the table as such. In the most general case, the entries in the dictionary are just a byte pattern.

Dictionary encoding works by maintaining a dictionary at both encoder and decoder. Encoding consists essentially of searching the dictionary for the longest matching phrase in the dictionary and transmitting that index to the decoder. Thus, the length field of LZ77 may be dispensed with, since the length is implicit in each entry. In searching for the longest matching phrase in the encoder, the match must be broken at some point. Because the decoder needs to build up an identical dictionary, it is necessary to send the byte encoding of the first character that breaks the match, similar to LZ77. Both encoder and decoder then build up another phrase, based on the phrase just encoded, followed by the new symbol. This then becomes a prefix for subsequent encoding. Nonmatching characters become the suffix that is appended to existing prefixes, and thus the dictionary is incrementally built up.

One potential for improvement would be the encoding of the nonmatching character. The Lempel–Ziv–Welch (LZW) variant solves this problem in an elegant way (Welch, 1984). It simply begins with a dictionary whose first entries

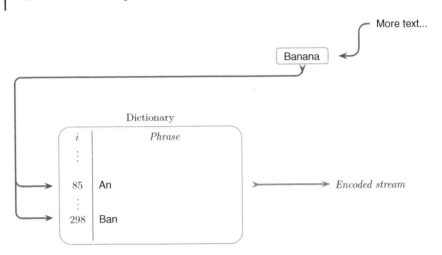

Figure 5.20 Lempel–Ziv dictionary-style compression. The longest match in the dictionary illustrated is "ban," followed by "an." The encoder and decoder could then add the phrase "bana" to their respective dictionaries. Future encodings of "banana" will then be more efficient, since the phrases "banan," then "banana" will be built up each time "banana" is encountered in the input stream.

are populated with all the standard characters to be encoded. Thus, if we were encoding text only, the dictionary would include "A", "B", "C", …, "Z." In this way, *any* previously unseen character could be encoded via a dictionary index. All that is needed to transmit is a sequence of indexes, with this covering both single symbols and any other strings as the dictionary is built up.

For example, if XABC and XYZ are in the dictionary, then the input XABCXYZ would result in output of a code for XABC, a code for XYZ, and an update to the dictionary of the new phrase XABCX. It might seem that updating the dictionary in this way is inefficient, but after some amount of data has been compressed, the dictionary will become populated with common phrases, and the encoding becomes more and more efficient.

An issue with LZ78 is that the dictionary will eventually fill up. This may take some time but invariably will occur at some point. Several strategies have been proposed for dealing with this eventuality. The simplest is to just erase the entire dictionary (apart from the single-symbol entries in LZW) and start again. But it may be a pity to throw away what is, in effect, the recent history of likely phrases. A count could be employed to indicate the frequency of use for each entry (so-called LFU, or *Least Frequently Used*, approach). But some patterns that are often used (and hence have a high usage count) may have occurred some time ago. Words used in the previous sections of this text, for example, may be used in subsequent sections, but the relative frequency might decrease to zero for some words eventually. Thus, the LRU or *Least Recently Used* approach may be preferable. Consider the example of encoding a list of phone numbers: Since the index is sorted in alphabetic order, then names (which are just phrases to the compression algorithm) would exhibit a very strong locality effect. Street names, though, would have a very weak locality pattern.

A subtle problem is related to repeated phrases; this is best explained by the example. Suppose XABC was in the dictionary of both encoder and decoder and that the input string is XABCXABCX. The encoder would output the code for XABC, add XABCX to its dictionary, and then start trying to find a match starting at the second X. It would find XABCX in the dictionary and output this code. But at that point, the decoder would not have XABCX in its dictionary – it is still waiting on the code for the symbol that follows the initial XABC. Such an exception must be handled so as to ensure correct decoding.

A large number of lossless data compression algorithms exist and are in use for many data transmission and storage applications. One example is LZO (Oberhumer, n.d.), used for compression on the Mars Rover. Another is the BWT (Burrows and Wheeler, 1994), used in some public domain data compression programs. The choice of algorithm in any particular situation depends on the data compression required, the complexity (and hence time taken to compress), and memory available.

5.6.3 Differential PCM

The previous sections dealt with lossless encoding – the original data is always transmitted such that the receiver can recreate an exact replica of the data. However, in many situations involving real-world sampled data such as speech and images, it is not necessary to ensure an exact reconstruction. This approach can lead to very large reductions in bit rate. The fact that the reconstruction is not exact leads to these methods being termed *lossy* – some information is lost.

The first and simplest method is *Differential PCM*. The term PCM stands for *Pulse Code Modulation* and refers to sampling a signal and transmitting the sampled value in binary form. The receiver can convert the binary-weighted code back into an analog voltage, which is an approximation of the original. It does not matter whether it is a pixel intensity, a sound sample, or some other analog information source. The number of levels (and hence bits) required for PCM depends on the data type and expected resolution or quality of the recon-struction, but as a rule between 8 and 16 bits are required. Since this number of bits must be transmitted with each sample, it effectively acts as a multiplier on the number of bits per second. That is, bits per second equates to samples per second, times bits per sample.

However, there is usually a significant similarity between successive samples, leading to some degree of redundancy. This in turn implies that there is some predictability in sample values: One or more previous samples may be used to predict the value of the next sample. At the simplest level, the value of one sample at instant n is a good predictor of the sample at instant $n + 1$.

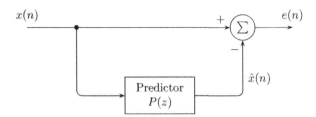

Figure 5.21 A simplified differential encoder, without quantization.

But how does this help reduce the bit rate? The basic premise is that, if the prediction is good, the amount of prediction error (the actual sample value, minus the prediction) will be a small number (ideally zero). This is the differential part in DPCM. If the prediction error is small, it is possible to use fewer bits to transmit the prediction error, rather than the actual sample value itself. Then, the decoder performs essentially the inverse operation: It forms its own prediction, accounts for the error, and thus reconstructs the original value.

5.6.3.1 Sample-by-sample Prediction
Let us first look at the larger picture, assuming that the prediction is, on average, fairly good. The situation shown in Figure 5.21 applies. A mathematical model is to imagine the true signal $x(n)$ as being an estimated signal $\hat{x}(n)$ plus some error amount,

$$x(n) = \hat{x}(n) + e(n) \tag{5.48}$$

The error $e(n)$ could be positive or negative at each sample instant n. The source signal $x(n)$ is used as an input to the predictor function in Figure 5.21. This prediction is then subtracted from the true value to form an error signal, so that

$$e(n) = x(n) - \hat{x}(n) \tag{5.49}$$

These two equations are, of course, equivalent. The receiver decodes the value by performing the inverse operation; this is shown in Figure 5.22. It forms the prediction based on the samples at the output (that is, the samples presented for reconstruction). The error signal is received from the channel, and this is added to the prediction. Mathematically, the receiver adds the prediction to the error:

$$x(n) = \hat{x}(n) + e(n) \tag{5.50}$$

There is, however, one slight complication with this approach – it neglects the effects of quantization. Recall that quantization reduces the representational accuracy from what is theoretically an infinite number of values, down to a finite number of values as dictated by the number of bits used. Thus, referring

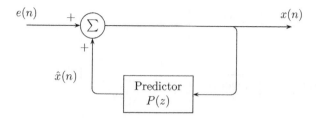

Figure 5.22 A differential decoder. The output is based on the prediction formed at the decoder, added to the difference (prediction error) values received over the channel.

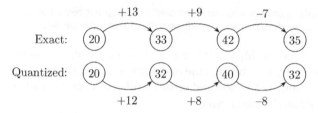

Figure 5.23 Prediction sequence, with and without quantization.

to Figure 5.22, the prediction at the decoder must be based on what it reconstructs as $\hat{x}(n)$, rather than the true value $x(n)$. After all, the receiver has no knowledge of the true signal that the encoder sees. Similarly, the error known at the decoder is $\hat{e}(n)$ rather than $e(n)$. Thus, we must modify our prediction to be based on these inexact samples.

If the calculated prediction is now denoted as $\tilde{x}(n)$ and the estimates available to the decoder as $\hat{x}(n)$, then based on the quantized error $\hat{e}(n)$,

$$\hat{x}(n) = \tilde{x}(n) + \hat{e}(n) \tag{5.51}$$

A simple example serves to illustrate this point. Suppose we have sample values 20, 33, 42, and 35 as shown in Figure 5.23. The differences are shown above each step to be +13, +9, −7. Now suppose we are constrained to quantize the errors to be a multiple of 4. That is, 0, ±4, and ±8 are the only allowable error values. Starting at 20, the quantized prediction errors are +12, +8, −8. As shown, the resulting sequence of 20, 32, 40, 32 does not precisely match the original samples. Of course, due to quantization, we will never precisely match the values. However, we want to be as close as possible, and certainly do not want the resulting values to diverge from the true values over many sample steps.

This may not appear to be a significant problem over a small run of samples, but over a larger number, the errors could accumulate. Since the decoder only has access to the quantized values and can only base its prediction on past output samples that it has itself calculated, the prediction should only be based on that premise. So we can modify the encoder block diagram as shown in

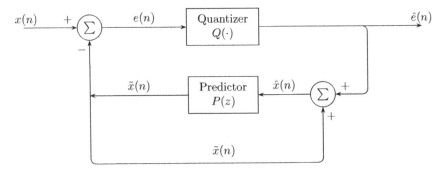

Figure 5.24 A DPCM encoder using quantization in the prediction loop. It is best if the prediction is based on what the decoder knows about, not what the encoder can see.

Figure 5.24 to take account of this fact. Careful comparison with the previous figures shows that we have embedded the operation of quantization in the loop, and although the same prediction and error calculation is performed, it is performed *based only on the quantized error*.

A reasonable question to ask is how to form the prediction, in the best possible way. Of course, we can only go on past samples. One idea is to form an average of previous samples. We can generalize that further, by introducing a *weighted linear sum*. Neglecting the quantization, for an order P predictor, the prediction formed as a weighted sum is

$$\hat{x}(n) = a_1 x(n-1) + a_2 x(n-2) + \cdots + a_p x(n-P)$$

$$= \sum_{k=1}^{P} a_k x(n-k) \tag{5.52}$$

where a_k is the kth prediction coefficient. This means that sample $x(n)$ is predicted using one or more previous samples. The simplest case, as used in the earlier numerical example, was just to use the previous sample, unaltered. In the above equation, this would correspond to a prediction of order $P = 1$ with the weighting coefficient $a_1 = 1$. As it turns out, this is not a bad prediction for image data. Several questions arise from this. First, in what sense do we mean "best" or "optimal" prediction? In that case, how do we calculate the best coefficient values a_k? And what is the best prediction order P?

The first questions – predictor optimization criteria and predictor coefficient values – are closely related and best addressed together. The predictor order is less clearcut – for an increasing order of prediction, we generally find that a better estimate results, but it is often a case of diminishing returns. Higher orders sometimes only yield marginally better prediction. It depends on the nature of the data to be encoded.

To address the problem of solving for the predictor coefficients, we first need to state the criteria. We want to have the lowest average error over some length of samples, and the error could be positive or negative. Using the squared error between predicted and actual values works well and also makes it possible to derive the solution mathematically. Consider a simple, first-order predictor again and the approach to determine the best value for the predictor coefficient a_1 (that is, the order is $P = 1$). The predicted value is formed as

$$\hat{x}(n) = a_1 x(n-1) \tag{5.53}$$

The prediction error in that case is

$$
\begin{aligned}
e(n) &= x(n) - \hat{x}(n) \\
&= x(n) - a_1 x(n-1)
\end{aligned}
\tag{5.54}
$$

To remove positive and negative errors, we can take the instantaneous squared error:

$$e^2(n) = [x(n) - a_1 x(n-1)]^2 \tag{5.55}$$

This gives the error at one sample instant. But the signal is changing over time. So the strategy is to average this error over a block of N samples. The average squared error is

$$
\begin{aligned}
\overline{e^2} &= \frac{1}{N} \sum_n e^2(n) \\
&= \frac{1}{N} \sum_n [x(n) - a_1 x(n-1)]^2 \\
&= \frac{1}{N} \sum_n [x^2(n) - 2x(n)\, a_1\, x(n-1) + a_1^2\, x^2(n-1)]
\end{aligned}
\tag{5.56}
$$

This looks complicated, but remember that we are using the known sample values $x(n), x(n-1), \ldots$, and we wish to determine the value of a_1. We want the minimum average error, and since we have effectively got a polynomial in x, we can take the derivative with respect to a_1:

$$\frac{d\,\overline{e^2}}{d\,a_1} = \frac{1}{N} \sum_n [0 - 2x(n)x(n-1) + 2a_1\, x^2(n-1)] \tag{5.57}$$

Then, as with any minimization problem, set the derivative equal to zero:

$$\frac{d\,\overline{e^2}}{d\,a_1} = 0 \tag{5.58}$$

This gives an equation where the predictor value a_1 is actually the optimal predictor value, which we denote as a_1^*:

$$\frac{1}{N} \sum_n x(n)x(n-1) = a_1^* \frac{1}{N} \sum_n x^2(n-1) \tag{5.59}$$

Solving for a_1^* gives

$$a_1^* = \frac{1/N \sum_n x(n)x(n-1)}{1/N \sum_n x^2(n-1)} \tag{5.60}$$

In theory, we need to sample the signal forever to do this in a mathematical sense. However, in practice, we can get away with updating the predictor periodically and use a finite-sized block of N samples. This makes sense because the signals actually change over time – consider the scanned pixels of an image or sampled speech data. The statistical characteristics are approximately constant over a small interval of time, not forever. So by taking the summation over a large number of samples, we may use *autocorrelations* defined as

$$R(0) = \frac{1}{N} \sum_n x^2(n) \tag{5.61}$$

$$R(1) = \frac{1}{N} \sum_n x(n)x(n-1) \tag{5.62}$$

So the solution becomes

$$a_1^* = \frac{R(1)}{R(0)} \tag{5.63}$$

For typical image data, the similarity means that $R(1)$ is only a little less than $R(0)$, and we find predictors of the order of 0.8–0.95 give good predictions with small errors (compare this with the value of one that we assumed earlier).

To make a better prediction, we could extend the weighted averaging over more previous samples. For a second-order predictor,

$$\hat{x}(n) = a_1 x(n-1) + a_2 x(n-2)$$
$$\therefore e(n) = x(n) - \hat{x}(n)$$
$$= x(n) - [a_1 x(n-1) + a_2 x(n-2)]$$
$$\therefore e^2(n) = \{x(n) - [a_1 x(n-1) + a_2 x(n-2)]\}^2 \tag{5.64}$$

Again, over many samples, the average square error is

$$\overline{e^2} = \frac{1}{N} \sum_n e^2(n)$$
$$= \frac{1}{N} \sum_n \{x(n) - [a_1 x(n-1) + a_2 x(n-2)]\}^2 \tag{5.65}$$

This time, however, we have an optimization problem involving two variables, a_1 and a_2. We then use partial derivatives to optimize each separately

$$\frac{\partial \overline{e^2}}{\partial a_1} = \frac{1}{N} \sum_n \{2[x(n) - (a_1 x(n-1) + a_2 x(n-2))] \times [-x(n-1)]\} \tag{5.66}$$

and again set to zero

$$\frac{\partial \overline{e^2}}{\partial a_1} = 0$$

The optimal predictors are again denoted by $*$, so $a_1 \to a_1^*$ and $a_2 \to a_2^*$, and we have

$$\frac{1}{N} \sum_n \{2[x(n) - (a_1^* x(n-1) + a_2^* x(n-2))] \times [-x(n-1)]\} = 0$$

Rearranging,

$$\frac{1}{N} \sum_n x(n)x(n-1) = a_1^* \frac{1}{N} \sum_n x(n-1)x(n-1)$$
$$+ a_2^* \frac{1}{N} \sum_n x(n-1)x(n-2)$$

Using the definition for autocorrelation $R(\cdot)$,

$$R(1) = a_1^* R(0) + a_2^* R(1) \tag{5.67}$$

So, we now have one equation but two unknowns. However, we can likewise find partial derivatives with respect to a_2^* to give

$$R(2) = a_1^* R(1) + a_2^* R(0) \tag{5.68}$$

Now, we have two equations in two unknowns. This may be easier to see if written in matrix form,

$$\begin{pmatrix} R(1) \\ R(2) \end{pmatrix} = \begin{pmatrix} R(0) & R(1) \\ R(1) & R(0) \end{pmatrix} \begin{pmatrix} a_1^* \\ a_2^* \end{pmatrix} \tag{5.69}$$

or simply

$$\mathbf{r} = \mathbf{R}\,\mathbf{a}^* \tag{5.70}$$

where $\mathbf{r}$ and $\mathbf{R}$ are calculated from the available data samples $x(n)$. Then, the vector of optimal predictor coefficients $\mathbf{a}^*$ may be determined by inverting matrix $\mathbf{R}$ and solving the resulting matrix equation. The error for each sample is then

$$e(n) = x(n) - \overbrace{\sum_{k=1}^{P} a_k x(n-k)}^{\hat{x}(n)} \tag{5.71}$$

An example of the application of this approach is shown in Figure 5.25. We start with a known test case, whose data is generated by the equation $x(n) = 1.71x(n-1) - 0.81x(n-2)$. For a random input to this system, the least-mean-square prediction approach implemented using the following MATLAB code yielded coefficients $a_1 = 1.62$ and $a_2 = -0.74$. Naturally,

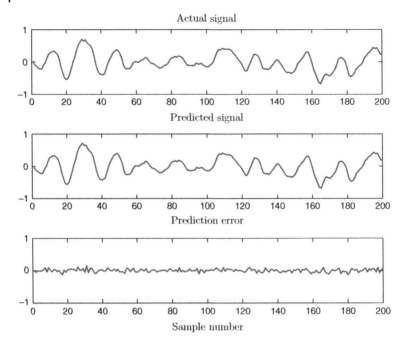

Figure 5.25 Linear prediction, showing the actual signal, the predicted signal, and the error. Calculating the autocorrelations over a larger block of samples will give a better prediction on average.

the coefficients are not precisely equal to that as expected, and as a result, the prediction error is not zero. However, as may be observed in Figure 5.25, the prediction aligns well with the input, and thus the error is generally quite small.

The MATLAB code below shows how it is possible to estimate the parameters of a system using this approach. It starts with a known system and uses a random input to that system in order to determine the output. The goal is to estimate the system parameters using the output alone – that is, without any knowledge of the system itself, except for an assumed order that determines the number of coefficients.

The input to the system is random noise. The calculation of the predictor parameters follows the autocorrelation and matrix inversion as outlined above. In order to predict the samples using the `filter()` function, it is necessary to form the prediction coefficients as an augmented vector am = [1 ; -a] ;. This is because `filter()` expects data of the form

$$a_1 y(n) = -a_2 y(n-1) - a_3 y(n-2) - \cdots + b_1 x(n) \tag{5.72}$$

with input x and output y, whereas the problem as formulated had

$$x(n) = a_1 x(n-1) + a_2 x(n-2) + \cdots + e(n) \tag{5.73}$$

with input being the error $e(n)$ and output $x(n)$ based on previous samples $x(n-1), x(n-2), \ldots$. The predicted value $\hat{x}(n)$ is denoted as xhat.

```
N = 200;

% poles at radius r  angle omega
r = 0.9;
omega = pi/10;
p = r*exp(j*omega);

a = poly([p conj(p)]);
roots(a)

% system input
e = 0.05*randn(N, 1);

% response to input
x = filter(1, a, e);

% calculate autocorrelations
R0 = sum(x .* x)/N;
R1 = sum(x(1:N-1) .* x(2:N))/(N);
R2 = sum(x(1:N-2) .* x(3:N))/(N);

% autocorrelation matrix & vector
R = [R0 R1 ; R1 R0 ];
r = [R1 ; R2];

% optimal predictor solution
a = inv(R)*r;

% optimal predictor parameters as a filter
am = [1 ; -a];

% estimated output
xhat = filter(1, am, e);
```

5.6.3.2 Adaptive Prediction

The method described in the previous section updates the predictors for each data block as it is encoded. To reconstruct the data at the receiver, it is necessary to have both the error samples as well as predictor parameters. So, either the predictor parameters have to be sent to the decoder separately or the decoder must calculate the predictor parameters using the previous block (which the decoder already has) and employ those for the current block. The former has the disadvantage of requiring extra bits to encode the predictor parameters, while the latter has the disadvantage of working with out-of-date information.

What about updating the predictor parameters on every sample, rather than after buffering a block of N samples? This is termed *adaptive prediction*, as opposed to the *blockwise prediction* discussed previously. Let the predictor again be defined by

$$e(n) = x(n) - \hat{x}(n)$$

$$= x(n) - \sum_{k=1}^{P} h_k x(n-k) \tag{5.74}$$

that is essentially the same as the blockwise predictor, although we use h_k for the coefficients to avoid confusion. If we allow the predictor to vary with each sample and write the coefficients as a vector $\mathbf{h}$, we have

$$e(n) = x(n) - \mathbf{h}^T(n)\mathbf{x}(n-1) \tag{5.75}$$

The vector of prediction coefficients varies over time and is

$$\mathbf{h}(n) = \begin{pmatrix} h_1 \\ h_2 \\ \vdots \\ h_P \end{pmatrix} \tag{5.76}$$

and the vector of previous samples, starting at the last one, is

$$\mathbf{x}(n-1) = \begin{pmatrix} x(n-1) \\ x(n-2) \\ \vdots \\ x(n-P) \end{pmatrix} \tag{5.77}$$

Again, this is a minimization problem. This time, however, instead of averaging over a block of N samples, we just take each sample and update the predictor. The estimate of the gradient ∇ of the predictor in the h_1 direction is the partial derivative:

$$\hat{\nabla}_{h_1} e^2(n) = \frac{\partial}{\partial h_1}[x(n) - \mathbf{h}^T(n)\mathbf{x}(n-1)]^2 \tag{5.78}$$

$$= \frac{\partial}{\partial h_1}\{x(n) - [h_1 x(n-1) + h_2 x(n-2) + \cdots]\}^2$$

$$= 2\{x(n) - [h_1 x(n-1) + h_2 x(n-2) + \cdots]\}$$

$$\times \frac{\partial}{\partial h_1}\{x(n) - [h_1 x(n-1) + h_2 x(n-2) + \cdots]\}$$

Figure 5.26 Adaptive linear prediction, illustrating how one predictor parameter converges.

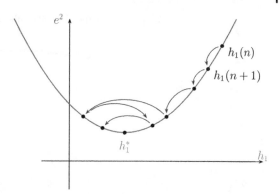

$$= 2 \ e(n) \ [-x(n-1)]$$

$$= -2 \ e(n) \ x(n-1) \tag{5.79}$$

This last line is quite easy to compute. A similar derivation gives the estimate of the gradient in the h_2 direction as

$$\hat{\nabla}_{h_2} e^2(n) = -2 \ e(n) \ x(n-2) \tag{5.80}$$

So, now we know the direction in which the error is heading, because we know the gradient. At each new sample, we aim to update the predictor **h** by a quantity proportional to the *negative* gradient of $e^2(n)$, because we want to seek the *minimum* error.

This is a critical point and is illustrated in Figure 5.26. The curve shown represents the squared error at each possible h_1 parameter setting, and the minimum error occurs at the lowest point of the curve corresponding to $h_1 = h_1^\star$. Suppose an update at step n is performed at point $h_1(n)$. The gradient or slope of the curve at this point is positive, but the value of h_1 must be reduced in order to get closer to the optimum point of minimum error at $h_1^\star$. Similarly, to the left of $h_1^\star$, the gradient of the curve is negative, but the optimum value is at a higher h value (moving to the right). This is why the update is performed in the negative gradient direction.

From the steps shown in the figure, it may be seen that the step size is also critical. As better parameter estimates occur over each iteration of the gradient algorithm, for points $h_1(n), h_1(n+1), \ldots$, the squared error reduces. The larger the step size, the faster the error reduces. However, it is possible to overshoot the minimum point as illustrated, resulting in the algorithm "hunting" either side of the minimum error at $h_1^\star$. A smaller step size would clearly be preferable in that case, but that also means that the initial convergence will be slower.

Although the preceding discussion is framed in terms of one parameter h_1, there are in fact several predictor parameters $h_1, h_2, h_3, \ldots$ and same arguments regarding convergence and step size apply to them all separately. The

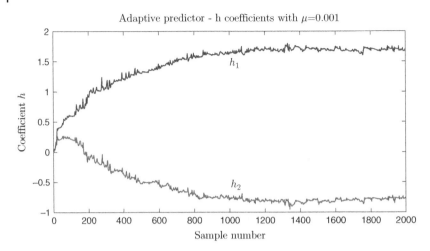

Figure 5.27 Adaptive linear prediction, showing the convergence of the predictor coefficients $\mathbf{h} = [h_1 \ h_2]^\mathsf{T}$ for a given step size parameter μ.

incremental adjustment of the predictor at each sample, for all the predictor parameters, is

$$\mathbf{h}(n+1) = \mathbf{h}(n) - \mu \ \hat{\nabla} e^2(n) \tag{5.81}$$

where μ is the adaptation rate parameter. Using the partial derivatives just found, this becomes

$$\mathbf{h}(n+1) = \mathbf{h}(n) + 2\mu \ e(n)\mathbf{x}(n-1)) \tag{5.82}$$

In expanded form this is

$$\begin{pmatrix} h_1(n+1) \\ h_2(n+1) \\ \vdots \\ h_P(n+1) \end{pmatrix} = \begin{pmatrix} h_1(n) \\ h_2(n) \\ \vdots \\ h_P(n) \end{pmatrix} + 2 \ \mu \ e(n) \begin{pmatrix} x(n-1) \\ x(n-2) \\ \vdots \\ x(n-P) \end{pmatrix} \tag{5.83}$$

This update step occurs on every sample, rather than every block of samples as in blockwise prediction. The convergence process is illustrated for a numerical example in Figure 5.27. Here, we have a second-order predictor with two parameters h_1, h_2. The convergence toward final values is clear, but it is also evident that there is some noise associated with the reestimation of the predictor parameters.

5.7 Image Coding

Digital images take up a lot of storage space and thus can take a long time to transmit. Consider an image of dimension 1000×1000 pixels (width $\times$ height), which is not an especially high resolution. To represent each of the three primary colors (red, green, blue) with one byte each would require $3 \times 1000 \times 1000$ bytes.[1] If both dimensions were doubled, it would need $2 \times 2 = 4$ times as much space again. The problem is even more acute with video. If such a still-image sequence were to be used as a video source, replayed at 50 frames per second, around 50 times this amount would required per second. A one hour video sequence would require $3 \times 50 \times 60 \times 60 \times 1000^2 \approx 540$ GB. This is a significant amount of storage space. To transmit such a video in real time would entail 1200 Mbps (millions of bits per second).

Thus, some form of compression of the data is almost always required in practice. In the case of storage, that may be *desirable* in order to reduce the costs. In the case of bandwidth-constrained channels, compression may be *essential* to enable video transmission in real time.

Fortunately, there is a significant amount of redundancy in visual information. We can exploit that fact and reduce the amount of data required in many instances. This is because the perception of image/video content is reliant on both the display device and the human visual system (HVS). There is of course a tradeoff: Reducing the amount of information by, effectively, throwing some away will reduce the image quality. The choice of compression algorithm may then be framed in terms of whether or not this loss of quality is perceptible, or whether the degradation is permissible. So, for movie content, we may want the degradation to be imperceptible, whereas for videoconferencing calls, some loss of quality may be an acceptable tradeoff. In some situations, such as medical imaging, no reduction in quality at all is acceptable. In that case, we must revert to lossless algorithms as discussed earlier and accept that the compression achievable will inevitably be much lower.

Many of the compression algorithms to be discussed are quite effective at reducing the bitrate, while maintaining acceptable quality, however, their computational complexity – and thus the speed of processor required to perform the compression – can be significant. Generally, increased computational speed comes at a higher cost. Furthermore, higher speed invariably means greater power consumption, which is a key consideration for mobile devices. Finally, the complexity aspect also impacts memory requirements, which may not be insignificant either. Usually, images are decomposed into smaller blocks (also termed sub-blocks) of size 8×8 pixels. The reason for this is twofold. First, the similarity of pixels in an image generally extends over a small range, and thus it makes sense to choose a small range of pixels that are likely to be similar. Second, such a modest block size allows the designer to tailor computational units to buffer these blocks and work on them independently.

---

1 In common usage, a megabyte (MB) is the multiplier of (1024^2), but some argue that a megabyte should be (1000^2) and that the term mebibyte (MiB) should be used to refer to a multiplier of (1024^2).

Original   Mean 16 × 16   BTC 16 × 16

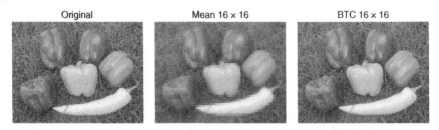

Figure 5.28 BTC example image. On the left is the original, then using the block mean only, and finally the BTC coded image for a 16 × 16 subblock. Note the blockiness evident in the mean-only image, although of course the average number of bits per pixel is quite small. With a better algorithm and transmitting more parameters, a substantially better image quality results.

5.7.1 Block Truncation Algorithm

The simplest approach to reducing the amount of data would be to average pixels over a neighborhood and only transmit the average level. There is a great deal of correlation between adjacent pixels and hence a great deal of redundancy that can be exploited. For a square block of N pixels on a side, the resulting $M = N^2$ pixels may be represented by one single value: the average. However, this leads to artificial "blockiness" in the reconstructed image. This may be seen by comparing the leftmost and middle images of Figure 5.28.

The *Block Truncation Coding* (BTC) algorithm is a simple extrapolation of the idea of using the mean of a block only. BTC, originally proposed in Delp and Mitchell (1979), preserves not only the mean but also the variance[2] about the mean, in the reconstructed block. Essentially, the mean contains the perceptually important level of average brightness, while the variance represents the average brightness variation about the mean. The aim is to determine the minimum set of parameters that can achieve equal mean and variance in the original and reconstructed subblocks. Although BTC does not provide high compression, it is instructive to consider its operation, which serves as a basis to help understand the better-performing DCT (Section 5.7.2).

In BTC, the block is reconstructed with only two levels of pixel: a if the current source pixel is less than the mean $\bar{x}$ and b if the source pixel is greater than the mean. Thus, a single bit per pixel is required to select either a or b, plus the values of a and b themselves. We must then show that it is possible to compute values for a and b that would satisfy this mean- and variance-preserving characteristic.

The image is decomposed into blocks of pixels, each represented as a matrix $\mathbf{X}$, with N pixels on a side, and thus $M = N^2$ pixels in total. Treating these

---

2 Strictly speaking, we are referring to the *sample variance*, and in MATLAB we would use `var(data, 1)`.

pixel values as a simple list, the mean $\bar{x}$, mean-square $\overline{x^2}$, and variance σ^2 are computed as

$$\bar{x} = \frac{1}{M} \sum_{i=0}^{N-1} \sum_{j=0}^{N-1} x(i,j) \tag{5.84}$$

$$\overline{x^2} = \frac{1}{M} \sum_{i=0}^{N-1} \sum_{j=0}^{N-1} x^2(i,j) \tag{5.85}$$

$$\sigma^2 = \frac{1}{M} \sum_{i=0}^{N-1} \sum_{j=0}^{N-1} [x(i,j) - \bar{x}]^2 \tag{5.86}$$

For M pixels in total and Q equal to the number of pixels greater than the mean, we can write the mean and mean square for both the original and reconstructed blocks. In order to preserve the mean and mean square in the reconstructed block, we match the mean and mean square using the equations

$$M\bar{x} = (M - Q)a + Qb \tag{5.87}$$

$$M\overline{x^2} = (M - Q)a^2 + Qb^2 \tag{5.88}$$

The left-hand side represents the original image, and the right is computed from the reconstructed image. Thus, we effectively have two equations with two unknown parameters a and b. All other values $(M, Q, \bar{x}, \overline{x^2})$ can be calculated from the source block. The solution is complicated a little because we have nonlinear terms such as a^2 and b^2. Solving the equation pair (5.87) and (5.88) yields the lower reconstructed pixel value as

$$a = \bar{x} - \sigma \sqrt{\frac{Q}{M - Q}} \tag{5.89}$$

and the higher reconstructed pixel value as

$$b = \bar{x} + \sigma \sqrt{\frac{M - Q}{Q}} \tag{5.90}$$

The following example shows the calculations involved. The pixel values are taken from a real grayscale image. A small block size of 4×4 has been utilized to illustrate, although, in practice, a larger size could be used so as to attain a lower bitrate. The original pixel (8-bit integer) block is

$$\mathbf{X} = \begin{pmatrix} 62 & 37 & 36 & 46 \\ 74 & 49 & 47 & 53 \\ 90 & 71 & 53 & 56 \\ 101 & 81 & 58 & 59 \end{pmatrix} \tag{5.91}$$

The mean of the $M = 16$ pixels is $\bar{x} = 60.81$ (with $\sigma^2 = 315.15$ and $\overline{x^2} = 4013.31$). The number of pixels Q above the mean is 6, and the bitmask is thus

$$\mathbf{B} = \begin{pmatrix} 1 & 0 & 0 & 0 \\ 1 & 0 & 0 & 0 \\ 1 & 1 & 0 & 0 \\ 1 & 1 & 0 & 0 \end{pmatrix} \tag{5.92}$$

with 1 representing values above the mean and 0 representing values below the mean.

From these numerical values, we can apply Equations (5.89) and (5.90) to calculate $a = 47.06$ and $b = 83.73$. The new subblock to reconstruct the mean and mean square is then

$$\hat{\mathbf{X}} = \begin{pmatrix} 83.73 & 47.06 & 47.06 & 47.06 \\ 83.73 & 47.06 & 47.06 & 47.06 \\ 83.73 & 83.73 & 47.06 & 47.06 \\ 83.73 & 83.73 & 47.06 & 47.06 \end{pmatrix} \tag{5.93}$$

The parameters associated with this block are $\bar{x} = 60.81$, $\sigma^2 = 315.15$, and $\overline{x^2} = 4013.31$. These are exactly the same as the original block, as expected. After rounding, the pixel values become

$$\hat{\mathbf{Y}} = \begin{pmatrix} 84 & 47 & 47 & 47 \\ 84 & 47 & 47 & 47 \\ 84 & 84 & 47 & 47 \\ 84 & 84 & 47 & 47 \end{pmatrix} \tag{5.94}$$

The reconstructed block has mean $\bar{x} = 60.88$, variance, $\sigma^2 = 320.86$, and mean square $\overline{x^2} = 4026.63$. These are close to, but not identical with, the theoretical values due to the rounding of the pixels in the last stage.

The result of using the BTC is illustrated in Figure 5.28 for a larger blocksize of 16 × 16. Using the mean of the block requires only one parameter for every 16 × 16 subblock and thus yields substantial compression (effectively a ratio of 256:1). However, the reconstructed image shows that the subblocks are quite evident and visually unappealing. Using the BTC algorithm, only two parameters plus the bitmask are required in order to encode the entire subblock. Thus, there is one bit per pixel for the bitmask, plus that required to transmit a, b. The result shows that the reconstructed image is markedly superior to the mean-only subblock reconstruction and in many respects is indistinguishable from the original, even though the bit rate is approximately 1.06 bpp (bits per pixel).

5.7.2 Discrete Cosine Transform

The BTC method examined in Section 5.7.1 appears to work well, giving a low bit rate and good quality. However, it requires a relatively large subblock to achieve low rates, leading to a greater loss of perceptual quality for the image overall. In addition, there is some loss of fidelity within each subblock.

A more advanced algorithm, the *Discrete Cosine Transform* (DCT), finds very widespread use in both image and video compression. The definition of the DCT may be traced to Ahmed et al. (1974), and subsequently it was found to perform well when used to compress images. It forms the basis of many low-rate algorithms in use today, such as MPEG for HDTV and JPEG for still images. The underlying idea is to remove as much of the statistical redundancy in image blocks as possible, and from that point of view, it is not substantially different to the BTC method discussed previously. However, the approach taken is quite different. Furthermore, the DCT is able to achieve fractional bitrates (less than 1 bit per pixel), which BTC cannot, as the latter is asymptotically limited by the requirement for the bitmask to be transmitted.

It is important to understand that the DCT itself does not produce any compression. The results of the mathematical transform, however, are able to represent the image subblocks more faithfully with fewer parameters.

Rather than starting with the mathematics, let us start with the result. We again form subblocks within the image, almost always 8×8 pixels, and work on those separately. In the following, we assume that the pixel values are simply intensities – imagine a grayscale picture. The method can be extended to color images, by processing the derived color components separately. This will be discussed later; however, for now note that there is even more redundancy in the color information, since the HVS is more sensitive to luminance variations rather than color exactness.

The first stage is to perform the transformation of the 8×8 pixel blocks. This results in the same number of coefficients. If we consider a large number of subblocks and generate the histogram over one or more images, we get something like that illustrated in Figure 5.29. This figure shows only nine coefficients, being the upper left portion of the entire 8×8 set of coefficients. The upper-left coefficient with index $(0, 0)$ is often called the "DC coefficient," since it is actually proportional to the average intensity over the block. The other coefficients – index $(0, 1)$, $(1, 0)$, $(1, 1)$, and so forth – clearly exhibit a peaked distribution. As we have seen, this type of distribution makes for better encoding, since some values are much more likely than others. The more likely values can be encoded with shorter codewords (such as Huffman codes) or more efficiently encoded with differential encoding (DPCM) between adjacent blocks. The peaked distribution also allows more efficient quantization – we can employ larger step sizes for the quantizer for less likely values, while having smaller step sizes for more likely values.

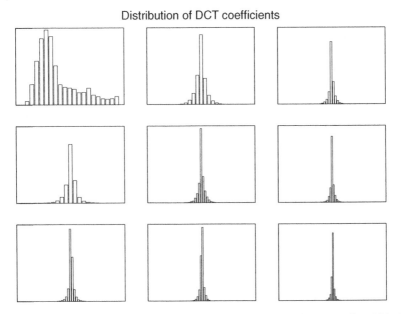

Distribution of DCT coefficients

Figure 5.29 Histograms of the DCT coefficients in the upper-left portion of a subblock. Note that for an 8 × 8 subblock, there would be an equivalent number of coefficient histograms. Only the upper-left 3 × 3 coefficient histograms are shown here.

What this means, in practice, is that instead of using typically 8 bits per pixel for each and every pixel in an 8 × 8 subblock, fewer bits may be allocated to the transformed coefficients. Even better, some coefficients are almost certain to be zero or very close to zero – which means we do not even need to encode them at all.

It is helpful to picture the process of DCT representation as a set of *basis images*. Figure 5.30 shows a single large block comprised of 8 × 8 smaller subblocks. Each subblock is composed of 8 × 8 pixels. We will refer to each of these as a "tile," and imagine the 8 × 8 image subblock to be made up of a combination of tiles. In fact, we really want a *weighted combination* of tiles. That is to say, the average value is represented by the upper-left or (0, 0) tile, so we encode this as a coefficient times the individual pixel values of the tile (which, in this case, are all equal as the figure shows). Next, consider tile (0, 1). This shows a vertical stripe, with higher intensity at the left and lower (darker) at the right. The particular image subblock under consideration may happen to have an underlying pattern similar to this, so we can approximate the true image subblock by this tile multiplied by a certain amount (a weighting coefficient value). Similarly, horizontal tiles – index (1, 0), (2, 0), and downward – are also weighted, as are the middle tiles, where various checkerboard pattern combinations exist. Overall, the

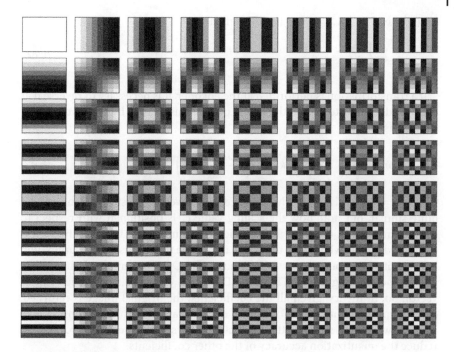

Figure 5.30 DCT basis images for an 8 × 8 transform. Each basis image is an 8 × 8 block of pixels, and there are 8 × 8 basis images in total.

image subblock is reconstructed by a combination of these tiles, each weighted separately.

This is where the HVS and the display device resolution both play an important part. Tiles toward the upper-left side represent more "coarse" image detail and are likely to be necessary. Tiles toward the lower-right, and along the diagonal, are less likely to be required. If we are prepared to sacrifice some of these more detailed tiles, then it is not necessary to encode the corresponding coefficients (they have an implicit value of zero).

What is needed, then, is a way to generate the subblock tile patterns and some way to work out the weighting of each tile in order to reconstitute the subblock. Clearly the patterns are regular, and it turns out that a cosine function is a very good choice for this task, if we wish to pack maximum energy into as few coefficients as possible.

The transformation is a two-dimensional one for images, though it is easier to start with one dimension. Given the input block (vector) of length N with pixels in vector $\mathbf{x}$, the transformed coefficients in vector $\mathbf{y}$ are determined by

$$\mathbf{y}(k) = \sqrt{\frac{2}{N}} \, c_k \sum_{n=0}^{N-1} \mathbf{x}(n) \cos\left(\frac{(2n+1)k\pi}{2N}\right) \qquad (5.95)$$

for indexes $k = 0, 1, \ldots, N - 1$ with coefficients

$$
c_k = \begin{cases} \frac{1}{\sqrt{2}} & : k = 0 \\ 1 & : k \neq 0 \end{cases} \tag{5.96}
$$

If a transformation is performed at the encoder, it is necessary to reverse it at the decoder, so as to obtain the image pixels. This is done by the *Inverse DCT* (IDCT) as follows:

$$
\mathbf{x}(n) = \sqrt{\frac{2}{N}} \sum_{k=0}^{N-1} c_k \, \mathbf{y}(k) \cos\left(\frac{(2n+1)k\pi}{2N}\right) \tag{5.97}
$$

where $n = 0, 1, \ldots, N - 1$. The forward and inverse transformations appear quite similar, but note how the scaling value c_k is positioned with respect to the summation. In the DCT it is outside the summation, because the sum is computed over n, whereas in the IDCT it must be moved inside the summation because it is computed over k. Of course, the DCT is completely invertible. That is, performing the IDCT on the DCT of a vector gets back the original vector. So, in theory, given all the coefficients, the original image is returned, exactly and without error. To achieve a much larger degree of compression, though, we forego some of the coefficients (by setting them to zero) and/or reduce the quantization accuracy of the other coefficients.

The tiles discussed previously are, in effect, the basis images. That is, the basic blocks from which an image is rebuilt. In the one-dimensional case, these *basis vectors* $\mathbf{a}_k$ are calculated as

$$
\mathbf{a}_k = \sqrt{\frac{2}{N}} \, c_k \cos\left(\frac{(2n+1)k\pi}{2N}\right) \tag{5.98}
$$

Each vector $\mathbf{a}_k$ (with index k ranging $0, 1, \ldots, N - 1$) has components $n = 0, 1, \ldots, N - 1$. It is also possible to formulate the DCT (and IDCT) as matrix multiplications. After all, in the 1D case, we are taking a column vector of $N \times 1$ and forming an output column vector also of dimension $N \times 1$. Matrix theory tells us that this could be done as a matrix–times–vector multiplication: We would need an $N \times N$ matrix, which is multiplied by the vector, to give the resulting output vector. That is, the forward transform is a multiplication:

$$
\mathbf{y} = \mathbf{Ax} \tag{5.99}
$$

For example, a 4×4 DCT kernel matrix is

$$
\mathbf{A} = \begin{pmatrix} a & a & a & a \\ b & c & -c & -b \\ a & -a & -a & a \\ c & -b & b & -c \end{pmatrix} \tag{5.100}
$$

where

$$a = \frac{1}{2}$$

$$b = \frac{1}{\sqrt{2}} \cos \frac{\pi}{8}$$

$$c = \frac{1}{\sqrt{2}} \cos \frac{3\pi}{8}$$

To simplify the matrix–vector approach, consider just a 2-pixel input (which would have a corresponding 2-valued output). Using the equations for the DCT, we could rewrite the computation in matrix form, and the 2×2 DCT transform matrix would then be

$$\mathbf{A}_{(2 \times 2)} = \frac{1}{\sqrt{2}} \begin{pmatrix} 1 & 1 \\ 1 & -1 \end{pmatrix} \tag{5.101}$$

If we examine the equations carefully, it is found that the basis vectors form the *rows* of the transform matrix. Similarly, the basis vectors also form the *columns* of the inverse transform matrix. The 2×2 result above can be interpreted intuitively: The first output coefficient is formed by multiplying the two input pixels by unity. Thus, it is an averaging of the two. The second output coefficient is formed by multiplying the pixels by $+1$ and -1, and thus is a differencing operation.

So, we can write the forward transform $\mathbf{y} = \mathbf{A}\mathbf{x}$ as being explicitly composed of row vectors

$$\begin{pmatrix} | \\ \mathbf{y} \\ | \end{pmatrix} = \begin{pmatrix} - & \mathbf{a_0} & - \\ - & \mathbf{a_1} & - \\ & \vdots & \end{pmatrix} \begin{pmatrix} | \\ \mathbf{x} \\ | \end{pmatrix} \tag{5.102}$$

We thus have the *basis vectors* as the *rows* of the transformation matrix. Each output coefficient in $\mathbf{y}$ is a vector dot product, of the form

$$y_0 = \begin{pmatrix} | \\ \mathbf{a_0} \\ | \end{pmatrix} \cdot \begin{pmatrix} | \\ \mathbf{x} \\ | \end{pmatrix} \tag{5.103}$$

with $y_1 = \mathbf{a_1} \cdot \mathbf{x}$, and so forth. In effect, we are multiplying and adding, or "weighting," the input vector by the basis vector.

Following a similar idea, the inverse transform is a multiplication:

$$\mathbf{x} = \mathbf{B}\mathbf{y} \tag{5.104}$$

Again, we can write the matrix in terms of vectors, but this time as column vectors:

$$\begin{pmatrix} | \\ \mathbf{x} \\ | \end{pmatrix} = \begin{pmatrix} | & | & | \\ \mathbf{b_0} & \mathbf{b_1} & \cdots \\ | & | & | \end{pmatrix} \begin{pmatrix} y_0 \\ y_1 \\ \vdots \end{pmatrix} \tag{5.105}$$

The *basis vectors* are the *columns*. The output may then be written a little differently, as

$$
\begin{pmatrix} | \\ \mathbf{x} \\ | \end{pmatrix} = y_0 \begin{pmatrix} | \\ \mathbf{b_0} \\ | \end{pmatrix} + y_1 \begin{pmatrix} | \\ \mathbf{b_1} \\ | \end{pmatrix} + \cdots \tag{5.106}
$$

So, it can be seen that the output sum $\mathbf{x}$ is a weighted sum of scalar–vector products. This, then, explains the forward transform as a process of determining how much of each basis vector we require and the inverse transform as a process of adding up those basis vectors in the right proportions.

The 1D case may be extrapolated to two dimensions. Now, instead of basis vectors, we have basis blocks (or matrices), and these are just the tiles discussed previously. The full two-dimensional DCT is

$$
Y(k,l) = \frac{2}{N} c_k c_l \sum_{m=0}^{N-1} \sum_{n=0}^{N-1} X(m,n) \tag{5.107}
$$
$$
\cos\left(\frac{(2m+1)k\pi}{2N} \right) \cos\left(\frac{(2n+1)l\pi}{2N} \right)
$$

where $k, l = 0, 1, \ldots, N-1$ with a corresponding inverse two-dimensional DCT

$$
X(m,n) = \frac{2}{N} \sum_{k=0}^{N-1} \sum_{l=0}^{N-1} c_k c_l Y(k,l) \tag{5.108}
$$
$$
\cos\left(\frac{(2m+1)k\pi}{2N} \right) \cos\left(\frac{(2n+1)l\pi}{2N} \right)
$$

where $m, n = 0, 1, \ldots, N-1$.

It should be noted that the amount of computation required is considerable. A direct implementation of the 2D DCT equation requires a summation over all $N \times N$ pixels in the input block, just to get *one* output coefficient. There are $N \times N$ output coefficients for each subblock. And, of course, a great many subblocks are required to form an image. It is no surprise, then, that there are several fast DCT algorithms, somewhat akin to the Fast Fourier Transform (FFT). Indeed, many of the fast DCT algorithms actually employ the FFT (for example, Narashima and Peterson, 1978).

5.7.3 Quadtree Decomposition

Fixed blocking of an image has some advantages: It is able to make the bitrate constant, and it simplifies the processing required to encode an image. A fixed bitrate is important for many types of channels, though that is not to imply that blocked encoding precludes variable bit rates. Overlaying of fixed blocks on an image is potentially somewhat unnatural, though. It would be better if more bits were allocated to "active" or detailed areas of an image, and fewer bits were

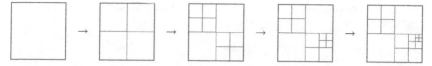

Figure 5.31 Quadtree decomposition. The recursive decomposition from left to right shows how some subblocks are subdivided further, while others are not.

allocated to less active areas. The definition of "active" may be something as simple as variance from the mean. This leads us to variable-blocksize encoding, one type of which is *quadtree decomposition.*

In quadtree decomposition, we start with an image as shown in the leftmost block of Figure 5.31. The block is subdivided into four equal-size quarter blocks. If the activity measure of these blocks indicates that they are substantially the same, we may not need to subdivide the starting block. If, on the other hand, the blocks have a degree of variance, we may decide to subdivide and continue. The act of dividing or not can be communicated with a single bit (0 = split, 1 = don't split).

Importantly, the process can be repeated on each of the four resulting sub-blocks. Thus, as we proceed in the figure from left to right, the upper-left and lower-right blocks are subdivided initially, but the upper-right and lower-left are not. This means that the blocks that are not subdivided may be represented by a constant value (grayscale luminance, or color if required). Continuing along to the right, it is seen that one of the subblocks is further subdivided, and so forth. In this way, the algorithm is selecting smaller and smaller block sizes for more active areas of the image. Such an algorithm is ideal for recursive implementation: At each stage, the input block is split into four, or it is left alone. If it is split into four smaller blocks, the same process is repeated on those smaller blocks.

Figure 5.32 shows an example image decomposed in his way, with the block boundaries made visible (in practice, the block boundaries are of course only hypothetical, and not actually coded in the image itself). Choosing the threshold at which to decompose blocks affects the resulting number of blocks. A larger threshold means fewer blocks will be subdivided.

5.7.4 Color Representation

Thus far, we have mainly considered images as comprising only the luminance or brightness. Color introduces a whole new set of problems, and a great deal of research and standardization effort has gone into efficient representation of color. Essentially, color as represented by an active (light-producing) display consists of the three additive primary components: red, green, and blue (RGB). As far as our perception is concerned, most of the "valuable" information is contained in the luminance, which is essentially the sum R + G + B. The color information may be overlaid, in a sense, to add color to a luminance-only image.

Original image   Quadtree partitioned image

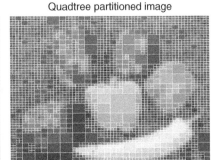

Figure 5.32 Example quadtree decomposition of a grayscale image. The block boundaries are made visible in this illustration in order to show the variable block sizes and how they correspond to the local activity of the image.

Sampling and display devices may be considered to work in the RGB domain. Colors are composed of the relative weights of the R, G, and B signals. Thus, for example, red plus green gives yellow, but red plus 50% green gives orange. The actual weightings given to the colors is highly subjective, and our visual system is somewhat nonlinear when it comes to color perception. That is, to say, a particular stimulus of one color does not appear as bright as another color. There are various evolutionary theories to account for this. Additionally, because image capture and display devices must be based on the physical chemistry of particular devices (notably, semiconductors), we are again constrained in terms of true color representation.

Since the HVS is more sensitive to luminance than to color, we can reduce the amount of data required to encode color. This has traditionally been done in analog television systems using so-called color difference signals, and this has continued into digital color sampling. The luminance is represented as a signal Y, and we need other signals to complement that in order to convey color. Since we have three color signals (R + G + B) to start with, it is reasonable that, in addition to Y, we ought to have two other signals. These are the color differences, and it is usual to work with the blue color difference (Cb) and red color difference (Cr) signals. The red color difference Cr is proportional to R minus Y, and the blue color difference Cb is proportional to B minus Y.

The most commonly employed weightings are the ITU-R Recommendation 601 (often referred to using the earlier definition of CCIR601). In matrix form, these may be written as (Acharya and Ray, 2005; ITU-R, n.d.)

$$\begin{pmatrix} Y \\ Cb \\ Cg \end{pmatrix} = \begin{pmatrix} 0.299 & 0.587 & 0.114 \\ -0.169 & 0.331 & 0.500 \\ 0.500 & -0.419 & -0.081 \end{pmatrix} \begin{pmatrix} R \\ G \\ B \end{pmatrix} \tag{5.109}$$

Y	Y	Y	Y		Y		Y		Y		Y
CrCb	CrCb	CrCb	CrCb		CrCb Y	CrCb Y		CrCb Y	CrCb Y		

Figure 5.33 Chrominance subsampling of a 4 × 4 block of pixels. Each pixel starts out as RGB, then is converted to luminance Y plus color differences CrCb. The color may be subsampled as shown, with little visual impact on the image itself.

This now allows us to represent the luminance with full resolution, but the color at reduced resolution. For every 4 pixels, we may have the so-called 4 : 4 : 4 representation, which includes Y, Cr, and Cb for each pixel. The 4 : 2 : 2 representation reduces the color resolution in the horizontal direction, as illustrated in Figure 5.33. Another representation, so-called 4 : 2 : 0, is also shown in the figure. This subsamples the color (or, more correctly, the color differences) in both horizontal and vertical directions.

5.8 Speech and Audio Coding

The most obvious approach to encoding speech (and audio) is to simply sample the waveform, and transmit (serially) the resultant bitstream. For example, telephone-quality speech sampled at 8 kHz and 16 bits/sample could be encoded with a bitstream rate of 128 kbps. This would allow the receiver to reconstruct the waveform as it was at the source up to a certain bandwidth.

However, the bitrate for direct sampling in this manner is large, and many situations where multiplexed transmission (Internet) and/or bandwidth-constrained channels (wireless) dictate that more users can be accommodated if the rate is reduced. Reduction in the bitrate is certainly possible, though generally it involves a tradeoff in terms of quality. The rates achievable also depend somewhat on the compression algorithm complexity.

In contrast to simple approaches to waveform-approximating coding, such as adaptive quantization and companding discussed earlier, is the parametric coder. As this name implies, it encodes not the waveform itself, but some parameters that represent the waveform. The key conceptual leap is to do away with exactness of reconstruction in the time domain and use the frequency domain to gage effectiveness. In addition, perceptual criteria are often employed – this means that the perception of how the waveform sounds, rather than the exact shape of the waveform, is used in the design.

5.8.1 Linear Prediction for Speech Coding

The largest class of low-rate parametric speech encoders employs the so-called analysis-by-synthesis approach (ABS). The key concept is to sample speech and process blocks or frames of around 10–20 ms and then derive the parameters that best represent that frame in the frequency domain. Those parameters are then coded and sent; the decoder reconstructs a waveform which approximates (according to some criteria) the frequency-domain characteristics of the original waveform. The exact sample-by-sample correspondence between the original and reconstructed speech no longer exists.

In this way, the compression stage at the encoder translates a large number of samples into a suitable parametric representation – that is, a set of parameters. The decompression stage reconstructs the speech (or audio) and, as it turns out, generally requires lower computational complexity.

To elaborate on the compression stage, consider the problem of prediction, which was introduced in Section 5.6.3.1. Here, we wish to predict a new sample at the encoder by way of a weighted linear sum of past samples. This works well for voiced speech (segments of speech produced with the lungs and vocal tract, typically vowel sounds such as "ay" or "ee") but less well for unvoiced speech (typically abrupt word endings and consonant-like sounds such as "k" and "ss"). The prediction for each sample of speech in the segment or frame is formed as a weighted linear sum

$$\hat{x}(n) = a_1 x(n-1) + a_2 x(n-2) + \cdots + a_P x(n-P)$$

$$= \sum_{k=1}^{P} a_k x(n-k) \tag{5.110}$$

and thus the true sample is the prediction plus an error term

$$x(n) = \hat{x}(n) + e(n) \tag{5.111}$$

where $x(n)$ is the speech signal sample at instant n, $\hat{x}(n)$ is the predicted value of the signal, and $e(n)$ is the estimation error inherent in the predictions.

The prediction $\hat{x}(n)$ may be imagined as a deterministic component – that is, described by model parameters. The error $e(n)$ is the error, which ideally is of course zero, but, in practice, would be a small, hopefully random, value. Rearranging the predictor equation gives

$$e(n) = x(n) - \overbrace{\sum_{k=1}^{P} a_k \hat{x}(n-k)}^{\hat{x}(n)} \tag{5.112}$$

The question, then, is how to determine the P parameters $a_k \colon k = 1, \dots, P$. For speech a value of $P = 10$ typically works sufficiently well; hence a 10th order

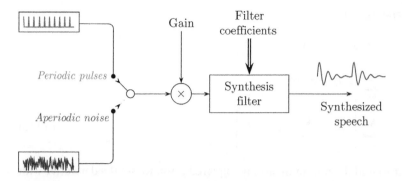

Figure 5.34 Linear predictive coder with switched pulse or noise excitation.

predictor is used. This will be addressed shortly. First, though, consider one method by which the linear prediction approach may be performed, assuming that we can find the predictor parameters. We could explicitly transmit the prediction error $e(n)$, but this would require an error corresponding to *each* sample. This would result in a bitrate that is less than the original (if the error is small) but in effect would still form a waveform-approximating coder. To convert it into a parametric coder, we recognize that this really is just a filtering operation (Section 5.3.2), and all that is required is the filter parameters. A filter, however, needs some sort of input. We further recognize that, for human speech, the excitation could be a simple pulse train for voiced speech and a random waveform for unvoiced speech. This type of approach is employed in the LPC10 codec and is shown in simplified form in Figure 5.34.

To derive the LPC parameters a_k, the autocorrelation function is useful. For our purposes, we can define it as

$$R_{xx}(k) = \sum_{n=0}^{N-1} x(n)x(n-k) \tag{5.113}$$

where k is the relative offset. Note that for a given data record, autocorrelation is symmetric, and hence

$$R_{xx}(k) = R_{xx}(-k) \tag{5.114}$$

The linear prediction estimate of order P is

$$\hat{x}(n) = \sum_{k=1}^{P} a_k x(n-k) \tag{5.115}$$

The average error is

$$\bar{e} = E\{[x(n) - \hat{x}(n)]^2\}$$

$$= \sum_{n=1}^{N-1} [x(n) - \hat{x}(n)]^2$$

$$= \sum_{n=1}^{N-1} \left[x(n) - \sum_{k=1}^{P} a_k x(n-k) \right]^2 \qquad (5.116)$$

In order to find the minimum average squared error, we set the derivative with respect to the predictor parameters equal to zero

$$\frac{\partial \bar{e}}{\partial a_m} = 0 \qquad (5.117)$$

Then, applying the chain rule for derivatives

$$\sum_{n=0}^{N-1} 2 \left[x(n) - \sum_{k=1}^{P} a_k x(n-k) \right] \left\{ \frac{\partial}{\partial a_m} \left[x(n) - \sum_{k=1}^{P} a_k x(n-k) \right] \right\} = 0 \qquad (5.118)$$

We then need the derivative term $\partial/\partial a_m$. This will be equal to zero for all a_k for which $k \neq m$. In the case of $k = m$, it will simplify to $-x(n-m)$. Mathematically, this may be expressed as

$$\frac{\partial}{\partial a_m} \left[x(n) - \sum_{k=1}^{P} a_k x(n-k) \right]$$

$$= \begin{cases} -x(n-m) : & k = m \\ 0 & : \text{ otherwise} \end{cases} \qquad \text{for} \quad m = 1, \dots, P \qquad (5.119)$$

Therefore, the expression that we have to minimize is

$$\sum_{n=0}^{N-1} \left[x(n) - \sum_{k=1}^{P} a_k x(n-k) \right] [-x(n-m)] = 0$$

$$\sum_{n=0}^{N-1} x(n)x(n-m) = \sum_{n=0}^{N-1} \left[\sum_{k=1}^{P} a_k x(n-k)x(n-m) \right]$$

$$= \sum_{k=1}^{P} a_k \left[\sum_{n=0}^{N-1} x(n-k)x(n-m) \right] \qquad (5.120)$$

The left-hand expression may be recognized as the autocorrelation at lag m, and the right-hand expanded, to give

$$
\begin{aligned}
R_{xx}(m) = & \sum_{k=1}^{P} a_k \left[\sum_{n=0}^{N-1} x(n-k)x(n-m) \right] \\
= & a_1 \sum_{n=0}^{N-1} x(n-1)x(n-m) \\
& + a_2 \sum_{n=0}^{N-1} x(n-2)x(n-m) \\
& \cdots + a_P \sum_{n=0}^{N-1} x(n-P)x(n-m)
\end{aligned}
$$

$$(5.121)$$

Dropping the xx subscript for convenience, we have for $m = 1$,

$$R(1) = a_1 R(0) + a_2 R(-1) + \cdots + a_P R(-(P-1)) \tag{5.122}$$

And for $m = 2$,

$$R(2) = a_1 R(1) + a_2 R(0) + \cdots + a_P R(-(P-2)) \tag{5.123}$$

The $R(\cdot)$ values are computed up to $m = P$,

$$R(P) = a_1 R(P) + a_2 R(P-1) + \cdots + a_P R(0) \tag{5.124}$$

Because we have a large number of terms in the predictor equation (typically $P = 10$ for speech), it is easier to write all these equations as a matrix formulation:

$$
\begin{pmatrix}
R(0) & R(-1) & \cdots & R(-(P-1)) \\
R(1) & R(0) & \cdots & R(-(P-2)) \\
\vdots & \vdots & \ddots & \vdots \\
R(P-1) & R(P-2) & \cdots & R(0)
\end{pmatrix}
\begin{pmatrix}
a_1 \\ a_2 \\ \vdots \\ a_P
\end{pmatrix}
=
\begin{pmatrix}
R(1) \\ R(2) \\ \vdots \\ R(P)
\end{pmatrix}
\tag{5.125}
$$

Recognizing that autocorrelation is symmetrical, $R(k) = R(-k)$, and so

$$
\begin{pmatrix}
R(0) & R(1) & \cdots & R(P-1) \\
R(1) & R(0) & \cdots & R(P-2) \\
\vdots & \vdots & \ddots & \vdots \\
R(P-1) & R(P-2) & \cdots & R(0)
\end{pmatrix}
\begin{pmatrix}
a_1 \\ a_2 \\ \vdots \\ a_P
\end{pmatrix}
=
\begin{pmatrix}
R(1) \\ R(2) \\ \vdots \\ R(P)
\end{pmatrix}
\tag{5.126}
$$

This representation may be more compactly written as a matrix equation

$$\mathbf{R\,a = r} \tag{5.127}$$

where

$$
\mathbf{R} = \begin{pmatrix}
R(0) & R(1) & R(2) & \cdots & R(P-1) \\
R(1) & R(0) & R(1) & \cdots & R(P-2) \\
\vdots & \vdots & \vdots & \ddots & \vdots \\
R(P-1) & R(P-2) & R(P-3) & \cdots & R(0)
\end{pmatrix} \tag{5.128}
$$

is the matrix of autocorrelation values, and

$$
\mathbf{r} = \begin{pmatrix}
R(1) \\
R(2) \\
\vdots \\
R(P)
\end{pmatrix} \tag{5.129}
$$

is a vector of autocorrelation values. This matrix equation may be solved for the desired linear prediction (LP) parameters

$$
\mathbf{a} = \begin{pmatrix}
a_1 \\
a_2 \\
\vdots \\
a_P
\end{pmatrix} \tag{5.130}
$$

This provides a means to solve for the optimal predictor parameters a_p. It would appear that this entails considerable computational complexity (a $P \times P$ matrix inversion). However, the symmetric nature of the autocorrelation matrix $\mathbf{R}$ facilitates efficient solution methods. The Levinson–Durbin recursion is most often used to simplify the computation.

As it stands, the preceding theory gives a good prediction of a signal, with correspondingly small error. The obvious approach would be to quantize and send the error signal. However, this would require one or more bits per sample, and we can do better than that. As mentioned at the start of the section, tracking the exact waveform on a sample-by-sample basis forms a waveform encoder. We can apply the parametric approach by dispensing with the exact reconstruction requirement, and instead sending only the predictor parameters. These then form the input to a filter, and the output approximates the short-term spectrum of the speech.

We must return, then, to the question of what to use for the input to the filter. To start with, a series of pulses to approximate the pitch of the speech signal over a short time frame gives a coarse representation of the speech spectrum. This may be improved upon by recognizing that this type of excitation is adequate for voiced speech sounds which have strong periodicity. When the periodicity is less pronounced (during so-called unvoiced speech), white noise excitation suffices. We then have the LP encoder discussed earlier, in Figure 5.34. This type of coder can produce acceptable speech at rates as low as 2 kbps. The speech thus produced is sometimes described as intelligible, but lacking naturalness. For some applications, this is acceptable (this type of coder was originally developed for military applications). However, for commercial voice communication services, some degree of naturalness is essential. This leads to the analysis-by-synthesis approach.

5.8.2 Analysis by Synthesis

To increase the perceptual quality and naturalness of the LP coder, several modifications have been proposed. The obvious candidate is to alter the type of excitation at the source, rather than the binary voiced/unvoiced excitation classification.

The use of a feedback loop also becomes desirable. In this way, the synthetic speech, which would otherwise only be generated at the decoder, is produced and analyzed in the encoder. The encoder adjusts the excitation parameters so as to produce the best speech, according to some defined criterion.

One possibility for the excitation is to use multiple pulses, giving a *multi-pulse excitation* (MPE) coder. This requires the placing of pulses within the excitation frame. Thus, as well as the placement of the pulses, the magnitude of the pulses must be determined. Another simpler variation is *regular pulse excitation* (RPE), wherein the pulse spacing is not variable but rather fixed.

Finally, another type of coding approach is termed CELP, or *Code Excited Linear Prediction* (Figure 5.35). In this method, rather than using known pulse-type excitations, random excitation vectors are stored in a codebook. Each excitation vector is tested for synthesis of speech at the encoder, and the best-performing one is selected. Thus, rather than pulse magnitudes (and possibly positions), a vector of samples is selected. Since the encoder and decoder have identical pre-stored codebooks, the encoding then consists of

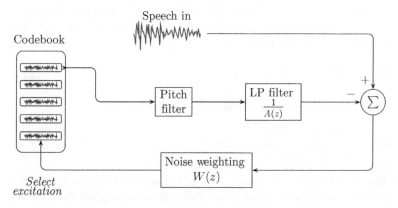

Figure 5.35 The essential arrangement of a code-excited linear predictive coder. The excitation is selected according to the match between the synthesized speech and the original.

transmitting the index of the best-matching codevector. This approach is able to give good quality speech at very low rates, however, this comes at the expense of substantial additional computational complexity. This is because each possible candidate must be processed and tested at the encoder. For a codebook of, say, 10 bits, this requires the testing of each of $2^{10} \approx 1000$ frames, if each were to be evaluated independently.

5.8.3 Spectral Response and Noise Weighting

The LPC process as described above typically models the lower frequency content better than higher frequency. In an attempt to compensate for this, a process called pre-emphases is applied. This entails a highpass filter of the form:

$$H_{\text{pre}}(z) = 1 - \lambda z^{-1} \tag{5.131}$$

with typically $\lambda = 0.95$, applied before the LPC analysis stage. A de-emphasis filter may be used to reverse the process, and it is just the reciprocal of this

$$H_{\text{de}}(z) = \frac{1}{1 - \eta z^{-1}} \tag{5.132}$$

Exact matching of the pre- and de-emphasis stages would occur if $\eta = \lambda$. However, in practice, setting $\eta = 0.75$ (a little less than λ) works well. This gives a "sharpening" of the decompressed speech.

A more advanced extension of this idea is to use a so-called *noise shaping* filter. Here, the premise is that the amount of noise perceived at a given frequency is less than its apparent power when that frequency has a higher energy content. That is, louder portions of speech mask the noise more effectively. The corollary of this is that more noise is perceived in a band when that frequency band has lower speech energy. Thus, it makes sense to base the noise weighting on the LPC filter itself. If the LPC filter is $A(z)$, then a noise-weighting filter following these principles is defined by

$$W(z) = \frac{A(z)}{A(z/\gamma)} \qquad 0 \le \gamma \le 1 \tag{5.133}$$

Expanding the $A(z)$ filter out, the noise-shaping filter becomes

$$W(z) = \frac{1 - \sum_{k=1}^{P} a_k z^{-k}}{1 - \sum_{k=1}^{P} a_k \gamma^k z^{-k}} \qquad 0 \le \gamma \le 1 \tag{5.134}$$

The coefficient γ is typically chosen to be of the order of 0.9–0.95, and this gives a frequency response which tracks slightly below the original. Because the noise-shaping filter is used in cascade with the LPC filter, the overall response of the system becomes $1/A(z/\gamma)$. It is worth investigating what this actually means, and how it affects the frequency response. Since we are replacing

$$z \rightarrow \left(\frac{z}{\gamma} \right) \tag{5.135}$$

each z term with exponent k becomes

$$\left(\frac{z}{\gamma} \right)^k = \gamma^{-k} z^k \tag{5.136}$$

Thus, the poles become

$$\left(\frac{z}{\gamma} + p \right) = \frac{1}{\gamma}(z + \gamma p) \tag{5.137}$$

which means the pole magnitude is reduced by a small factor γ. It is important to realize that the magnitude is affected, but not the pole angle. Since the pole angle determines the frequency, the overall frequency peak locations are unchanged.

To illustrate this, consider Figure 5.36 which shows the poles of an LPC function, derived using the algorithms discussed earlier for a particular frame of speech. The frequency response corresponding to this filter is also shown.

Now suppose we apply the above noise-weighting principle, with $\gamma = 0.85$. The poles are altered as shown in Figure 5.37. It is clear that the pole angles are unchanged, but the radial distance is reduced. The corresponding frequency response has peaks at the same locations, however, the peaks are somewhat reduced in size and spread out due to the poles moving inward. This results in a slightly muted speech signal as compared with the unaltered LPC.

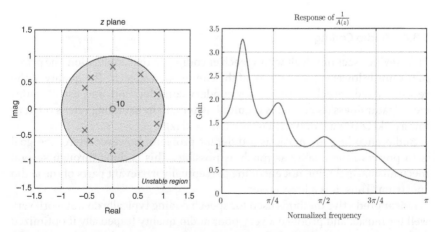

Figure 5.36 The poles of a linear predictor, and the corresponding frequency response. The resonances model the vocal tract, so as to produce "synthetic" speech.

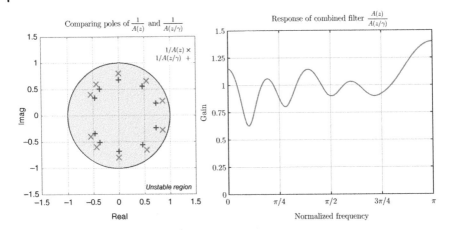

Figure 5.37 The poles of a linear predictor (×), and the corresponding noise-weighted poles (+) for a frame of speech. For the indicated value of $\gamma = 0.85$, the poles move inwards and flatten the spectrum shown in Figure 5.36. The resulting noise-weighting filter $W(z)$ is shown on the right.

Considering the location of the noise-weighting filter $W(z)$ in the feedback loop of Figure 5.35, less gain is given to areas around the LPC spectral peaks. This means that more quantization noise is allocated to the stronger spectral areas where it is less objectionable, and as a result, less quantization noise is present in the less-strong areas of the LPC spectrum. Hence, the term *noise-shaping* – the quantization noise is "shaped" according to the frequency spectrum of the signal to be encoded.

5.8.4 Audio Coding

The previous sections dealt with speech encoding. In that application, the prime motivator is lowering the bit rate of the compressed speech. The quality of the reconstructed speech is usually a secondary consideration, and usually sufficient "naturalness" is all that is required. After all, the sampling rate is often as low as 8 kHz (for a bandwidth of less than 4 kHz).

Audio encoding for music is a different problem, however. Here, the goal is to preserve the quality as much as possible. This is done through several mechanisms, including not encoding perceptually irrelevant parts of the audio spectrum. Thus, it is a lossy coder.

Linear predictive methods used for speech coding typically do not work very well for music, and produce a very poor audio quality (especially if optimized for speech signals). Thus, different techniques are applied. Audio coding (as opposed to speech coding) is often used in an off-line (not-real-time) mode,

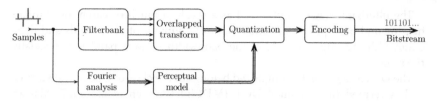

Figure 5.38 Audio encoding using sub-band coding (filterbanks), an overlapped transform, perceptual weighting, and finally, entropy coding.

meaning that the signal does not have to be compressed immediately, and the very tight constraints for real-time speech encoding can be relaxed. The result, in practice, is that buffering periods may be longer, leading to greater compression. Furthermore, the computational requirements for compression are not as critical, thus allowing more sophistication in the algorithms.

Audio coders have developed through research over a number of years, and continue to evolve. It is a complex field, and here we present only the outline of the basic principles. A more complete summary, with extensive references is given in Brandenburg (1999).

Perceptual coding aims to reduce the resolution in such a way that it would not greatly affect the perceived audio quality, thus saving on parameters which must be coded and transmitted. Naturally, the fidelity of typical sound reproduction electronics (the amplifiers and speakers), and the listening environment itself, all have an effect on the audio in some way.

Figure 5.38 shows a block diagram of the generic audio encoder, as exemplified in the MP3 family of coders. In speech coders, it is assumed that a pulse-like excitation is sufficient. This provides an encoding advantage because the time waveform itself is not coded, but rather a frequency-domain representation. In audio coders, it is best to split the source code into frequency bands and code each of those separately, according to their perceptual relevance. This is done by the filterbank stage. A transformation stage (a modified DCT) is then performed on each band, so as to reduce the correlation between samples, in much the same way that an image coder uses the DCT (except now in one dimension). Overlapped frames are utilized to reduce blocking artifacts, which may be audible. The quantization stage then assigns appropriate binary representations to the transformed filterbank coefficients. This quantization stage is informed by the perceptual model, which in turn uses Fourier analysis of the original signal. Finally, entropy-efficient bit allocation is performed to produce the coded bitstream. It is evident that the encoder is generally more complex than the decoder, because it must perform the filtering, quantization, and perceptual analysis.

The filterbank stage is optimized as a result of much research into human hearing and perception of sounds. Both the number of filter bands and their bandwidth are important. Also, the bandwidth of each band is not usually the same.

The DCT stage which is employed belongs to the class of lapped transforms, and is termed the Modified DCT (MDCT) (Princen et al., 1987; Malvar, 1990). The MDCT is unusual for transforms, in that it has a different number of outputs as compared to inputs. The MDCT takes $2N$ inputs and has N outputs,

$$X(k) = \sum_{n=0}^{2N-1} x(n) \cos\left[\left(\frac{\pi}{N}\right)\left(n + \frac{1}{2} + \frac{N}{2}\right)\left(k + \frac{1}{2}\right)\right] \tag{5.138}$$

This may also be written in terms of the full double-length input block size, with $M = 2N$, as

$$X(k) = \sum_{n=0}^{M-1} x(n) \cos\left[\left(\frac{\pi}{2M}\right)(2n + 1 + N)(2k + 1)\right] \tag{5.139}$$

The inverse MDCT takes N inputs and has $2N$ outputs

$$x(n) = \frac{1}{N} \sum_{k=0}^{N-1} X(k) \cos\left[\left(\frac{\pi}{N}\right)\left(n + \frac{1}{2} + \frac{N}{2}\right)\left(k + \frac{1}{2}\right)\right] \tag{5.140}$$

Note the forward and inverse are of the same form, except for the scaling constant.

The use of overlapping blocks serves to reduce the blocking effects at block boundaries, which may be audible. Overlap and addition of successive blocks reconstructs the output sequence exactly. The overlapping process is illustrated in Figure 5.39. A simplified case with a block size of $M = 4$ is illustrated, in order to provide a concrete example of the principle of perfect reconstruction. Each of these blocks produces an output block of length $N = 2$, which is exactly half the size of the input. The input parameters shown in the code below produce the output as illustrated in Figure 5.39. Of course, in practice, the block size is substantially larger. Since the number of outputs of the inverse transform is larger than the input, aliasing occurs, and the outputs of the inverse transform, if taken directly, do not produce the original sequence. It is necessary to add successive blocks as illustrated in the diagram. As shown, the first and last sub-blocks are not overlapped, and thus are not reconstructed exactly.

```
% The mdct and imdct functions are generic for any length,
% but the example below shows a 4-point block with 2-point
% outputs.

% input vector
x = [6 8 7 4 9 2 3  7 1 4];
disp(x);

% partition input vector into 4-sample overlapping blocks
x1 = x(1:4);
x2 = x(3:6);
x3 = x(5:8);
x4 = x(7:10);

disp(x1);
disp(x2);
disp(x3);
disp(x4);

% take MDCT of each of these blocks 4->2
X1 = mdct(x1);
X2 = mdct(x2);
X3 = mdct(x3);
X4 = mdct(x4);

% take inverse MDCT of each transformed block 2->4
y1 = imdct(X1);
y2 = imdct(X2);
y3 = imdct(X3);
y4 = imdct(X4);

disp(y1');
disp(y2');
disp(y3');
disp(y4');

% combine output blocks with overlap
y = zeros(1, length(x));

y(1:2)  = y1(1:2)';    % will not be correct
y(3:4)  = y1(3:4)' + y2(1:2)';
y(5:6)  = y2(3:4)' + y3(1:2)';
y(7:8)  = y3(3:4)' + y4(1:2)';
y(9:10) = y4(3:4);    % will also not be correct
```

The following function shows the general Modified DCT for $M = 2N$ inputs and N outputs.

```
function [X] = mdct(x)
    x = x(:);
    M = length(x);
    N = round(M/2);

    X = zeros(N, 1);
    for k = 0:N-1
        s = 0;
        for n = 0:2*N-1 % M-1
            % these are equivalent
            % s = s + x(n+1)*cos( (pi/N)*(n + 1/2 + N/2)...
            %*(k + 1/2) );
            s = s + x (n +1) * cos ( ( pi/(2*M) )
            *(2* n + 1 + N) *(2* k + 1) ) ;
        end
        X(k+1) = s;
    end
end
```

The complement is the Inverse Modified DCT for N inputs with $M = 2N$ outputs.

```
function [y] = imdct(X)
    X = X(:);
    N = length(X);

    y = zeros(2*N,1);
    for n = 0:2*N-1 % M-1
        s = 0;
        for k = 0:N-1
            s = s + X(k+1)*cos( (pi/N) *(n + 1/2 + N/2)...
            *(k + 1/2) ) ;
        end
        y(n+1) = s/N;
    end
end
```

The perceptual coding aspect relies heavily on perception research (Schroeder et al., 1979). Not only is the nonlinear sensitivity of the ear over different frequency bands exploited, but also the ability to perceive tones. The phenomenon of *masking* refers to the fact that one tone may be obscured or

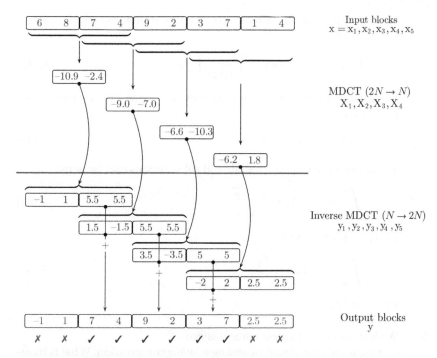

Figure 5.39 Overlapping blocks for the Modified DCT, with numerical values shown for a block size of *M* = 4 and *N* = 2 in order to illustrate the overlap and addition of successive blocks to yield perfect reconstruction.

masked by adjacent tones. If that is the case, then it need not be coded. Perceptual aspects of audio and speech coding are dealt with extensively in Painter and Spanias (1999, 2000).

5.9 Chapter Summary

The following are the key elements covered in this chapter:

- Scalar quantization, including fixed and variable-step characteristics.
- Vector quantization – training and search methods.
- Image encoding using block transforms.
- Speech encoding using ABS, and music encoding using transforms.

Problems

5.1 The Shannon bound gives the capacity of a channel for a given SNR. In deriving it, the entropy of a Gaussian signal is required.

Starting with the definition of entropy for a continuous variable x drawn from a set X as

$$H(X) = \int_{-\infty}^{\infty} f_X(x)\log_2\left(\frac{1}{f_X(x)}\right) dx \qquad (5.141)$$

Show that entropy (information content) of a signal with a Gaussian distribution is

$$H(X) = \frac{1}{2}\log_2 2\pi e\sigma^2 \qquad (5.142)$$

The following standard integrals and substitutions may be of use:

$$\int_{-\infty}^{\infty} x^2 e^{-ax^2} dx = \frac{1}{2}\sqrt{\frac{\pi}{a^3}} \qquad (5.143)$$

with $a = 1/(2\sigma^2)$, and

$$\int_{-\infty}^{\infty} e^{-ax^2} dx = \sqrt{\frac{\pi}{a}} \qquad (5.144)$$

with $a = 1/(2\sigma^2)$.

5.2 With reference to lossless compression:
a) Explain what is meant by *sliding window* compression. What is transmitted from encoder to decoder? What are the weaknesses of this approach?
b) Explain what is meant by *dictionary-based* compression. What is transmitted at each stage? What are the weaknesses of this approach?

5.3 Investigate the relationship between theoretical entropy and lossless compression as follows:
a) Write and test MATLAB code to calculate the entropy of a data file (in bits per symbol). Test using a grayscale image file in uncompressed format such as bmp or lossless jpeg.
b) Compress the data files used in the previous question part using a standard lossless compression such as zip, bzip, gzip, or similar. What is the size of the resulting file, and the compression ratio? What is the entropy of the compressed files according to your program from the previous question?

5.4 Huffman codes are one type of lossless code.
a) Construct a Huffman code for the symbols A, B, C, D using the probabilities 0.2, 0.4, 0.25, 0.15, using a tree that does *not* maintain the "sibling property." Calculate the entropy and the expected average bit rate.

b) Repeat using a tree that *does* maintain the "sibling property." Calculate the expected average bit rate. Compare with the results obtained in the previous question, for the code tree constructed without the sibling property.

5.5 Create a Huffman coding tree similar to that shown in Figure 5.14 using symbols A, B, C, D, E with probabilities 0.30, 0.25, 0.20, 0.18, 0.07. Using the MATLAB code provided for generating, encoding, and decoding Huffman codes:

a) Create each codeword and determine the average codeword length of the set.

b) Use a loop to encode and then decode each codeword. Check that the decoding is correct.

5.6 Show that the two-dimensional DCT is *separable*. That is, the DCT of a matrix can be performed by taking the one-dimensional transform of the rows, followed by the one-dimensional transform of columns of the resulting matrix.

5.7 For a DCT of dimension $N = 4$, use the following code for the forward 2D DCT to implement an *inverse* 2D DCT. Verify that the output after applying the DCT followed by the IDCT is equal to the original input vector **X**, subject to arithmetic precision errors.

```
N = 4;              % block size
X = rand(N, N);     % input block
Y = zeros(N, N);    % output block

% each output coefficient
for k = 0:N-1
    for el = 0:N-1

        % calculate one DCT coefficient
        s = 0;
        for n = 0:N-1
            for m = 0:N-1
                s = s + X(m+1, n+1)* ...
                    cos( (2*m+1)*k*pi/(2*N) ) * ...
                    cos( (2*n+1)*el*pi/(2*N) );
            end
        end

        if ( k == 0 )
            ck = 1/sqrt(2);
```

```
        else
            ck = 1;
        end

        if ( el == 0 )
            cl = 1/ sqrt (2) ;
        else
            cl = 1;
        end

        Y(k+1, el +1) = 2/N*ck*cl*s;
    end
end
```

5.8 The variance of a data sequence affects how predictable it is.

    a) Calculate the correlation coefficient of a typical grayscale image using MATLAB.

    b) An image source $\{x(n)\}$ has a mean of zero, unit variance, and correlation coefficient $\rho = 0.95$. Show that the difference signal $d(n) = x(n) - x(n - 1)$ has a significantly lower variance than the original signal.

    c) The ratio σ_x^2/σ_d^2 may be used to gage how good a prediction actually is for a set of data. Determine the ratio σ_x^2/σ_d^2 for a simple first-order differential predictor $\hat{x}(n) = a_1 x(n - 1)$.

5.9 Image prediction may be enhanced by using both horizontal and vertical prediction.

    a) A two-dimensional image coding system is implemented as shown below. Each pixel is predicted by the average of the previous pixel and the pixel in the previous row immediately above it.

$$\cdot \;\; \cdot \;\; x_c \quad \text{line } n - 1$$
$$\cdot \; x_b \; x_a \quad \text{line } n$$

The prediction equation is

$$\hat{x}_a = \frac{x_b + x_c}{2}$$

Assume that the mean of all samples is zero, that the correlation coefficient is identical across and down ($\rho_{ac} = \rho_{ab} = \rho$), and that the diagonal correlation coefficient may be approximated as $\rho_{bc} = \rho^2$. Find an expression for the signal variance to the prediction variance, σ_x^2/σ_e^2.

    b) For $\rho = 0.95$, find σ_x^2/σ_e^2, and compare with the one-dimensional predictor of the previous question.

5.10 The Laplacian distribution is often used as a statistical model for the distribution of image pixels. Since the dynamic range of the pixels is critical to allocating the quantization range and hence step size, it is useful to know the likelihood of pixels at extreme values being truncated. Given a zero-mean Laplacian distribution

$$f(x) = \frac{1}{\sigma\sqrt{2}} e^{-\sqrt{2}\,|x|/\sigma}$$

find the probability that $|x| > 4\sigma$.

5.11 A uniformly distributed random variable is to be quantized using a four-level uniform mid-rise quantizer. Sketch the appropriate decision-reconstruction characteristic. By minimizing the error variance, show that the optimal step size is $(\sigma\sqrt{3})/2$. Would this change for a different distribution?

5.12 A video coder for videoconferencing encodes images of size 512 (width) by 512 (height) at 10 frames per second. Direct vector quantization of 8×8 block is used, at a rate of 0.5 bits per pixel.
   a) Determine the number of blocks per image, and the number of blocks per second which the codec (coder–decoder) must encode.
   b) If each block is encoded as a mean value using 8 bits, and the remaining bits for the vector shape, determine the size of the VQ codebook.
   c) Determine the number of vector comparisons per second.
   d) Determine the number of *arithmetic* operations per second, and thus the search time if each operation takes 1 ns.

5.13 Generate 100 samples of a random Gaussian waveform.
   a) Feed this into the filter transfer function

$$\hat{y}(n) = b_0 x(n) + a_1 y(n-1) + a_2 y(n-2)$$

   with $a_1 = 0.9$, $a_2 = -0.8$, and $b_0 = 1$. Plot the vector $y(n), n = 0, \ldots, N-1$.
   b) Use MATLAB to estimate the order-2 autocorrelation matrix $\mathbf{R}$ and autocorrelation vector $\mathbf{r}$.
   c) From these, calculate the filter transfer function parameters $\hat{a}_k$. How close does the estimate $\hat{a}_k$ come to the true values of a_k?

6

Data Transmission and Integrity

6.1 Chapter Objectives

On completion of this chapter, the reader should be able to:

1) Distinguish between error detection and error correction.
2) Calculate the bit error probabilities for a simple channel coding scheme.
3) Understand the working of algorithms for block error detection and block error correction.
4) Explain the operation of convolutional coding, including path-search algorithms.
5) Explain private (secret) key encryption, key-exchange methods, and public-key encryption.

6.2 Introduction

Digital communications clearly depends on getting the correct bit sequence from the sender to the receiver. In a real communication system, there is no guarantee that the stream of bits emanating from the encoder will be received correctly at the decoder. Some bits may be subject to random errors due to noise and other factors such as clock synchronization. An important distinction is between error *detection* and error *correction*. Detecting errors is easier than correcting them. Not only do we usually require the correct bit sequence; the data must also be delivered in the correct order. Although this may sound strange, it is possible in packet-switched systems for the correct blocks of data to be delivered, but out of order.

In addition to checking data integrity due to random events, it is often important to ensure that there is no compromise to data integrity due to deliberate manipulation or theft of confidential data. The Internet represents an essentially insecure network – a sender does not have control over who

Communication Systems Principles Using MATLAB®, First Edition. John W. Leis.
© 2018 John Wiley & Sons, Inc. Published 2018 by John Wiley & Sons, Inc.
Companion website: www.wiley.com/go/Leis/communications-principles-using-matlab

may view the data in transit. Furthermore, wireless networks represent an even easier means to intercept and steal data, due to the fact that the radio signal may be received within a given range from the transmitter. By analogy with conventional lock-and-key systems, data can be encrypted using a digital key, which is really just a particular bit pattern. However, the problem then arises as to how to communicate that binary key to both parties (sender and recipient), given that the communication channel is assumed to be insecure. Such private or secret key systems require a separate channel to send the key from one party to another, but this is not always feasible. One interesting approach involves using a *different* key for decryption to the one used for encryption. This is termed *public-key encryption*. The idea is that the decryption key is only known to the intended recipient but that the encryption key can be in the public domain without any threat to security. Clearly, then, there should be no way to derive the secret decryption key from the public encryption key.

6.3 Useful Preliminaries

This section briefly reviews two useful concepts in modeling data transmission integrity. These are probability concepts (used for mathematical models of noise) and integer arithmetic (used for various encryption functions).

6.3.1 Probability Error Functions

Errors in digital systems – mistaking a 1 for a 0 or vice versa – occur due to noise in the process of transmission and reception. Electronic, electromagnetic, or optical noise comes from various sources, and one key concept is that of Additive White Gaussian Noise (AWGN). Such random noise is characterized by a probability density function (PDF) – essentially, how likely certain values of noise are, and thus how likely it is that errors will occur under given conditions.

Figure 6.1 shows a representative random noise waveform. Two aspects characterize this type of waveform: firstly, the *mean* or *average* value and secondly, the relative spread of values. The spread of values is the mathematical *variance*, which in turn is equivalent to the power in the noise signal. This is a positive quantity, whereas the mean or average is often zero.

To picture the spread of values and their likelihood, the PDF is employed. The PDF of the noise in Figure 6.1 is shown in Figure 6.2. Values around the mean are more likely, while larger amplitudes (positive or negative) are less likely. The curve does not directly provide the likelihood or probability though, since it is a *density* function. Rather, the area under the curve, between one amplitude x_1 and another x_2, is the probability that the signal lies between those levels. As a result, the total area under this curve must equal to one.

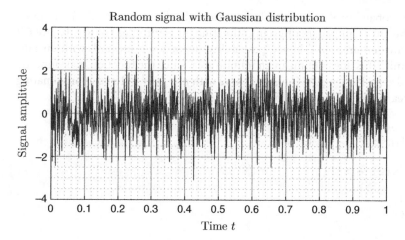

Figure 6.1 Additive noise is characterized by random values with a certain mean and variance.

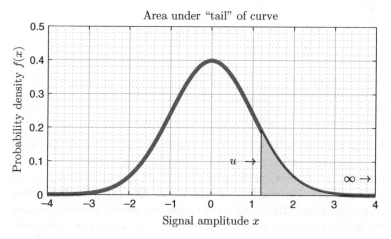

Figure 6.2 The probability density function is used to tell the likelihood of a signal amplitude falling between two levels. In the case illustrated, this is between $x = u$ and any higher value (that is, $x \to \infty$).

AWGN is commonly assumed in modeling communication systems. The characteristic curve is

$$f_X(x) = \frac{1}{\sigma\sqrt{2\pi}} e^{-(x-\mu)^2/2\sigma^2} \tag{6.1}$$

where $f_X(x)$ is the PDF of random variable X, indicating the density at point x. The mean is μ and variance σ^2.

The probability of some event is determined by the area under the PDF curve, and this problem occurs often in estimating bit error rates (BERs). Figure 6.2 illustrates an area (and hence probability) from $x = u$ upward, with an infinite limit. This does not mean an infinite area, since the function is asymptotic toward zero. Plotting the Gaussian curve with $\mu = 0, \sigma^2 = 1$ and approximating the area is accomplished as follows.

```
N = 400;                      % number of points
x = linspace(-4, 4, N);       % uniformly spaced points over
                              % the range
dx = x(2) - x(1);             % delta x

v = 1;                        % variance
m = 0;                        % mean

% Gaussian function
g = (1/sqrt(2*pi*v))*exp( -(x-m).^2/(2*v) );

plot(x, g);
fprintf(1, 'total area = %f\n', sum(g.*dx));

% find the area greater than u
u = 1.2;
i = min(find(x >= u));
area = sum(g(i:end).*dx);
fprintf(1, 'area under tail from u = %f\n', area);
```

Note that the accuracy achievable with this approach is limited by the size of the step dx, which corresponds to a small increment δx.

Since the small increment approach as above is an approximation, a better approach is necessary. Two methods commonly employed are the Q function and the erf function (van Etten, 2006). The Q function simply provides the area under a normalized Gaussian with zero mean and unit variance, which can be scaled according to the desired mean and variance. As might be expected, it is really the integral of Equation (6.1) with $\mu = 0$ and $\sigma^2 = 1$ and is expressed as

$$Q(u) = \frac{1}{\sqrt{2\pi}} \int_u^\infty e^{-x^2/2} \, \mathrm{d}x \tag{6.2}$$

A closely related (but not identical) function is the complementary error function erfc, defined as

$$\mathrm{erfc}(u) = \frac{2}{\sqrt{\pi}} \int_u^\infty e^{-x^2} \, \mathrm{d}x \tag{6.3}$$

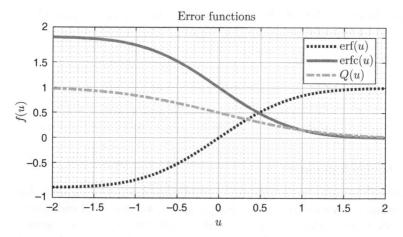

Figure 6.3 Comparing the error function, complementary error function, and the Q function.

Although similar, it is not identical to the Q function. The complementary error function is related to the error function:

$$\text{erf}(u) = \frac{2}{\sqrt{\pi}} \int_0^u e^{-x^2} \, dx \tag{6.4}$$

$$\text{erfc}(u) = 1 - \text{erf}(u) \tag{6.5}$$

Notice the limits (0 to u) of the error function compared with the complementary error function (0 to ∞). Effectively, these complement each other in the PDF area curve. For comparison, these three functions are shown in Figure 6.3.

Since the Q function gives the area under Gaussian *tail*, the relationship between Q and erfc is useful:

$$Q(u) = \frac{1}{2}\text{erfc}\left(\frac{u}{\sqrt{2}}\right) \tag{6.6}$$

$$\text{erfc}(u) = 2\,Q(u\sqrt{2}) \tag{6.7}$$

MATLAB includes the `erf()` and `erfc()` functions. If we know that a certain signal has zero mean and unit variance, then the likelihood of a random variable having a value greater than zero, for example, is computed with

```
0.5*erfc(0/sqrt(2))
ans =
    0.5000
```

This is the area under half the PDF curve. The shaded area of Figure 6.2, for which $u = 1.2$, is

```
0.5 * erfc (1.2/ sqrt (2))
ans =
       0.1151
```

The error function is used in the following sections for BER calculations.

6.3.2   Integer Arithmetic

Calculations for data transfer integrity checking, and newer approaches to data security and encryption, depend on integer arithmetic in general and *modulo arithmetic* in particular. Modulo arithmetic simply means counting numbers up to some limit, after which the count is reset to zero. More formally, counting $0, 1, 2, \ldots$ up to some number $N - 1$ is termed counting *modulo-N*. Thus modulo-7 counting would be as follows, where $\equiv$ means "equivalent to:"

$$
\begin{array}{llll}
0 \bmod 7 & \equiv & 0 \qquad & 7 \bmod 7 & \equiv & 0 \\
1 \bmod 7 & \equiv & 1 & 8 \bmod 7 & \equiv & 1 \\
2 \bmod 7 & \equiv & 2 & 9 \bmod 7 & \equiv & 2 \\
3 \bmod 7 & \equiv & 3 & 10 \bmod 7 & \equiv & 3 \\
4 \bmod 7 & \equiv & 4 & 11 \bmod 7 & \equiv & 4 \\
5 \bmod 7 & \equiv & 5 & 12 \bmod 7 & \equiv & 5 \\
6 \bmod 7 & \equiv & 6 & 13 \bmod 7 & \equiv & 6
\end{array}
$$

The notion of modulo arithmetic is related to factors and remainders. A number *modulo* another is simply the remainder after division:

$$14 \ \bmod \ 3 = 2$$

One way to calculate the modulo remainder is to use floating-point (fractional) calculations and the *floor* operator, where $\lfloor x \rfloor$ means rounding down x to the next lowest integer:

$$
\begin{aligned}
14 \ \bmod \ 3 &= \left(\frac{14}{3} - \left\lfloor \frac{14}{3} \right\rfloor \right) \cdot 3 \\
&= (4.6667 - 4) \cdot 3 \\
&= 2
\end{aligned}
$$

where $\lfloor x \rfloor$ is an integer that results from rounding down the fractional number x. The `floor` and `round` operators are built-in to MATLAB, so we can calculate the modulo result using either of these methods:

```
14/3
ans =
      4.6667
```

```
floor(14/3)
ans =
     4
14/3 - floor(14/3)

ans =
     0.6667

(14/3 - floor(14/3))*3
ans =
     2.0000

mod(14,3)
ans =
     2
```

Generalizing this,

$$a \bmod N = \left(\frac{a}{N} - \left\lfloor \frac{a}{N} \right\rfloor \right) \cdot N \tag{6.8}$$

This is a useful method for calculating modulo-division remainders in public-key encryption and decryption if floating-point arithmetic computations are available.

For any integers (counting numbers) a and b, we can write the larger number as a product of an integer and the smaller number, plus a remainder

$$a = qb + r \tag{6.9}$$

Here q is the quotient and r is the remainder. Of course, in the case where b divides into a perfectly, the remainder will be zero, but that is just a special case with $r = 0$. Then we usually term b a *factor* of a. Equivalently, we may write

$$\frac{a}{b} = q \text{ rem } r \tag{6.10}$$

For example,

$$14 = 3 \cdot 4 + 2$$

or

$$\frac{14}{3} = 4 \text{ rem } 2$$

A *prime number* has only itself and one as factors. That is, no other number divides into it with zero remainder. This is a well-known definition. A less well-known definition is that of *relatively prime* numbers. Two numbers are said to be *relatively prime* if they share no common factors. For example, 20 and 8 have common factors of 2 and 4, whereas 20 and 7 have no common factors and thus are relatively prime.

If two numbers do have common factors, there may be several of them. Sometimes, we are interested in the *greatest common factor* (GCF). A procedure for this is known as Euclid's algorithm, which dates to antiquity. To illustrate with a numerical example, suppose we wish to find the GCF of 867 and 1802. Using the quotient and remainder definitions as discussed, we could write

$$1802 = 2 \cdot 867 + 68$$

If there is a common factor between 867 and 1802, then it must be able to divide into both with zero remainder. But if this is the case, then the above equation shows that it must also divide into 68 evenly. This implies that the GCF of 867 and 1802 must also be the GCF of 867 and 68. So we can in turn write

$$867 = 12 \cdot 68 + 51$$

Effectively, we have a remainder after division (here 51) that must be less than the divisor 68. Repeating the same logic, we have in turn

$$
\begin{aligned}
1802 &= 2{\cdot}867 + 68 \\
867 &= 12{\cdot}68 + 51 \\
68 &= 1{\cdot}51 + 17 \\
51 &= 3{\cdot}17 + 0
\end{aligned}
$$

Now that the (final) remainder is to zero, the last divisor (17) must be the GCF. That is, 17 goes into 1802 and 867 evenly and is the largest number to do so. Compare this to an exhaustive search, which tries all possible numbers.

```
for  k = 1:867
    rem1 = abs(floor(867/k) - 867/k);
    if( rem1 < eps )
        fprintf(1, '%d is a factor of 867\n', k);
    end

    rem2 = abs(floor(1802/k) - 1802/k);
    if( rem1 < eps )
        fprintf(1, '%d is a factor of 1802\n', k);
    end

    if( (rem1 < eps) && (rem2 < eps) )
        fprintf(1, '%d is a common factor\n', k);
    end
end
```

Prime number calculations such as outlined above pave the way for so-called one-way trapdoor functions for data encryption. Multiplying two numbers together is relatively easy, but determining the factors of a given number is a more difficult problem. If the numbers are exceedingly large, and very few factors exist (ideally only the two numbers in question), the problem of factorization is much more difficult. Prime numbers and modulo arithmetic may be utilized in the transfer of symmetric encryption keys as well as the related method of public-key encryption.

6.4 Bit Errors in Digital Systems

The *rate* of errors encountered in a digital transmission system is clearly something we wish to minimize, and/or compensate for. This section introduces the key concept of the BER and relates it to the system overall, the transmitted signal power, and the external noise encountered.

6.4.1 Basic Concepts

The BER of a given system or link is the number of bits received in error divided by the total number of bits received. A given digital communication system may have a seemingly very low error rate, for example, 1 in 10^6. But consider that at a rate of 100 Mbps, there are 100×10^6 bits per second, and thus there would be about 100 errors per second on average at this BER. This would be considered a very poor channel at this data rate.

Furthermore, many systems are comprised of cascades of individual systems, and each will contribute their own error rate. For example, Figure 6.4 illustrates three telecommunication blocks, each with their own bit error rate BER_n.

To determine the overall error rate, it is necessary to remember that in a series cascade as shown, just one error in one of the systems will manifest itself as an error in the overall block. So we must first convert the error rate into a success rate, which is $1 - BER_n$. Since the likelihood of successful transmission overall dependent upon *all* of the subsystems being error-free, it is calculated using the product of all of the independent success likelihoods. Finally, the likelihood of an error in the overall system is the complement of this, or one minus the success rate. So we have

$$BER = 1 - \prod_n (1 - BER_n) \tag{6.11}$$

where $\prod_n$ means "product over n values." For example, if errors of three system blocks are 1 in 100, 1 in 1000, and 1 in 10000, the overall error is calculated as

$\mathrm{BER}_n = $ Bit error rate of stage n

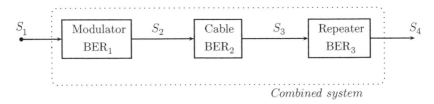

Figure 6.4 Errors in cascading systems.

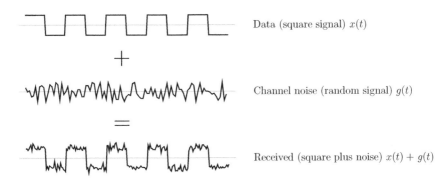

Figure 6.5 A binary 1/0 sequence with noise added.

```
ber = [1/100 1/1000 1/10000];
bertotal = 1 - prod(1-ber)
ans =
    0.0111
```

Thus, the highest rate (0.01) dominates the overall rate of 0.0111. In terms of reliability, adding more reliable systems in cascade with an unreliable one does not really help the overall reliability. In terms of communication systems, the weakest link is the one with the highest error probability.

Why do bit errors occur? Figure 6.5 illustrates a square wave, indicative of a binary data transmission with two levels: $+A$ and $-A$. Consider a square wave as an alternating 1/0 sequence (which of course could be any transmitted sequence of 1s and 0s). In the process of transmission and reception, noise is added. The resulting signal amplitude must be compared at the receiver against a threshold to decide whether a 1 or 0 was transmitted originally. Clearly, a significant amount of noise may cause an incorrect decision to be made.

The binary decision may be analyzed by first creating the desired points $\pm A$ on the plane as in Figure 6.6. The noise works to move the received point away

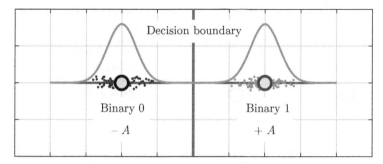

Figure 6.6 Two polar values +*A* and −*A*, with additive noise. The probability density shows the likelihood for each level at the receiver.

from the desired point. The random nature of the noise is captured in the probability density, which is superimposed above each expected point. In other words, the received signal is modeled as two PDFs. Importantly, an incorrect decision may be made when the PDF curves overlap. The extent of the overlap determines how often, on average, such incorrect decisions are made.

It is not necessary to limit attention to one bit per symbol interval transmitted. The *IQ* techniques discussed in Section 3.9.4, with, for example, 4 quadrant points, may result in the situation shown in Figure 6.7 when noise is added. Once again, provided the noise does not cross the decision boundary, the correct decision will always be made. But if the received signal does cross the decision boundary, an erroneous bit decision will be made.

It should be evident that in order to reduce the likelihood of an incorrect decision, larger signal amplitudes could be used, thus moving the −*A* point further away from the +*A* point. This approach increases the required signal power, which is usually only possible up to a certain level. The amount of power increase is relative to the noise present in the channel, so we conclude that the ratio of signal power to noise power is the critical quantity.

6.4.2 Analyzing Bit Errors

Two things influence the BER – the signal power and the noise power. The amount of noise present in the channel is certainly one factor. However we can, to some extent, mitigate this by increasing the power of the transmitted signal. This section aims to introduce a conceptual understanding of how signal power and noise influence the BER for a given type of modulation.

The channel noise itself is usually characterized as being random or nondeterministic. This is distinct to deterministic influences, which are predictable. Doppler shift in a carrier frequency, for example, is predictable. One common model for noise, as already discussed, is AWGN. Although the influence of noise can be reduced by using good design approaches, it is not possible to

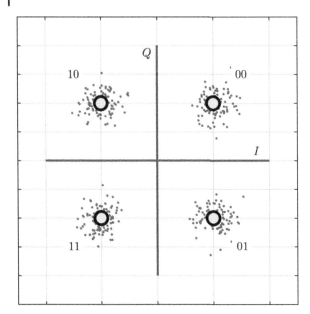

Figure 6.7 Extending the concept of received points to two orthogonal axes. Two bits are transmitted at a time, with the decision boundary being the axes themselves.

influence the external noise itself. Thus it is necessary to consider increasing the transmitter power. This is usually not desirable. Increased power results in shorter battery life for mobile devices, which is an undesirable characteristic. Additionally, there may be practical limits such as the creation of unwanted interference or even regulatory (legal) requirements regarding the maximum permissible transmission power.

Consider a simple case of a bipolar baseband digital waveform, where a binary value of 0 is represented as a voltage of $-A$, and a binary 1 uses a voltage of $+A$. Figure 6.8 shows a short sequence of bits recovered under such a scenario. The received signal amplitude provides the binary value for the assumed bit value. Errors where this differs from the transmitted data bitstream are indicated.

In order to analyze this scenario, Figure 6.9 shows the statistical distribution of received data levels as compared to their expected values of $\pm A$. The decision point (for two levels) is clearly midway between each. The shaded area indicates the likelihood that, if a binary 0 was transmitted, a binary 1 is decided upon by the receiver. Similarly, an argument could be made for the likelihood of a binary 0, if in fact a binary 1 was transmitted, by examining the complementary tail under the curve centered on $+A$.

The probability of a 1 being decided as the most likely bit value, given that 0 was transmitted, is written as $\Pr(1|0)$ – read as "probability of 1 received, given that 0 was sent." This corresponds to the area under the tail of the curve

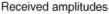

Received amplitudes

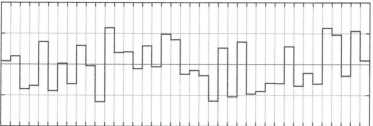

Data transmitted and received

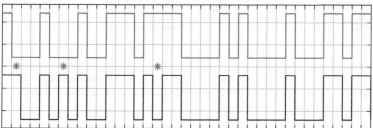

Figure 6.8 A small section of received data, together with the resulting bit stream. The bit errors are indicated as ∗. In some of these cases, the received amplitude is only just on the wrong side of the decision boundary, but incorrect nevertheless.

illustrated, and using the Gaussian PDF equation (6.1),

$$\Pr(1|0) = \frac{1}{\sigma\sqrt{2\pi}} \int_0^\infty e^{-[(x-\mu)/\sigma\sqrt{2}]^2} \, dx \tag{6.12}$$

The average is $\mu = -A$ for this case, so to simplify this we let

$$u = \frac{x - (-A)}{\sigma\sqrt{2}} \tag{6.13}$$

The derivative of this is

$$dx = \sigma\sqrt{2} \, du \tag{6.14}$$

Changing the variable of integration requires changing the integration limits

$$x \to \infty \Rightarrow u \to \infty$$

$$x = 0 \Rightarrow u = \frac{A}{\sigma\sqrt{2}}$$

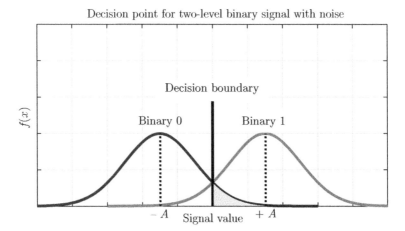

Decision point for two-level binary signal with noise

Figure 6.9 Two possible levels sent (±A) and received (PDFs centered on ±A). The shaded area indicates the probability of 0 being sent, yet with a decision made in favor of 1 at the receiver. Both the signal amplitudes and the statistical distribution of the noise influence whether the correct decision is made for each bit.

So the new expression to be evaluated is

$$Pr(1|0) = \frac{1}{\sigma\sqrt{2\pi}} \int_{+A/\sigma\sqrt{2}}^{\infty} e^{-u^2} \sigma\sqrt{2} \, du$$

$$= \frac{1}{\sqrt{\pi}} \int_{u}^{\infty} e^{-u^2} \, du \tag{6.15}$$

where the lower limit is $u = A/\sigma\sqrt{2}$. Comparing this expression to the definition of erfc in Equation (6.3), we have

$$Pr(1|0) = \frac{1}{2}\text{erfc}\left(\frac{A}{\sigma\sqrt{2}}\right) \tag{6.16}$$

Similar steps may be followed to find the probability of deciding upon 0 given that 1 was actually sent. The symmetry of the figure shows that this will be the same as before, and so we just use the same expression for Pr(0|1).

The total probability of error P_e is equal to the likelihood of selecting 0 when a 1 was transmitted or selecting 1 when a 0 was transmitted. Assuming that 1 and 0 are equally likely, each of these probabilities must then be weighted by 50%, with the overall probability of error then being

$$P_e = 0.5 \, Pr(0|1) + 0.5 \, Pr(1|0) \tag{6.17}$$

As deduced above, $\Pr(1|0) = \Pr(0|1)$ and the overall probability of error simplifies to

$$P_e = \frac{1}{2}\text{erfc}\left(\frac{A}{\sigma\sqrt{2}}\right) \tag{6.18}$$

As noted earlier, power is the key issue, so we arrange to have the squares of A and σ:

$$P_e = \frac{1}{2}\text{erfc}\left(\sqrt{\frac{A^2}{2\sigma^2}}\right) \tag{6.19}$$

While this gives the desired result, it is usual to incorporate the required bandwidth as well as signal energy. This is in order to make a fair comparison between different modulation schemes. We thus define the signal energy per bit as E_b, which is the total power integrated over one symbol interval T_s. As a result, the peak power per symbol interval is $E_b = A^2 T_s$. The σ^2 term represents the noise power. In practical terms, this power cannot extend over an infinite bandwidth (otherwise the noise would have infinite power). So it is more usual to incorporate the noise energy per unit bandwidth, $N_o = \sigma^2/B$, where B is the bandwidth. Thus we have energy (in Joules) per unit bandwidth (in Hz). The bandwidth is $B = (1/2)(1/T_s)$. Combining these equations yields $N_o = 2\sigma^2 T_s$.

The ratio of signal energy to noise per bit E_b/N_o is then

$$\frac{E_b}{N_o} = \frac{A^2 T_s}{2\sigma^2 T_s}$$
$$= \frac{A^2}{2\sigma^2} \tag{6.20}$$

Substituting this in Equation (6.19),

$$P_e = \frac{1}{2}\text{erfc}\left(\sqrt{\frac{E_b}{N_o}}\right) \tag{6.21}$$

This gives a relationship between probability of error P_e and the ratio of peak energy per symbol to the noise power per unit bandwidth. This expression will be different depending on where $\pm A$ are placed, and it follows that for different types of modulation, different P_e expressions result.

Equation (6.21) is shown as the theoretical curve in Figure 6.10, which is calculated using the code below.

```
SNRbdB = linspace(0, 12, 400);
SNRbAct = 10.^(SNRbdB/10);
Pe = (1/2)*(erfc(sqrt(SNRbAct)));
plot(SNRbdB, Pe);
set(gca, 'yscale', 'log');
```

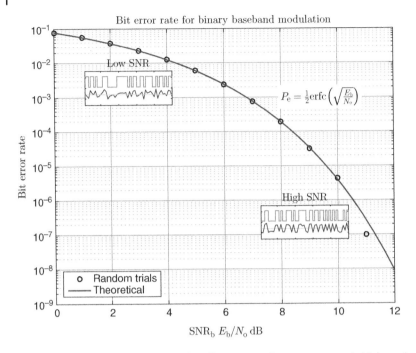

Figure 6.10 The theoretical and simulated bit error performance curves. At higher values of SNR per bit, increasing the signal power (or reducing the noise) results in a much greater reduction of the BER.

Note that increasing E_b/N_o (also called the SNR per bit, SNR_b) decreases the BER, but not in a linear fashion. At low SNR_b, an increase of 3 dB from 0 to 3 dB results in a reduction in BER from approximately 0.078 (about 1 error in 13) to 0.022 (about 1 error in 44). At higher SNR_b, an increase of 3 dB from 9 to 12 dB results in a reduction in BER from approximately 3.5×10^{-5} (about 3 in 10^4) to 9×10^{-9} (about 1 in 10^8). The proportional change is quite different.

To help verify the above theoretical approach, we can code a simulation that makes no assumptions, as follows. The approach is to generate a random bitstream, encode 1/0 as amplitudes $\pm A$, add noise according to the desired signal energy to noise ratio, and then set a 1/0 decision threshold on the resulting block of received samples. These are then compared with the original bitstream and the number of errors noted in proportion to the total number of bits in the simulation. Naturally, this simulation of a channel will not result in a precise value of BER and is dependent on the number of bits tested in the simulated data stream (variable N below). For the set of desired SNR values, the actual ratio is calculated using `EbNoRatio = 10 ^ (SNRbdB(snrnum)/10)`, which is effectively the reverse of the decibel conversion using $10\log_{10}(\cdot)$.

```matlab
N = 1000000;                      % number of bits in simulated
                                  % channel
SNRbdB = [0:1:12];                % SNR values to test
Ntest = length(SNRbdB);
A = 1;
BERsim = zeros(Ntest, 1);

for snrnum = 1:Ntest
    % desired simulated SNR
    EbNoRatio = 10^(SNRbdB(snrnum)/10);

    % simulated bit stream and data stream
    A = 1;
    tb = randi([0 1], [N, 1]);  % transmitted bit
    td = A*(2*tb - 1);          % transmitted data

    % Gaussian random noise
    g = randn(N, 1);
    varg = var(g);
    EbNoAct = (A^2)/(2*varg);

    % scale noise up so that resulting EbNo is one
    g = g*sqrt(EbNoAct);
    EbNo1 = (A^2)/(2*var(g));
    fprintf(1, 'Scaling EbNo to unity. EbNo1 = %f \n', ...
            EbNo1);

    % scale noise down to match desired EbNo
    g = g/sqrt(EbNoRatio);
    EbNo2 = (A^2)/(2*var(g));
    fprintf(1, 'Scaling EbNo to desired. EbNo2 = %f
                        Desired=%f\n', ...
                        EbNo2, EbNoRatio);

    % add noise to received data
    rd = td + g;

    % retrieve bitstream from data
    rb = zeros(N, 1);
    i = find( rd >= 0 );
    rb(i) = 1;

    % compare transmitted bitstream with received bitstream
    be = (rb ~= tb);
    ne = length(find(be == 1));
    BERest = ne/N;
    BERsim(snrnum) = BERest;
```

```
      BERtheory(snrnum)  =  1/2*erfc(sqrt(EbNoRatio));
end

plot(SNRbdB, BERsim, 's', SNRbdB, BERtheory, 'd', ...
     'linewidth', 2);
set(gca, 'yscale', 'log');
grid('on');
grid('minor');
xlabel('Eb/No');
ylabel('Pe');
```

The random data stream is generated using the random integer function randi(), which is then scaled according to the desired signal amplitude. Random Gaussian noise with unit variance is then generated to be added to the data stream. However, it must be scaled such that the actual E_b/N_o matches as closely as possible the desired value. First, the data vector is normalized such that $E_b/N_o = 1$. Then, the scaling could be accomplished by either multiplying the signal samples by the desired E_b/N_o, or dividing the noise by the desired E_b/N_o. Either achieves the same result.

Finally, the received bitstream is recovered by comparison of the noisy received signal to a decision threshold (in this case, zero) and determining the number of bit positions that differ. Figure 6.10 shows both the simulated data channel (circles) and the theoretically derived BER (lines) using Equation (6.21).

6.5 Approaches to Block Error Detection

In general, to add a degree of error tolerance to a data stream, we must add some additional, redundant information. It is of course desirable to minimize the amount of extra information added but at the same time the additional information ought to be able to detect any errors that might have occurred.

The exclusive-OR (XOR) binary function features in many error detection methods and is defined as

$$A \oplus B \triangleq A \cdot \overline{B} + \overline{A} \cdot B \tag{6.22}$$

By taking all possible combinations of binary variables (bits) A and B, we can write the truth table as shown in Table 6.1. For the case of two input bits, this function effectively acts as a detector of differences in the logic levels of the input – that is, the output is true (binary 1) when the inputs are different. This may be extended to multiple input bits if necessary, in which case it is effectively a detector that is true (binary 1) for an odd number of input 1s.

Table 6.1 The XOR function truth table.

A	B	$A \oplus B$
0	0	0
0	1	1
1	0	1
1	1	0

Conventionally, 1 is logically "true" and 0 is "false."

Table 6.2 Examples of computation of an even parity bit.

b_7	b_6	b_5	b_4	b_3	b_2	b_1	b_0	b_{parity}
1	0	1	1	1	0	1	1	0
0	0	1	1	0	0	1	0	1
1	1	0	1	1	0	0	1	1
1	0	1	1	1	1	0	1	0

Some of the earliest approaches to error detection require a single parity bit to be generated and sent along with the data (typically an 8-bit byte). The decoder also generates a parity bit from the data received and compares that with the received parity bit. This is illustrated in Table 6.2, which shows the generation of an even parity bit, so that the total number of 1 bits is even. This is easily generated using the XOR function in cascade. In effect, the XOR of two bits produces a single bit that represents the even parity bit.

At first sight, it might appear that this approach would detect any errors in the 8-bit pattern with a high degree of certainty. After all, if one bit was in error, it effectively becomes inverted, and this would be flagged as having the incorrect parity. But if a second bit were also inverted, the data byte would have the same parity as the original, and thus the error would not be detected. In reality, errors often occur in bursts, meaning that several sequential bits tend to be the subject of errors. Additionally, we must consider the case where the parity bit itself might be corrupted. Furthermore, the efficiency of this approach is not particularly good – for every 8 data bits, an extra bit must be transmitted just to detect errors. Finally, this approach does not even attempt to *correct* errors.

This scheme could be extended by buffering several bits, and calculating more than one parity bit. Suppose 16 bits are arranged in a 4×4 matrix as shown in Figure 6.11. Horizontal parity bits are calculated along each row, and vertical

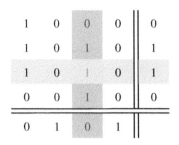

Figure 6.11 Computation of two-dimensional parity. There are 16 data or message bits in the 4 × 4 block, and 4 + 4 parity bits in the final row and column.

Table 6.3 A naïve repetition code.

Original bits		Coded bit-stream								
0	0	0	0	0	0	0	0	0	0	0
0	1	0	0	0	0	0	0	1	1	1
1	0	0	0	0	1	1	1	1	1	1
1	1	1	1	1	1	1	1	1	1	1

Nine bits are sent for every two real data bits.

parity bits down each column. Of course, this scheme is not particularly efficient as shown: 8 check bits are needed for 16 data bits, so there is considerable overhead.

6.5.1 Hamming Codes

At this point, it's not unusual to ask why we don't just repeat data bits, and use a "majority rule." The short answer is that such a scheme would be terribly inefficient. However, to seek a better scheme, suppose we use a simple repetition code as defined in Table 6.3.

Suppose the two bits 00 are to be encoded; the corresponding bit pattern to be transmitted is 000 000 000. Upon receiving this 9-bit pattern, the receiver translates it back to 00. Suppose, however, that in transmission it gets corrupted to the bit pattern 000 000 010. To the decoder, it is fairly clear that this bit pattern should have been 000 000 000, and hence it corresponds to the original bit sequence 00. This is based on the observation that only valid codewords should exist, and what was received was not a valid codeword. The codeword that appears to be the closest match is selected.

Before we formalize the definition of "closest match," consider what would happen if two bits were corrupted, such that the received codeword was 000 000 011. This would be declared as an invalid codeword (a correct assumption), but under the assumptions of "closeness," we might erroneously assume that the codeword transmitted was actually 000 000 111, and so the original bits were 01.

Thus, this system is not infallible. In fact, we can characterize all such error detection schemes in this way, in terms of the probability of detecting an error.

To formalize this notion of closeness of bit patterns, we can define the concept of an *error distance*, commonly called the Hamming distance (Hamming, 1950). Let $d(\mathbf{x}, \mathbf{y})$ be the number of locations in which codewords $\mathbf{x}$ and $\mathbf{y}$ differ. If two of these codewords are defined as

$$\mathbf{x} = 0\ 1\ 1\ 0 \quad 1\ 1\ 0\ 1$$
$$\mathbf{y} = 0\ 1\ 1\ 1 \quad 0\ 1\ 1\ 1$$

then the distance $d(\mathbf{x}, \mathbf{y}) = 3$. That is, they differ in three bit positions as illustrated. This notion may be extended to any set of codewords. Clearly, between any pair in the set of valid codewords, the Hamming distance may be determined. Within this set of codewords, there will be a minimum distance between one (or more) of the pairs of codewords. It is this minimum distance that governs the worst-case performance of an error detection system. As a result, the terminology *minimum distance* or d_{min} of a codeword is used to mean the *smallest* Hamming distance between any two codewords in the codeword set.

To give a concrete example, suppose the codewords are

$$\mathbf{x} = 0\ 1\ 1\ 0$$
$$\mathbf{y} = 0\ 1\ 0\ 0$$
$$\mathbf{z} = 1\ 0\ 0\ 1$$

Then there are three distances to be found: between $\mathbf{x}$ and $\mathbf{y}$, between $\mathbf{x}$ and $\mathbf{z}$, and between $\mathbf{y}$ and $\mathbf{z}$. These are

$$d(\mathbf{x}, \mathbf{y}) = 1$$
$$d(\mathbf{x}, \mathbf{z}) = 4$$
$$d(\mathbf{y}, \mathbf{z}) = 3$$

So the smallest of the distances is $d_{min} = 1$.

The reason why we define the minimum distance between codewords is that it determines the overall performance in terms of error detecting and correcting capability. Using the repetition code of Table 6.3 as an example, if two codewords are a distance d apart, it would require d single-bit errors to occur on the communication channel in order to convert one valid codeword into another. If this occurs, one valid codeword would be converted into another valid codeword, resulting in false decoding, since the error would not be detected as such. Mathematically, this means that in order to *detect* d errors, it is necessary that

$$d_{min} = d + 1 \tag{6.23}$$

Furthermore, if we wish to *correct* for d errors, we would need a distance $2d + 1$ code set. This is because even if d changes occurred, the closest original codeword could still be deduced. Mathematically, this means that in order to *correct d* errors, it is necessary that

$$d_{min} = 2d + 1 \tag{6.24}$$

Returning to the repetition code of Table 6.3, the minimum distance d_{min} is 3. This would permit detection of either 1 or 2 bits in error. Three bits in error, however, may be missed. Mathematically, the Hamming distance equation is $3 = d + 1$ and so $d = 2$ bit errors would be detected. However, it would only permit *correction* of one bit error (since $3 = 2d + 1$ and so $d = 1$). Both of these cases may be demonstrated by changing one and then two bits in the codewords of Table 6.3.

These arguments provide a performance bound for error detection and correction, given a set of codewords. But how are the codewords themselves determined? Clearly, they would need to be as small as possible so as to minimize the amount of additional information necessary in order to gain the advantage of error detection. How many additional bits are required, at a minimum? This problem was succinctly addressed by Hamming (1950). The following parameters are defined to be clear about what defines the message (data), and what defines the check bits:

M = The number of message or data bits to begin with.

N = The number of bits in total in the codeword.

C = The number of "redundant" (or "check") bits.

In our simple and somewhat arbitrary repetition code, there were $N = 9$ bits in each codeword, with $M = 2$ data bits effectively transmitted for each codeword and C check bits. Effectively, the number of redundant check bits is $N - M = 9 - 2 = 7$. Note that these check bits are not explicitly identified as such.

The number of valid states in the set of codewords is 4, calculated as $2^M = 2^2$. The total number of possible received codewords is $2^9 = 2^N$ (512 for this example). Thus the number of *invalid* states is $2^N - 2^M = 512 - 4$. Now, since the total number of bits in a codeword is the number of message bits plus the number of check bits, $N = M + C$. So the number of *invalid* codewords is

$$N_i = 2^N - 2^M$$
$$= 2^{M+C} - 2^M$$
$$= 2^M(2^C - 1) \tag{6.25}$$

Since there are $N_v = 2^M$ valid codewords, the ratio of invalid codewords to valid codewords is

$$\frac{N_i}{N_v} = \frac{2^M(2^C - 1)}{2^M}$$
$$= 2^C - 1 \tag{6.26}$$

Now take the case of a single-bit error occurring. If a single-bit error occurs, it could occur in any one of the N bit positions, and so the number of possible *single-bit error* patterns is $N_o = N \cdot 2^M$ (this is $9 \cdot 2^2$ in the example).

To ensure unambiguous decoding, we don't want two single-bit error patterns to be equally close in Hamming distance (otherwise the decoding would be ambiguous – which would we pick as the closest codeword?). In order to detect and correct single-bit errors, the number of invalid codewords, which we derived as $N_i = 2^M(2^C - 1)$, must be greater than (or at the very least, equal to) the number of single-bit errors. The latter was derived to be $N_o = N \cdot 2^M$. Thus the condition is

$$N_i \geq N_o$$
$$\therefore \quad 2^M(2^C - 1) \geq N \cdot 2^M$$
$$(2^C - 1) \geq N$$
$$(2^C - 1) \geq M + C \tag{6.27}$$

Finally, it is possible to relate the number of check bits required for a given number of data bits. Table 6.4 tabulates the values for data (message) bits and the required number of check bits as determined by the inequality given in Equation (6.27).

This provides a key to the number of check bits required, but it does not tell us how to actually calculate the check bits from a given set of data bits. So how are the check bits themselves derived? Fortunately, an ingenious procedure devised by Hamming (1950) provides the solution. It is able to construct codes, which can both detect and correct single-bit errors and detect double-bit errors.

A single-error correcting Hamming code is constructed as follows. The usual nomenclature is to define a Hamming $H(N, M)$ code, meaning a Hamming code of total length N bits, with M message bits. This implies $C = N - M$ check bits are necessary. Consider the simple case of a $H(7, 4)$ code. We first number the bit positions from 1 (LSB) to 7 (MSB), and define the four message databits as $m_3 m_2 m_1 m_0$, and the $7 - 4 = 3$ check bits as $c_2 c_1 c_0$. Then C check bits are placed in power-of-two positions (in this case, 1, 2 and 4). Thus we have at this point

$$
\begin{array}{ccccccc}
7 & 6 & 5 & 4 & 3 & 2 & 1 \\
& & & c_2 & & c_1 & c_0
\end{array}
$$

Then the M data bits are allocated to the remaining positions, to fill up the N positions in total. The code becomes

$$
\begin{array}{ccccccc}
7 & 6 & 5 & 4 & 3 & 2 & 1 \\
m_3 & m_2 & m_1 & c_2 & m_0 & c_1 & c_0
\end{array}
$$

Table 6.4 Calculating the required number of check bits to satisfy $(2^C - 1) \geq M + C$. Only selected values of M are tabulated for Hamming (N, M) codes.

M	C	$2^C - 1$	$N = M + C$
4	1	1	5
4	2	3	6
4	3	7	7
4	4	15	8
7	3	7	10
7	4	15	11
8	1	1	9
8	2	3	10
8	3	7	11
8	4	15	12
8	5	31	13
16	4	15	20
16	5	31	21
32	6	63	38

Next, write below each bit position the indexes' binary code ($1 = 001, 2 = 010, 3 = 011, \ldots$).

```
          7   6   5   4   3   2   1
         m3  m2  m1  c2  m0  c1  c0
MSB  1   1   1   1   0   0   0
     1   1   0   0   1   1   0
LSB  1   0   1   0   1   0   1
```

It is significant that each check bit has a single 1 below it. The check equations are written by XORing ($\oplus$) along each row corresponding to a 1 in the check bit position:

$$c_0 = m_3 \oplus m_1 \oplus m_0 \tag{6.28}$$

$$c_1 = m_3 \oplus m_2 \oplus m_0 \tag{6.29}$$

$$c_2 = m_3 \oplus m_2 \oplus m_1 \tag{6.30}$$

Thus, it may be seen that c_0 checks m_0, m_1, m_3, and so forth for the other check bits. The set of check bits are placed in the data word to define the entire codeword – that is, the message bits and check bits are interleaved.

At the decoder, it is necessary to use both the message bits and check bits to determine if an error occurred. Recall from the earlier discussion that a value that is XORed with itself equals zero. So we take each check equation in turn, and XOR both sides to get the error syndrome bit s_0,

$$s_0 = c_0 \oplus c_0$$
$$= c_0 \oplus m_3 \oplus m_1 \oplus m_0 \tag{6.31}$$

and similarly for s_1 and s_2. Note that the first line $c_0 \oplus c_0$ defines a quantity that will be zero by definition, and thus with no errors the second line will also equal zero. If a syndrome bit does not equal zero, it indicates that an error has occurred. Importantly, *the binary value of the syndrome points to the bit error position.*

To give a concrete example, consider that the data (message) to be sent is $m_3m_2m_1m_0 = 0101$. Using the check bit equations above, we have in this case

$$c_0 = 0 \oplus 0 \oplus 1 = 1$$
$$c_1 = 0 \oplus 1 \oplus 1 = 0$$
$$c_2 = 0 \oplus 1 \oplus 0 = 1$$

Assuming that the data and check bits are received without error, the error syndrome would be

$$s_0 = 1 \oplus 0 \oplus 0 \oplus 1 = 0$$
$$s_1 = 0 \oplus 0 \oplus 1 \oplus 1 = 0$$
$$s_2 = 1 \oplus 0 \oplus 1 \oplus 0 = 0$$

Thus, the error syndrome bits are all zero, indicating that no error has occurred.

Now suppose there is an error in bit m_2, and hence m_2 is inverted from 1 to 0. Repeating the above calculations for check bits and syndrome bits, the syndrome is

$$s_0 = 1 \oplus 0 \oplus 0 \oplus 1 = 0$$
$$s_1 = 0 \oplus 0 \oplus 0 \oplus 1 = 1$$
$$s_2 = 1 \oplus 0 \oplus 0 \oplus 0 = 1$$

The bit pattern $s_2s_1s_0 = 110$ in binary, or 6 decimal. Therefore, bit position 6 is in error, which points to m_2. Correcting the error is simply a matter of inverting this bit.

Clearly, the code also needs to check the check bits themselves. Suppose bit c_1 is inverted in transmission. The syndrome then is

$$s_0 = 1 \oplus 0 \oplus 0 \oplus 1 = 0$$
$$s_1 = 1 \oplus 0 \oplus 1 \oplus 1 = 1$$
$$s_2 = 1 \oplus 0 \oplus 1 \oplus 0 = 0$$

The bit pattern $s_2 s_1 s_0 = 010$ in binary or 2 decimal. Therefore, bit position 2 is in error, which points to c_2. Correcting the error is simply a matter of inverting this bit. Of course, it is not strictly necessary to do this because the check bit is not part of the useful data for the end receiver. However, since an error occurred, it is important that it defines which bit is erroneous, since the non-zero bit pattern indicates that an error did in fact occur.

All of the above works quite well for correcting single-bit errors, provided we group the data in blocks of 4, which is a significant disadvantage if we wish to correct longer bursts of errors. One way of mitigating this is to interleave blocks of data, as illustrated in Figure 6.12. In this case, the bits comprising each Hamming codeword are spread across several blocks. Thus, a burst of errors would need to last as long as one (vertical) block if it were to affect more than one (horizontal) codeword. Because Hamming codes can correct single-bit errors, each Hamming codeword then corrects its own block, and the net result is that the longer burst is also corrected. Of course, this approach requires a longer buffering delay, and hence total transmission time.

6.5.2 Checksums

In many cases, it is not necessary to correct for data errors, but merely to detect the fact that an error has occurred. This is a common requirement for Internet Protocol (IP) data transfer. A common method used is the *checksum*, which requires a form of summation of the data itself. The exact form of the checksum has many variants, involving various error detection capabilities (which should be maximized) and computational complexity (which should ideally be minimized). Checksums are suitable for computation in software, and as such have found widespread use.

The core idea of the checksum is to partition the data into a convenient size such as 8-bit bytes or 16-bit words. The data is then treated as a sequence of integers, and a summation formed. The end result is a single integer quantity that represents the uniqueness of that data block. The checksum is computed by the sender, and a similar computation performed at the receiver; a different result indicates an error in transmission.

Several questions arise from this basic description. For any moderate-sized data block, the summation is likely to overflow. That is, it cannot be represented in the 8 or 16 bits available for the checksum value. This is handled

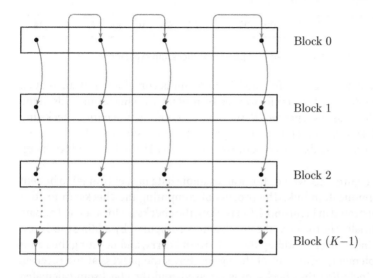

Figure 6.12 Block interleaving of single-bit correcting codes. All the data blocks are buffered in memory, and a Hamming code is computed for each horizontal block. The transmission is then ordered vertically, taking one bit from each block in turn, in the order indicated.

in various ways in different checksum algorithms. One approach is simply to discard the overflow; for 8-bit accumulation of the sum, this is effectively a modulo-256 summation. A second, improved approach is to wrap the overflow around and add it back in to the checksum. This is termed an *end-around carry*. For example, suppose we add the hexadecimal values E7 and 46. The result is 012D, which is clearly greater than can be contained in an 8-bit accumulator. End-around carry simply takes the overflow of 01 and adds it to the value 2D (that is, the value without the overflow), to form the result 2E.

A more subtle issue is the method of addition for the checksum. While a simple addition may have the advantage of lower complexity, if some bytes in a data block are reversed, then the same checksum would result. This may be addressed by forming a summation that is weighted by the position of each byte in the block.

It is necessary to embed the checksum value in the data stream, either at the end of the data message or in the header prepended to the data block. If the checksum value is included within the data header, as occurs with IP, the decoding steps are more complicated as it requires an explicit exception to the addition at the point where the checksum is stored. In practice, this problem is usually overcome by transmitting a modified checksum value, such that the checksum calculated over all the data, including the modified checksum, should be zero.

| 08 | 00 | *cc* | *cc* | 08 | 01 | 02 | 00 | B6 | EA | 95 | 10 | AA | AA | AA | AA |

Figure 6.13 A portion of a captured data packet for checksum evaluation.

The original specification for the IP checksum is contained in Braden et al. (1988). As IP data packets are forwarded by routers, the checksum fields in the packet header must be updated. Rather than recompute the entire checksum, it may be desirable to incrementally update the checksum based only on the changes in certain fields. This is addressed in RFC1141 (Mallory and Kullberg, 1990).

Consider Figure 6.13, which shows an example data packet (frame) to be sent over a communication link. The process of computing the checksum for this packet is illustrated in Figure 6.14. On the left, the checksum location in the data packet is initially set to zero. The checksum is computed by adding the 16-bit data words in the packet, adding the end-around carry, and negating the result (one's complement). The right of the figure shows the process at the receiver. The packet including the checksum is received, and the checksum calculated on the entire packet (making no special provision for the checksum bytes). The computation is exactly the same – add the 16-bit data words, add the overflow carry in an end-around fashion, and finally inverting the result. However, the end result will be all zero if there were no errors in the packet or the checksum bytes themselves.

This type of checksum computation – for both sending and upon receipt – is routinely performed for IP datagrams as well as TCP segment. One real-world problem that occurs in that case is the so-called endian ordering of the host processors at each end. In little-endian processors, the least-significant byte of a 2-byte quantity is stored in the lower memory address, as illustrated in the mapping from the data packet (Figure 6.13) to the in-memory checksum computation (Figure 6.14). Many processors store data in the opposite order – so-called big-endian ordering, with the lower memory byte address storing the *most significant* byte. This is shown in Figure 6.15. The lowest memory address still stores the first byte of the data packet (08 hexadecimal in the example), however if the data bytes taken sequentially are taken as 16-bit quantities, the 08 becomes the higher byte. Put another way, given that the first bytes of the packet are 08 00, the little-endian ordering would interpret this as 0008, whereas the big-endian ordering interprets this as 0800.

Clearly, this different ordering has significant implications for arithmetic operations on a block of data. If the results are to stay within the memory of one processor, then no problems occur. However, if the data is sent to (or received from) a machine with the opposite endian ordering, the checksum bytes may become reversed. Figure 6.15 shows that, even though the byte order is reversed due to the endian ordering, the net result is the same. In this case, the sum using little-endian ordering is AE4C, and using big-endian ordering it

	Sender					Receiver		
	$n+1$	n				$n+1$	n	
	AA	AA	High memory			AA	AA	High memory
	AA	AA				AA	AA	
	10	95				10	95	
	EA	B6				EA	B6	
	00	02				00	02	
	01	08				01	08	
	CC	CC				AE	4C	
	00	08	Low memory			00	08	Low memory
2	51	B1	Sum		2	FF	FD	Sum
	51	B3	End carry			FF	FF	End carry
	AE	4C	Complement			00	00	Complement

Figure 6.14 Checksum computation with the low-order byte of a 16-bit quantity first (little-endian ordering).

	Sender					Receiver		
		n	$n+1$				n	$n+1$
High memory		AA	AA		High memory		AA	AA
		AA	AA				AA	AA
		95	10				95	10
		B6	EA				B6	EA
		02	00				02	00
		08	01				08	01
		CC	CC				4C	AE
Low memory		08	00		Low memory		08	00
Sum	2	B3	4F		Sum	2	FF	FD
End carry		B3	51		End carry		FF	FF
Complement		4C	AE		Complement		00	00

Figure 6.15 Checksum computation with the high-order byte of a 16-bit quantity first (big-endian ordering).

is 4CAE. However, inserting the bytes into memory in the same order as the endian order of the processor results in the same sequence – 08 00 4C AE … using little-endian ordering, and 08 00 4C AE … using big-endian ordering.

Other variants of the basic checksum have been defined. The Fletcher checksum (Fletcher, 1982) computes a checksum based on the position of bytes within the data. A useful summary of checksum algorithms for data communications and their performance in terms of error detection may be found in Maxino and Koopman (2009). Checksums also find application in other areas, such as the Luhn checksum for credit card number verification.

6.5.3 Cyclic Redundancy Checks

The checksum discussed in the previous section is one type of data sequence check that may be used to check a frame (or block) of data. It is one class of Frame Check Sequence (FCS) used to ensure data integrity. Another class is the Cyclic Redundancy Check (CRC), which is introduced in this section.

Generally speaking, checksums are more suited to software computation, whereas CRCs are more suited to calculation via hardware. Whereas the checksum is computed with addition operations, the CRC may be computed using shift registers and XOR gates. For this reason, CRCs are more often encountered as check sequences at the end of a data frame (the trailer), typically in Ethernet local area networks.

Before introducing the formalities of the CRC, we start with an example. The calculations are bit oriented rather than byte or word oriented as in the checksum. Rather than basing the calculation on addition (possibly with enhancements, such as end-around carry and/or modulo arithmetic), the CRC is based on division. The bit sequence to be transmitted is considered as an (very long) integer, and it is divided by another integer, called the *generator*. In any integer division, there will be a result (termed the quotient) and a remainder (which may or may not be zero). The remainder is used as the integrity check, or CRC.

It might appear odd that division would be used as a method to check bit transmission. After all, division is a time-consuming operation – think of division of successive subtraction of one number from another. Furthermore, error checking is usually performed at the data-link layer, implying speeds of Mbps or Gbps. However, as will be shown, the division operation can be recast as an XOR operation. This is not true division in the conventional sense, but rather a type of binary-field division. Consider Figure 6.16, which shows how we might set out a long division to compute 682/7. We first multiply 9×7 and write the result 63 as indicated, immediately below the dividend. Subtracting this from the corresponding digits of the dividend leaves a remainder of 5. Bringing down the next digit of the dividend gives 2 after the 5. Noting that 7×7 is less than 52, the final remainder is 3. Of course, larger numbers would simply require

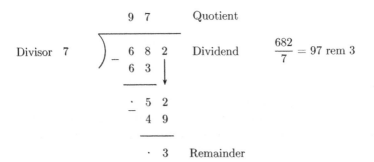

Figure 6.16 Division as a precursor to the CRC calculation.

iteration of the process over more steps. Note the salient aspects at each step: multiplication in order to obtain a number that has a final remainder less than the divisor, subtraction, and then finally bringing down the next digit.

Now consider an analogous process using binary arithmetic. Some observations are in order. Firstly, multiplication of an N-bit binary number by a binary digit 0 or 1 simply yields either zero, or the original number. Secondly, the shifting of the partial quotient at each stage in the decimal example is easily accomplished by a binary shift operation. Thirdly, and perhaps less obvious, is the fact that the remainder will always be less than the divisor. So in the decimal example, the divisor of 7 means that the remainder will range from 0 to 6.

The subtraction is replaced with a binary XOR operation. In a similar way to the observation that the remainder was always less than the divisor in decimal, in binary the number of bits occupied by the remainder will be one less than the number of bits in the divisor. So for an N bit divisor, the remainder will require $N - 1$ bits.

In the following example, we will use the binary sequence 1101 0011 as the message, which becomes the dividend. The divisor is chosen as 1011. When performing CRCs, this quantity is called the *generator* or *generator polynomial* (the polynomial representation will be explained subsequently).

The first step in the binary process is to write the data message with zeros appended. The number of zeros is one less than the size of the divisor as deduced above, hence three 0 bits are appended. Writing the divisor to the left results in

$$
1\ 0\ 1\ 1\ \overline{)\ 1\ 1\ 0\ 1\ 0\ 0\ 1\ 1\ 0\ 0\ 0}\qquad
\begin{array}{l}\text{Message}\\ +\ \text{three 0 bits}\end{array}
$$

Generator
polynomial

The next step begins the iteration of multiplication and reduction. We multiply the generator by either 1 or 0, according to whether the leftmost bit of the dividend (or partial result in subsequent stages) is 1 or 0. The 1 or 0 is multiplied by the generator, with the result written below as a partial product. In this example, it is a 1

$$
\begin{array}{l}\qquad\qquad\qquad 1\\[4pt]
1\ 0\ 1\ 1\ \overline{)\ 1\ 1\ 0\ 1\ 0\ 0\ 1\ 1\ 0\ 0\ 0}\qquad \begin{array}{l}\text{Message}\\ +\ \text{three 0 bits}\end{array}\\[2pt]
\qquad\qquad\quad\ 1\ 0\ 1\ 1\end{array}
$$

Generator
polynomial

The third step is to perform the XOR of the partial product with those bits immediately above it (it may help to remember that the XOR of two bits is true [binary 1] if the input bits are different). The leftmost XOR result will always be zero, but it is shown as a dot here, as it is ignored from now on:

```
                    1
                 ┌─────────────────────
        1 0 1 1  )  1 1 0 1 0 0 1 1 0 0 0      Message
                 ⊕  1 0 1 1                     + three 0 bits
    Generator       ─────────
    polynomial       · 1 1 0
```

The next step is to bring down the next bit from the message, to form a 4-bit number as a partial result:

```
                    1
                 ┌─────────────────────
        1 0 1 1  )  1 1 0 1 0 0 1 1 0 0 0      Message
                 ⊕  1 0 1 1 ↓                   + three 0 bits
    Generator       ─────────
    polynomial       · 1 1 0 0
```

Subsequent steps repeat the above, multiplying and producing a partial product, which we XOR to get a result with one fewer bits, and then bring the next message bit down:

```
                    1 1
                 ┌─────────────────────
        1 0 1 1  )  1 1 0 1 0 0 1 1 0 0 0      Message
                 ⊕  1 0 1 1 ↓                   + three 0 bits
    Generator       ─────────
    polynomial       · 1 1 0 0
                       1 0 1 1 ↓
                       ─────────
                        · 1 1 1 0
                       ⊕  1 0 1 1
                       ─────────
```

At the point shown below, we have to multiply the generator by a zero, in order to cancel the leftmost bit as in earlier steps:

```
                    1 1 1 1 0
                 ┌─────────────────────
        1 0 1 1  )  1 1 0 1 0 0 1 1 0 0 0      Message
                 ⊕  1 0 1 1 ↓                   + three 0 bits
    Generator       ─────────
    polynomial       · 1 1 0 0
                       1 0 1 1 ↓
                       ─────────
                        · 1 1 1 0
                       ⊕  1 0 1 1 ↓
                        ─────────
                         · 1 0 1 1
                        ⊕  1 0 1 1 ↓
                         ─────────
                          · 0 0 0 1
                         ⊕  0 0 0 0
                          ─────────
```

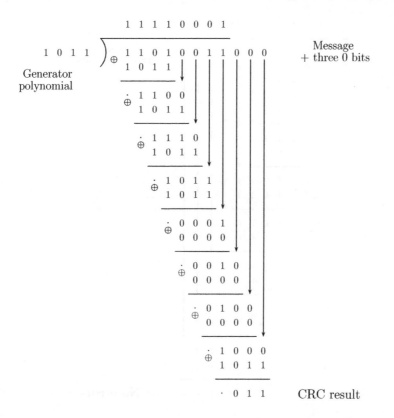

Figure 6.17 All the steps involved in the CRC calculation at the sender, with the final result shown.

This process continues until all of the input bits, including the three appended zeros, are exhausted. At that point, three remainder bits are left. The full process to completion is shown in Figure 6.17, where the remainder is shown to be 011.

The bitstream is then transmitted with the message first, followed by the calculated remainder bits. Thus, the receiver sees a data sequence with the remainder appended to it. Now, recall the earlier decimal example where we had 682/7, which gave a remainder of 3 (decimal). If we had subtracted the remainder from the "data" of 682, the remainder would be 0. That is, 679 is evenly divisibly by 7. A parallel may be drawn in binary, except that we do not subtract the remainder, but rather append it. The remainder simply takes the place of the three 0 bits, which were appended.

The calculation at the receiver, incorporating the calculated remainder bits, is shown in Figure 6.18. Identical steps are employed as in the transmitter, except that the zero padding is replaced with the sender-calculated remainder. If the

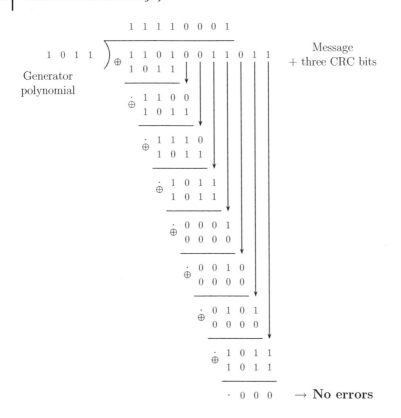

Figure 6.18 The steps involved in CRC calculation at the receiver, assuming no errors have occurred in transit.

message and CRC bits have been received without error, the remainder will be zero. This indicates to the receiver that there were no errors in transit:

The goal of the CRC is to detect errors, so we now examine the same example with an error embedded. Suppose the two bits shown boxed in Figure 6.19 were inverted, resulting in a 2-bit error burst. The figure shows the same calculation steps that are undertaken upon receipt of data to yield a final remainder of 110. This is not zero, and it indicates the presence of an error.

There are some circumstances under which error detection schemes can fail, and the CRC is no exception. Figure 6.20 shows the same data sequence and generator once again, except that a 4-bit error burst occurred as indicated by the boxed set of bits. Iteration of the CRC process will show that the final remainder is zero, which under normal circumstances would indicate that no errors have occurred. However, this is not the case.

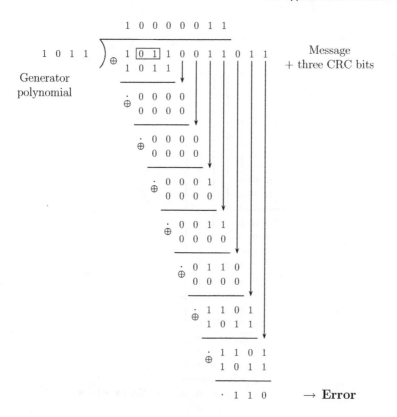

Figure 6.19 The steps involved in CRC calculation at the receiver, when an error has occurred in transit. The error is detected in this case.

Under what circumstances can the CRC fail – that is, not detect an error? Careful examination of Figure 6.20 will show that the 4-bit error pattern is contrived to be the XOR of the original data pattern 1001 with the generator pattern 1011, as shown below:

$$
\oplus\ \frac{\begin{array}{cccc} 1 & 0 & 0 & 1 \\ 1 & 0 & 1 & 1 \end{array}}{\begin{array}{cccc} 0 & 0 & 1 & 0 \end{array}}\ \ \begin{array}{l} \text{Original data} \\ \text{Generator bits} \\ \text{Actual bits in error} \end{array}
$$

The resulting error pattern of 0010 is what is replaced in the data stream. An analogous case in the decimal example is where the divisor (7) was subtracted from the message. That, of course, would not change the decimal remainder. The likelihood of this occurring is exceedingly small, and both the length and composition of the generator bit sequence determine the probability of missing an error. Thus, the choice of generator is critical.

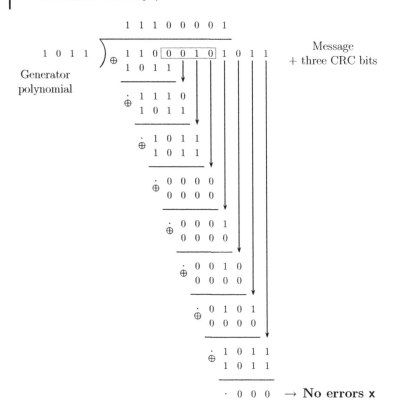

Figure 6.20 The steps involved in CRC calculation at the receiver, when an error has occurred in transit. The error is *not* detected in this case.

It should be noted that the data sequence protected by the CRC is generally much, much longer than that used in the simple example above. For example, an Ethernet frame of 1500 bytes would be protected by a 32-bit CRC.

The field of CRC error detection is quite extensive, and much literature is devoted to their performance in terms of error detection, the selection of the generator polynomial, and their complexity. They are employed in many data communication applications such as Bluetooth and Ethernet.

It was noted above that the generator bit sequence is also termed a *generator polynomial*. The reason for this is that the analysis of CRCs considers them as a division process, and polynomials are an ideal structure for capturing the division process. CRC polynomials are then written in the form of aX^n where a is the binary value of the coefficient and n is the bit position. For example, the previous divisor of 1011 may be written as

$$g(X) = 1X^3 + 0X^2 + 1X^1 + 1X^0 \tag{6.32}$$

This is usually abbreviated to

$$g(X) = X^3 + X + 1 \tag{6.33}$$

following the usual rules of exponents and neglecting the zero terms. Note that a generator polynomial will always have its MSB and LSB set and have a bit length one more than the desired CRC value.

6.5.4 Convolutional Coding for Error Correction

Previous methods described for error detection and correction employ a fixed block of data, for which a parity check is computed in some way. The original data block is transmitted to the receiver unchanged, with the error check digits appended (or sometimes prepended) to the block. The class of error detection and correction termed *convolutional coding* adopts a different approach. Essentially, the raw data itself is not sent. Rather, a sequence of check symbols is sent (the *codewords*), from which both error checking information and the original data may be determined. Unlike methods such as checksums, convolutional codes are typically used where error *correction* is required, and they operate on a continuous data stream rather than a defined block of input data. The notion of convolutional codes was first proposed some time ago (Elias, 1954), but the approach has undergone considerable evolution since then. In this section, we describe the fundamental underpinnings of convolutional codes, leading to the Viterbi algorithm (Viterbi, 1967) for efficiently decoding the encoded bitstream, which is very widely used.

To begin with, suppose we employed a simple 3-bit Hamming code, such that a 1 bit was sent as 111 and a 0 bit as 000. It would require inverting three bits consecutively in order to convert one valid codeword into another. But if just one bit was inverted, then we might (for example) receive the codeword 001 if 000 was actually sent. This could be corrected if we assumed that the least number of bits were inverted. If we received 011, then it might be corrected to 111 as the closest valid codeword. This type of decoding is *instantaneous*, since each corrected bit is available immediately once the 3-bit symbol is received; it is not necessary to wait any longer.

The essence of convolutional coding, though, is to imagine the source plus channel as a type of finite-state machine. At any time, the machine exists in a known state, and that state in turn governs what codewords are possible. Not all codes are possible in each state. If the receiver believes that the sender is in a certain state with certain allowable codewords able to be sent on the channel, then any different code received not only denotes a channel error but also implies an incorrect assumption as to the sender's state (either now or in the recent past). Essentially, it is this extra information regarding allowable state (and state transitions permitted) that gives a performance advantage.

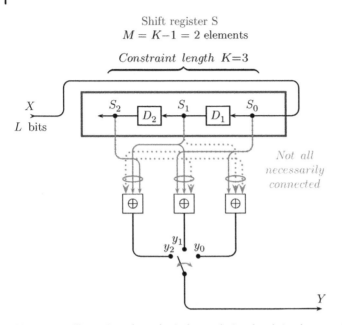

Figure 6.21 Illustrating a hypothetical convolutional code implementation. One-bit delay elements are represented as *D*, with the "convoluted" channel codeword produced by XORing a combination of the input and delayed inputs. The dotted lines are not connected in this example design. Of course, such a structure is not unique, and many permutations of this type of layout are possible.

If an incorrect bit sequence is received, then both the received bit sequence and the expected state are used to correct the incorrect bit(s). Thus, the receiver has to track the "expected" state continuously, so as to facilitate the decoding process. Furthermore, since the sequence of states is not known in advance, the receiver may require several received codewords in order to determine what the most likely state was over recent steps. As a result, this type of approach is termed *delayed-decision* decoding, as opposed to *instantaneous* decoding.

To introduce the idea of a state-driven convolutional code, consider the block diagram shown in Figure 6.21. Here, we continue with the use of a 3-bit output codeword for each input (or message data) bit. Such a design is clearly inefficient, but its simplicity helps to illustrate several important aspects of convolutional codes. In the diagram, message bits are shifted in from the right, and these are delayed by successive delay elements D_1 and D_2. The outputs of each of these forms the *state bits*; thus, there are four possible states in this particular design. The output bits of the channel codeword are formed by XORing various combinations of the state bits and the input bit. As each bit is shifted in, three bits are produced at the output, which are transmitted sequentially.

Table 6.5 State table for the simple illustrative example of convolutional code operation.

Current state			Input X	Output			Next state	
S_2	S_1	S_0		y_2	y_1	y_0	S_2	S_1
0	0	0		0	0	0	0	0
0	0	1		0	1	1	0	1
0	1	0		1	1	0	1	0
0	1	1		1	0	1	1	1
1	0	0		1	0	0	0	0
1	0	1		1	1	1	0	1
1	1	0		0	1	0	1	0
1	1	1		0	0	1	1	1

To analyze this arrangement, a *state table* may be constructed as shown in Table 6.5. The three output bits are defined by combinations of the input bit and the current state bits, and these are output serially. In this example, the input bit X is passed through to output bit y_0, but in general this need not be the case. Output y_1 is formed by the XOR of S_0 and S_1, so $y_1 = S_1 \oplus S_0$. Similarly, output y_2 is formed by the XOR of S_1 and S_2, so $y_2 = S_2 \oplus S_1$. Note that the number of XOR gates is not required to be the same as the number of delays. The "Next State" column defines the transition of the encoder after the output bits have been transmitted serially. In the absence of any channel errors, the decoder follows the encoder state exactly; in the presence of channel errors, it may deviate to incorrect states for some number of input bits. Ideally, it will eventually return to track the correct sequence of states at some later time.

From the state table, the *state diagram* of Figure 6.22 may be deduced. This shows the states themselves, the permissible transitions due to each input bit, and the corresponding output bits that comprise a codeword. In this depiction, the states are shown as circles, and the transitions between states as directed lines. Only certain transitions are permitted, according to a given design (which in turn determines the state table). The input bit for each transition is shown, together with the 3-bit output pattern.

As data bits are received, the state transitions are followed, and as a result the corresponding output bits may be determined. The system cycles through consecutive states as per the state diagram, emitting codewords as it goes. The tree diagram of Figure 6.23 shows the time evolution of states, starting from

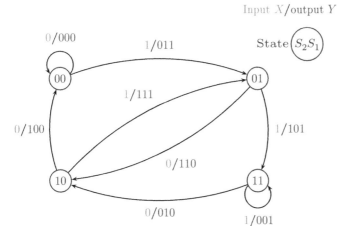

Figure 6.22 Convolutional code state transition diagram. The output codeword (consisting of three bits in this case) is determined by the present state and the current input bit. The present state, in turn, is determined by the recent history of states visited.

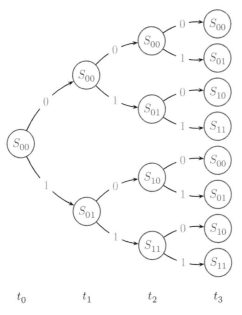

Figure 6.23 Convolutional coding tree diagram. Each time step represents one new bit to be encoded.

an assumed initial state. Each input bit leads to one of two new states. A problem becomes evident almost immediately: the number of branches and nodes grows exponentially, such that after even a small number of input bits, the tree becomes unmanageably large. However, it must be recognized that there are

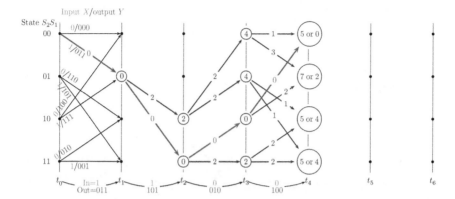

Figure 6.24 Convolutional code trellis diagram. The state transitions are shown in the block from t_0 to t_1 for convenience. At a later time t_4 we can choose between two possible paths, according to the lowest cost to get there.

only four possible states (in this example), and so the tree may be "folded" into a *trellis diagram* as shown in Figure 6.24. All of the input and output bits are shown here in the transition between time instants t_0 to t_1 for convenience, but this is not repeated in subsequent states for clarity.

In the trellis of Figure 6.24, the sender (encoder) state transitions from S_{00} to S_{01} for an input bit of 1, and remains in state S_{00} for an input bit of 0. The value inside each node circle is used to keep track of the total cost per path, which is derived from the Hamming distance between the received data and that expected according to the known state transitions. At the receiver, the starting state transitions to state S_{00} at t_1 if the symbol 000 were to be received (and a 0 bit decoded), whereas it transitions to state S_{01} at t_1 if the symbol 011 were received (and a 1 bit decoded).

At the receiver, if 000 was received, the state transitions to S_{00}, whereas receiving 011 forces the state transition to S_{01}. This of course is exactly the same as the sender. The choice as to which transition to take is governed by the Hamming distance between the received symbol and the permissible symbols in the current state. In this case, the Hamming distance between 011 (received) and 000 (upper path out of state S_{00}) is 2, whereas the Hamming distance between 011 (received) and 011 (lower path out of state S_{00}) is zero, and thus the choice of best (lowest cost) state transition is clearly to take the path to S_{01}. The cost of 0 is shown in the directed line from S_{00} to S_{01}.

At each subsequent time, each state has two possible outgoing paths. The cumulative sum of Hamming distances is calculated at each state for a given time, and thus there are multiple possible cumulative costs. This may be seen in Figure 6.24, since by the time we reach time t_4, there are several possible paths that could have been taken, depending on the branch decision at each node. As shown, the cumulative path costs for state S_{00} at time t_4 could be either 5 or 0,

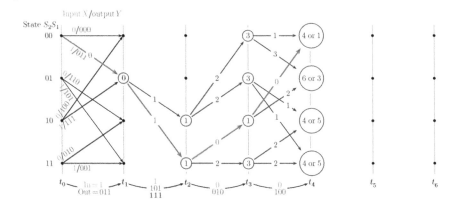

Figure 6.25 Convolutional code trellis diagram when a single bit is in error in the transition from t_1 to t_2.

depending on the path taken through the trellis, which determines the branch we arrived from at the last step. This observation will prove to be important in showing how errors may be corrected.

Now suppose that a one-bit error occurs between t_1 and t_2 as depicted in the trellis of Figure 6.25. Recall that a one-bit error could have been instantaneously corrected with a 3-bit codeword, but now we wish to see what happens with a convolutional code. At time t_2, both of the states indicated have a Hamming distance of 1 from the previous state. Note that the decoder can only calculate a Hamming distance as the distance between the received codeword of 111 and the two possible paths emanating from the previously assumed state S_{01}, since the true codeword of 101 is, of course, unknown to the decoder. At this point, it is uncertain which was the best path, and hence what the original bit should be. So, the decision is delayed.

Proceeding to the next time step t_3, the minimum path cost is shown for each time/state node. Finally, at time t_4, we can see that the lowest cumulative cost at time t_4 is 1, occurring in state S_{00}. This, in turn, was arrived at from state S_{10} at time t_3. Reversing the trellis steps at each node according to the lowest cumulative cost reveals the overall lowest-cost path (in reverse), which is $S_{00}(t_4) \rightarrow S_{10}(t_3) \rightarrow S_{11}(t_2) \rightarrow S_{01}(t_1)$. This phase is termed *backtracking*. By reversing this recovered path, the most likely original path of $S_{01}(t_1) \rightarrow S_{11}(t_2) \rightarrow S_{10}(t_3) \rightarrow S_{00}(t_4)$ is recovered.

Since it appears that a single-bit error can be corrected, a two-bit error pattern is examined as shown in Figure 6.26. Recall that such an error could not be corrected by the 3-bit instantaneous Hamming code approach. At step t_4, we find that the lowest path cost of 2 occurs at state S_{00}. Backtracking from each node to the optimal (total lowest-cost) predecessor reveals the true path once again. Thus, a two-bit error could be corrected by this code, provided that

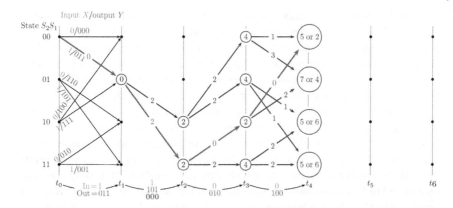

Figure 6.26 Convolutional code trellis diagram with two erroneous bits received between t_1 and t_2.

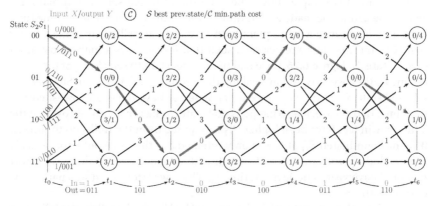

Figure 6.27 Convolutional code complete trellis diagram, showing all possible state transitions. Even for a relatively small number of steps, the number of possible paths becomes prohibitively large.

the decoding is delayed beyond where the initial error occurs. In fact, this is a characteristic of convolutional codes in general – they are *delayed-decision* codes.

The complete trellis from t_0 to t_6 is now shown in Figure 6.27. The correct path through the trellis is shown. The best previous state S and corresponding minimum path cost C are shown for each node. This information is retained for each node, for the backtracking phase, which finds the lowest cost through the trellis. This lowest-cost search is one of the generic family of shortest-path problems, which are solved by the class of algorithms called *dynamic programming*. Such problems also occur in other areas of telecommunications, such as finding the best route for data packets in a packet-switched network.

It would seem to be necessary to continually try numerous possible paths through the trellis as codewords are received, in order to correct for errors. The above approach would need to be tried for each new codeword, and so the problem of keeping track of multiple possible paths would seem to be challenging. Not only that, the number of paths would grow for longer delays, which would occur for more states and/or bursts of errors in the channel.

The Viterbi algorithm solves the optimal trellis path problem in an elegant way. The key observation is that if two paths merge at a single node, the higher cost path can always be eliminated from consideration. This is because subsequent steps in time cannot decrease the path cost, and so the observed minimum path cost at a node must be the minimum thereafter. Thus, only two path costs need to be maintained for each state, and these paths can be culled to just one by selecting the predecessor corresponding to the lowest cumulative cost at each node.

This does *not* mean that the entire lowest path cost is determined immediately – as in the example above, the decision on which state is optimal for a given time cannot be made at that time instant, but is deferred until some time later. This is because a subsequent path through the trellis may make a lower cost path increase at a greater rate, such that it overtakes a higher cost at an earlier state. In Figure 6.26, if we had made a decision to select one of the paths at time t_2, then we may end up making the wrong overall decision. Suppose we selected state S_{10} at time t_2, since it happened to have the equal lowest cost at that time. In this case, it also happens that state S_{11} has an identical cost, but the latter may in fact have been higher. As we can see from the trellis up to time t_4, if we had eliminated S_{11} at t_2 and only retained S_{10}, then we could not possibly have ended up with the correct overall path at time t_4, since there is no way we could have visited node S_{10} at time t_3, which we later found was part of the correct path.

To see how the optimal path is determined by Viterbi decoding, the following code shows the development of a convolutional coder trellis. We first need a state table that stores the number of states as follows.

```
% StateTable.m - state transition and codeword generation

classdef StateTable
    properties
        NumStates         % number of possible states
        NumStateBits      % bits for each state
        NumOutBits        % bits for each output
    end

    methods (Access = public)
        function TheStateTable = ...
        StateTable(NumStates, NumStateBits, NumOutBits)
```

```
                    TheStateTable.NumStates = NumStates;
                    TheStateTable.NumStateBits = NumStateBits;
                    TheStateTable.NumOutBits = NumOutBits;
            end

            function ShowTable(TheStateTable)
                    disp(TheStateTable);
            end
        end % end methods

        % add other methods   helper functions here

end
```

This state table must help the trellis generate the sequence of states. Given the current state, it must be able to determine the next possible state, as well as the output codeword for a given input bit. This is accomplished with Map-State().

```
% StateTable.m

methods (Access = public)
    % map state and input to output and next state
    function [Y, State, NextStateIndex, NextState] = ...
                    MapState(TheStateTable, StateIndex, x)

        % bitstring – LSB: highest index, MSB: lowest index
        State = TheStateTable.StateBits(StateIndex);
        Y = dec2bin(0, TheStateTable.NumOutBits);

        Y(3) = StateTable.xorstr([x]);
        Y(2) = StateTable.xorstr([x State(2)]);
        Y(1) = StateTable.xorstr([State(2) State(1)]);

        % next state = current state shifted left, input bit
        % shifted into LSB
        NextState = [State(2:end) x];

        % convert binary string to actual index
        NextStateIndex = bin2dec(NextState)+1;
    end
end

methods (Access = private)
    % Converts the current state (integer) into a character
```

```
    % bitstring
    function str = StateBits(TheStateTable, StateIndex)
        str = dec2bin(StateIndex-1, TheStateTable.
        NumStateBits);
    end
end

methods (Static)
    % multi-bit XOR, which is true if an odd number of 1's
    function resbit = xorstr(str)
        isodd = false;
        if( mod(length(find(str == '1')), 2) == 1 )
            isodd = true;
        end

        resbit = '0';
        if( isodd )
            resbit = '1';
        end
    end

    function d = HammingDist(x, y)
        d = length(find(x ~= y));
    end
end
```

The trellis must be able to encapsulate both the state and state transitions, as well as facilitate the Viterbi backtracking procedure. This can be done by defining a trellis node as follows.

```
% CNode.m - node class for convolutional coding

classdef CNode < handle

    properties
        StateIdx        % index of this node's state
        TimeIdx         % time index of this node

        NextStateIdx    % index of next node
        Bit             % input bit for transition
        Code            % output code for transition
        PrevStateIdx    % index of previous state

        PathCost        % cumulative path cost
        PathIdx         % best backtrack path index
    end
```

```
methods   (Access = public)
    function [hNode] =  CNode(StateIdx , TimeIdx)
        hNode. StateIdx = double.empty;
        hNode. TimeIdx = double.empty;
        hNode. NextStateIdx = double.empty;
        hNode. PrevStateIdx = double.empty;
        hNode. Bit = char.empty;
        hNode. Code = char.empty;

        hNode. PathCost = inf;
        hNode. PathIdx = 0;

        if( nargin == 2 )
            hNode. StateIdx = StateIdx;
            hNode. TimeIdx = TimeIdx;
        end
    end

    function SetNext(hNode, Idx , StateIdx , Bit , Code)
        hNode. NextStateIdx(Idx) = StateIdx;
        hNode. Bit(Idx) = Bit;
        hNode. Code(Idx, :) = Code;
    end

    function ClearNext(hNode)
        hNode. NextStateIdx = double.empty;
        hNode. Bit = char.empty;
        hNode. Code = char.empty;
    end

    function AddPrev(hNode, StateIdx)
        hNode. PrevStateIdx = [hNode. PrevStateIdx
        StateIdx];
    end

    function ClearPrev(hNode)
        hNode. PrevStateIdx = double.empty;
    end
end % end methods
end
```

Each trellis node must store the index of the next state, as well as the index of predecessor states. The latter is to facilitate the backtracking phase. The trellis itself is comprised of a regular lattice of nodes. These nodes are indexed by the combination of state and time. The state table object, created earlier, is also stored within the trellis itself.

```
% CTrells.m - class for trellis for convolutional coding

classdef CTrellis  < handle

    properties
        NumStates          % number of possible states
        MaxTimeIdx         % number of time intervals

        NumStateBits       % bits for each state
        NumOutBits         % bits for each output

        StateTable         % state mapping table
        hNodeTable         % array of all nodes, linked in a
                           % trellis
    end

    methods
        function [hTrellis] = CTrellis(NumStates,
        MaxTimeIdx, ... NumStateBits, NumOutBits)

            % save dimensions of the trellis
            hTrellis.NumStates = NumStates;
            hTrellis.MaxTimeIdx = MaxTimeIdx;

            % save parameters of the trellis
            hTrellis.NumStateBits = NumStateBits;
            hTrellis.NumOutBits = NumOutBits;

            % create a state table for the trellis
            hTrellis.StateTable = StateTable(NumStates,
            NumStateBits, NumOutBits);

            % create pointers to initial state nodes
            hTrellis.hNodeTable = CNode();
            for TimeIdx = 1:hTrellis.MaxTimeIdx
                for StateIdx = 1:hTrellis.NumStates
                    hTrellis.hNodeTable(StateIdx,
                    TimeIdx) = ...
                            CNode(StateIdx, TimeIdx);
                end
            end
        end
```

```
%------------------------------------------------
% add other methods here:
% PopulateNodes()
% EmitCodewordSeq()
% ForwardPass()
% Backtrack()
% ShowPathCosts()
%------------------------------------------------

    end % end methods

end   % end class
```

As well as the trellis constructor, the nodes within the trellis itself must be created. This is accomplished with `PopulateNodes()`, which creates forward and backward pointers according to the defined state table. This is done for each possible input bit (1 or 0).

```
% CTrellis.m - CTrellis class

function PopulateNodes(hTrellis)

    % set branch transitions for each node in trellis
    for TimeIdx = 1:hTrellis.MaxTimeIdx
        for StateIdx = 1:hTrellis.NumStates

            X = '01';
            for k = 1:2
                % current input bit
                x = X(k);
                [Y, State, NextStateIdx, NextState] =
                hTrellis.StateTable.MapState(StateIdx, x);

                % forward pointer to state at next time
                % index
                if( TimeIdx == hTrellis.MaxTimeIdx )
                    hTrellis.hNodeTable(StateIdx, TimeIdx).
                    ClearNext();
                else
                    hTrellis.hNodeTable(StateIdx, TimeIdx).
                    SetNext(k, NextStateIdx, x, Y);
                end

                % backwards pointer
                if( TimeIdx == 1 )
                    hTrellis.hNodeTable(StateIdx, TimeIdx).
                    ClearPrev();
```

```
                        end

                        if ( TimeIdx < hTrellis.MaxTimeIdx )
                            hTrellis.hNodeTable(NextStateIdx,
                            TimeIdx+1).AddPrev(StateIdx);
                        end
                    end
                end
            end
        end
    end
```

At this point, the trellis has been initialized and is able to encode a bitstream. Given a sequence of bits, the trellis is traversed from start to end, emitting a codeword for each time step.

```
% CTrellis.m - CTrellis class
function [CodeSeq] = EmitCodewordSeq(hTrellis, BitSeq)
    BitSeqLen = length(BitSeq);
    TimeIdx = 1;
    StateIdx = 1;

    % returned digit sequence
    CodeSeq = char.empty;

    % traverse the list of nodes, forward direction
    for n = 1:BitSeqLen
        b = BitSeq(n);    % current bit

        % find this bit in forward table
        bits = { hTrellis.hNodeTable(StateIdx, TimeIdx).
        Bit };
        bits = char(bits);
        ibit = find(b == bits);

        Code = hTrellis.hNodeTable(StateIdx, TimeIdx).
        Code(ibit,:);
        CodeSeq(TimeIdx, :) = Code;

        % link to next node in trellis
        StateIdx = hTrellis.hNodeTable(StateIdx, TimeIdx).
        NextStateIdx(ibit);
        TimeIdx = TimeIdx + 1;
    end
end
```

The receiver must first take the codeword sequence and save the cumulative path cost at each node. This is accomplished in the `ForwardPass()` method, which takes the previously constructed trells and applies the received codeword sequence to it. The Hamming distance is calculated for each node transition, and the cumulative Hamming distances along the paths are calculated. These are compared to the current cost in traversing to the next node, and if the new cost is lower than the current cost, it is saved along with the previous-node pointer for backtracking.

```
% CTrellis.m - CTrellis class

function ForwardPass(hTrellis, CodewordSeq)
    NumDigits = length(CodewordSeq);

    % initialize path costs
    for TimeIdx = 1:hTrellis.MaxTimeIdx
        for StateIdx = 1:hTrellis.NumStates

            if( TimeIdx == 1 )
                hTrellis.hNodeTable(StateIdx, TimeIdx).
                PathCost = 0;
                hTrellis.hNodeTable(StateIdx, TimeIdx).
                PathIdx = 0;
            else
                hTrellis.hNodeTable(StateIdx, TimeIdx).
                PathCost = Inf;
                hTrellis.hNodeTable(StateIdx, TimeIdx).
                PathIdx = 0;
            end
        end
    end

    for TimeIdx = 1:hTrellis.MaxTimeIdx-1

        RxCode = CodewordSeq(TimeIdx, :);
        for StateIdx = 1:hTrellis.NumStates
            for k = 1:2

                Code = hTrellis.hNodeTable(StateIdx,
                TimeIdx).Code(k,:);
                Bit = hTrellis.hNodeTable(StateIdx,
                TimeIdx).Bit(k);

                NextStateIdx = hTrellis.hNodeTable(StateIdx,
                TimeIdx).NextStateIdx(k);
```

```
                         HamDist = hTrellis.StateTable.HammingDist
                         (RxCode, Code);

                         PathCost = hTrellis.hNodeTable(StateIdx,
                         TimeIdx).PathCost;
                         NewCost = PathCost + HamDist;
                         CurrCost = hTrellis.hNodeTable(NextStateIdx,
                         TimeIdx+1).PathCost;

                         if( NewCost < CurrCost )
                             hTrellis.hNodeTable(NextStateIdx,
                             TimeIdx+1).PathCost = NewCost;
                             hTrellis.hNodeTable(NextStateIdx,
                             TimeIdx+1).PathIdx = StateIdx;
                         end
                     end
                 end
             end

    end
```

It is useful to be able to display the path costs and state index transitions, and this is performed for the entire trellis as follows.

```
% CTrellis.m - CTrellis class

function ShowPathCosts(hTrellis)

    for StateIdx = 1:hTrellis.NumStates
        for TimeIdx = 1:hTrellis.MaxTimeIdx
        PathCost = hTrellis.hNodeTable(StateIdx,
        TimeIdx).PathCost;
        PrevStateIdx = hTrellis.hNodeTable(StateIdx,
        TimeIdx).PathIdx;

        fprintf(1, '%d / %d \t', PrevStateIdx, PathCost);
        end
        fprintf(1, '\n');
    end
end
```

Finally, the backtracking function starts at the end time and determines the lowest cumulative cost across all states at that time. From there, the previous index is used to backtrack one time step. This is repeated until the start of the trellis.

```
% CTrellis.m - CTrellis class

function [StateSeq, BitSeq, CodeSeq] = Backtrack(hTrellis)

    RevStateSeq = zeros(1, hTrellis.MaxTimeIdx);
    CurrTimeIdx = hTrellis.MaxTimeIdx;

    TermCosts = cell2mat( { hTrellis.hNodeTable(:,
      CurrTimeIdx).PathCost } );
    [CurrCost, CurrStateIdx] = min(TermCosts);
    FwdTimeIdx = 1;

    while( CurrStateIdx > 0 )
        RevStateSeq(FwdTimeIdx) = CurrStateIdx;
        CurrCost = hTrellis.hNodeTable(CurrStateIdx,
          CurrTimeIdx).PathCost;
        fprintf(1, 'curr state/time %d/%d  cost %d\n',
          CurrStateIdx, CurrTimeIdx, CurrCost);

        NextStateIdx = hTrellis.hNodeTable(CurrStateIdx,
          CurrTimeIdx).PathIdx;
        CurrStateIdx = NextStateIdx;
        CurrTimeIdx = CurrTimeIdx - 1;
        FwdTimeIdx = FwdTimeIdx + 1;
    end

    % reverse the bit-string order
    StateSeq = fliplr(RevStateSeq);
    BitSeq = char('x'*ones(1, hTrellis.MaxTimeIdx-1));
    CodeSeq = char('x'*ones(hTrellis.MaxTimeIdx-1, hTrellis.
      NumOutBits));

    for TimeIdx = 1:hTrellis.MaxTimeIdx-1
        StateIdx = StateSeq(TimeIdx);
        NextStateIdx = StateSeq(TimeIdx+1);

        % find this bit in forward table
        NextStates = hTrellis.hNodeTable(StateIdx, TimeIdx).
          NextStateIdx;
        istate = find(NextStateIdx == NextStates);
        Digits = hTrellis.hNodeTable(StateIdx, TimeIdx).
          Code(istate,:);
        DigitSeq(TimeIdx, :) = Digits;
        Bit = hTrellis.hNodeTable(StateIdx, TimeIdx).
          Bit(istate);
        BitSeq(TimeIdx) = Bit;
```

```
            Code = hTrellis.hNodeTable(StateIdx, TimeIdx).
              Code(istate, :);
            CodeSeq(TimeIdx, :) = Code;
      end
end
```

The following shows how the example trellis in this section was created.

```
% TestCTrellis.m

clear classes

NumStates = 4;
MaxTimeIdx = 7;
NumStateBits = 2;
NumOutBits = 3;

hTrellis = CTrellis(NumStates, MaxTimeIdx, NumStateBits,
   NumOutBits);
hTrellis.PopulateNodes();

BitSeq = '110010';
[CodeSeq] = hTrellis.EmitCodewordSeq(BitSeq);

% this is the transmitted code sequence
CodeSeq

% to insert an error in transmission
%CodeSeq(2,:) = '111';

% the receiver performs the following processing
hTrellis.ForwardPass(CodeSeq);
hTrellis.ShowPathCosts();

[StateSeq, BitSeq, CodeSeq] = hTrellis.Backtrack();
StateSeq
BitSeq
CodeSeq
```

To insert an error in transmission, the received code sequence may be altered once the correct bit sequence is known at the sender. Using CodeSeq(2,:) = '111' sets an incorrect received bit sequence at time step 2. The forward–backward procedure must then repeated as above to recover the correct bit sequence.

6.6   Encryption and Security

Security of information may imply access to physical resources where that information is stored. Accessing a wireless server, for example, may be restricted based on the credentials (password or token) given by the user of a system. Since IP networks are packet-switched and process each data packet independently, any additional security steps must also operate at high speed, so as to introduce minimal delay.

As well as physical resources, security also involves knowledge transfer. In transmitting any message, especially over a wireless communication channel, the possibility always exists that a third party may intercept the message. For many types of communication, this may not be critical. For others, it is important and perhaps even essential that the communication remains private between the sender and receiver. Financial transactions, for example, require a high level of confidence by all parties in the transaction. This concept extends much further than is often realized. As well as keeping communication (a "message") secret, it may be desirable to ensure the *integrity* of the message – that it has not been altered, and that it came from the person who claims to have sent it. It is often held that there are three key aspects to securing a communication system:

  i) *Confidentiality* implies the privacy of a message and requires encryption of the message contents.
 ii) *Integrity* implies that a message has not been tampered with in transit.
iii) *Authentication* means that it is possible to verify the sender of a message.

The "traditional" security system is one of physical security and requires a physical key. The key matches one or more specific locks. But it also has short-comings, which have parallels in electronic security: a key can be lost or stolen, and a key can be copied. In the electronic realm, we have an additional problem: that of key distribution, or how to transmit the electronic key in the first place. However, there are also additional capabilities: we can authenticate the person who presents a message. Ideally, we would base such authentication on:

  i) Something you *have* – for example, an identity card.
 ii) Something you *know* – for example, a password or Personal Identification Number (PIN).
iii) Something you *are* – using biometrics such as the user's fingerprint.

A useful listing of terminology associated with secure systems may be found in the NIST Digital Identity Guidelines Standard (NIST, 2017). The following sections examine some of the underpinnings of digital security, including both encryption and authentication of user identity as well as electronic documents.

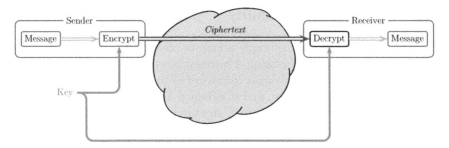

Figure 6.28 Information flow in private (or secret) key encryption. The network cloud could be comprised of many intermediate hops, which may be able to be accessed by a third party, hence it is assumed to be insecure. Ideally, the key is somehow sent to the receiver via a separate, secure channel.

6.6.1 Cipher Algorithms

In order to encrypt (or encipher) a message, an encryption algorithm is necessary, which alters the original message stream in such a way as to make the original message unintelligible to someone who may intercept some or all of the message (an *eavesdropper*). Of course, a corresponding decryption method is also required. This is shown diagrammatically in Figure 6.28.

The sender of the message encrypts the *cleartext* or *plaintext* message using an *encryption key* and transmits the *ciphertext*. The receiver performs the reverse operation, and so it must have knowledge of the decryption key. In this case, the encryption key and decryption key are the same. The reverse operation may in fact be similar to the encryption, but there are some systems (to be discussed) in which the decryption is not identical to the encryption. The cloud in the middle indicates the possibly insecure path that the message must traverse. This could be a physical connection (such as a wireless path) or intermediate routers in the Internet, over which the sender has no control. It is assumed that, in the worst case, someone may be able to intercept some or all of the ciphertext. If they have the key, they could decrypt the message. Even if they don't have the key, they could attempt to determine the key, though guessing or other more sophisticated means.

The overall security is based on the electronic key itself – knowledge of the key permits decryption of the received message. The underlying assumption is that the message is accessible to all, and thus there is no assumption of physical security. This of course is the case in wireless networks in particular and the Internet in general.

Note in Figure 6.28 that the key itself is shown traversing from sender to receiver *outside* the hostile cloud area. Of course, if someone intercepted the key, they could decrypt all of the messages. Thus the key must be sent separately via a secure channel. This is the so-called key distribution problem, which we will return to in Section 6.6.3.

Some security systems are based on the secrecy of the algorithm itself: the particular steps involved in encrypting a message. It is generally agreed that this is a bad idea, for two reasons. First, it is quite possible that someone may reverse-engineer an encryption device (including software), and if this is the only secret part, then the secrecy is destroyed forever. Second, perhaps less obvious, is the open nature of an algorithm. If the steps of an encryption algorithm are available for all to scrutinize, it is likely that any weak points will be revealed. This enables the designers to deploy a stronger encryption system. Thus, the secrecy of the key is a critical element.

6.6.2 Simple Encipherment Systems

A very simple, often-used encryption system is based on the XOR function. Consider the binary XOR function, defined as

$$A \oplus B \triangleq A \cdot \overline{B} + \overline{A} \cdot B \qquad (6.34)$$

By taking all possible combinations of binary variables (bits) A and B, we can write the truth table as shown in Table 6.6.

A simple encryption function may be formed by taking the message in blocks of N bits and XORing with an N bit key. The process is repeated for all blocks in a message, reusing the same key. This system is invertible, in that applying the exact same XOR operation between the ciphertext and the key, the original plaintext may be restored. This is illustrated in Figure 6.29.

It is not difficult to prove the invertibility of XOR encryption. If the message is M and key K, then the encrypted message E is

$$E = M \oplus K \qquad (6.35)$$

Table 6.6 The digital XOR function truth table.

A	B	$A \oplus B$
0	0	0
0	1	1
1	0	1
1	1	0

The output is true (binary 1) only if the input bits A and B are different.

To encrypt one byte of the plaintext message

	Binary								ASCII	Hex
Message byte	0	1	1	0	0	0	0	1	"a"	61H
Key	0	1	0	1	1	0	0	0	"X"	58H
Result	0	0	1	1	1	0	0	1	"9"	39H

To decrypt one byte of the ciphertext

	Binary								ASCII	Hex
Ciphertext	0	0	1	1	1	0	0	1	"9"	39H
Key	0	1	0	1	1	0	0	0	"X"	58H
Decrypted byte	0	1	1	0	0	0	0	1	"a"	61H

Figure 6.29 A simple example of XOR encryption and decryption.

Performing the XOR operation on the encrypted message with the same key gives

$$M' = E \oplus K$$
$$= (M \oplus K) \oplus K \tag{6.36}$$

upon substitution of the encryption expression (6.35). Examination of the truth table of Table 6.6 reveals that any binary value XORed with itself ($A = B$) gives zero, so

$$M' = M \oplus (K \oplus K)$$
$$= M \oplus 0 \tag{6.37}$$

Once again, the truth table of Table 6.6 is used to determine that XORing of a value with zero (input B, $A = 0$) is itself, so

$$M' = M \oplus 0$$
$$M' = M \tag{6.38}$$

Thus, the decrypted message is exactly the same as the original message.

Two other simple types of encryption are also commonly employed. Because they were historically employed with human-readable text, we describe these in terms of "letters," however they could be used with any arbitrary data. The first of these is a simple *substitution cipher*, wherein one letter is substituted with another from a table (or, even simpler, with a table formed by a cyclic rotation of the alphabet of all possible symbols in order). This type of cipher has been used since antiquity. The other approach is to take the letters in blocks and rearrange the order. This is the essence of a *transposition cipher*. The inverse operation requires knowledge of the substitution table for a substitution cipher and the permutation order for the transposition cipher. Of course, we can use these methods in combination, and Figure 6.30 shows just that for a simple two-letter encoding.

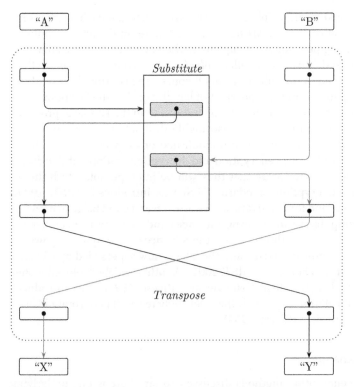

Figure 6.30 Using substitution and transposition to create a more complex cipher system.

Each of these methods has their weaknesses. The simple substitution cipher is vulnerable to statistical attacks. If it is known that the plaintext represents a particular type of conversation, for example, English text, then the frequency of occurrence of letters in the alphabet is known (or at least, to a degree of approximation). Thus with a simple substitution, the relative frequencies will be retained, albeit for different letters. This would form a clue as to the actual substitution used. This type of *monoalphabetic cipher* thus provides very weak encryption.

Apart from statistical analysis, a simple exhaustive search is possible, though not always feasible. If the substitution were just a simple rotation of the alphabet, it could be easily cracked by brute force. However a completely random permutation of the alphabet would be more difficult. If there are 26 letters available, then there are 26 possible substitutions for the first letter, 25 for the second, and so forth, and hence 26! (26 factorial) possible permutations for the entire alphabet (less one, since like-for-like substitution would not encrypt the message).

Alternatively, a keyword or phrase may be used such that each letter represents a different substitution starting point. Each letter in the phrase could be used as the starting point of a new alphabet. The key is reused once all letters in it have been used. This is a so-called *polyalphabetic cipher*, since more than one alphabet is used in the substitution. Of course, the continual reuse of the key may create patterns in the ciphertext of length equal to the key phrase.

In practice, much stronger ciphering operations than the simple approaches described above are required. It is also highly desirable that an encryption process be standardized rather than unpublished or proprietary. This allows multiple vendors to produce encryption and decryption equipment (hardware and software) in the knowledge that they will be interoperable. With this in mind, the Data Encryption Standard (DES) was introduced (NIST, 1999). When introduced, and for some time afterward, it was a good balance between difficulty of computation in securing messages and inability to decipher via brute-force exhaustive methods. Given the advances in computing power, a stronger cipher became necessary, and the DES is now superseded by AES, the Advanced Encryption Standard (NIST, 2001). Another widely deployed cipher is RC5, described in RFC2040 (Baldwin and Rivest, 1996). This introduces data-dependent rotations, in which the intermediate results are rotated based on other bits in the data (Rivest, 1994).

6.6.3 Key Exchange

In each of the encryption methods discussed so far, there is one underlying problem: how to make both (or all) parties aware of the secret key, without exposing it to others. If the underlying communication channel is insecure, this would seem to be an unsolvable problem – and for a long time, it was.

The key exchange (or, more generally, key distribution) problem has received a great deal of research over many years, and perhaps the best-known approach is the so-called Diffie–Hellman–Merkle algorithm (Diffie and Hellman, 1976). It requires the use of modulo arithmetic, and the raising of integers to a power. Modulo arithmetic defines counting up to a specified value (the modulus), then resetting the count to zero. The numerical properties of modulo arithmetic have been found to be essential both to decrypting an encrypted message back to its original form and to ensuring at least some degree of secrecy. The secrecy is based on the difficulty of factorizing very large numbers (perhaps hundreds of digits), and as a result, the key must be considered to be a number.

To understand the motivation as a one-way mathematical operation, consider the result $r = a \cdot b$. If a and b are given, this reduces to a simple calculation. However if only r is given, it is more difficult (though of course not impossible) to determine the factors a, b – especially if there is only one possible solution. Next, suppose the calculation involved $r = a^p$. This is easy given a, p only, but if only r, a is given, the problem of finding p is somewhat more difficult. Various

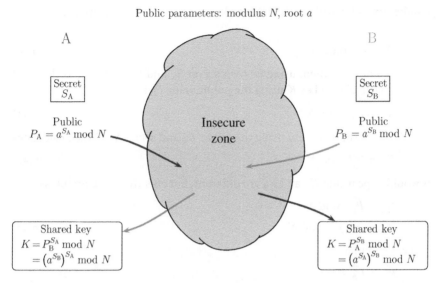

Public parameters: modulus N, root a

Figure 6.31 Key exchange across an insecure transmission channel. Computed parameters P_A, P_B may be made public, but S_A, S_B must be key secret.

exponents p could be tried iteratively. A direct mathematical solution is provided if we employ logarithms, since $p = \log_a r$. To make the inverse calculation more difficult, suppose we required $r = a^p \bmod N$ using modulus N. This does not have a direct solution using logarithms (due to the mod N operator), and so determining p if the other parameters are given is far more difficult.

Consider Figure 6.31, which shows two parties A and B who wish to exchange a secret key for the purposes of encrypting messages. The communication channel is assumed to be insecure. Both parties agree to use an integer base a and a modulus N, which must be a prime number. The requirements on the choice of a and N, as well as the desirability of N being prime, are discussed after the present example.

The first step is for A to generate a random integer S_A, which is kept secret. Likewise, B generates a random integer S_B. Although these have to be kept secret, somehow we have to exchange a key (or, more precisely, compute a value for the key, which is guaranteed to be identical at both sides of the communication). To exchange a key secretly, A computes

$$P_A = a^{S_A} \bmod N \tag{6.39}$$

and sends it to B. Since S_A is assumed to be secret to A, then ideally nobody else should be able to work out P_A except, perhaps, by guessing – and such guessing ought to be unlikely to yield the correct answer. The quantity P_A is then in the public domain, and it is assumed that an interceptor could obtain this number (possibly without A or B even knowing). Similarly, B computes a public quantity

$$P_B = a^{S_B} \bmod N \tag{6.40}$$

and sends it to A. Then, using its own secret S_A and the public knowledge of P_B, A can calculate a key K using the public value P_B from

$$K_A = P_B^{S_A} \bmod N \tag{6.41}$$

In a similar fashion, except using the public P_A and secret S_B, B then calculates

$$K_B = P_A^{S_B} \bmod N \tag{6.42}$$

It would appear that K_A and K_B are different, but effectively A computes

$$K_A = P_B^{S_A} \bmod N$$
$$= (a^{S_B})^{S_A} \bmod N \tag{6.43}$$

while B computes

$$K_B = P_A^{S_B} \bmod N$$
$$= (a^{S_A})^{S_B} \bmod N \tag{6.44}$$

Thus $K_A = K_B = K$, and *both secret keys K are equal*. Of course, K must be kept private to A and B and not divulged, in order to maintain the secrecy.

The reason behind using the modulus in calculating the secret key is to thwart direct calculation of K. Suppose the calculation did not use modulo-N arithmetic. In that case, the initial computation at A would be $P_A = a^{S_A}$, with P_A and a known. Thus $S_A = \log_a P_A$, and the secret value S_A could be directly calculated. Using modulo arithmetic, the logarithm operation does not exist in the conventional form, and thus direct inversion is infeasible.

A simplified numerical is presented in Figure 6.32. To simplify the computations, very small numbers are chosen in order to illustrate the calculations required. In practice, the numbers employed will be much, much larger (of the order of 100 digits or more).

Suppose the base a is 5 and the modulus N is 23. Side A generates a random secret integer $S_A = 3$, then calculates the public quantity $P_A = 5^3 \bmod 23$, and sends it to B. Side B generates a random secret integer $S_B = 5$, then calculates the public quantity $P_B = 5^5 \bmod 23$, and sends it to A. Thus far, both sides have used public information ($N = 23$ and $a = 5$), as well as their own secret random numbers. Next, it is necessary for each side to derive a number using both public information and their own individual secrets to arrive at the same answer (the shared secret key K).

A then calculates $K = 20^3 \bmod 23 \equiv 19$, which uses the public value $P_B = 20$ together with the private value $S_A = 3$. B calculates in a similar fashion with different parameters: $K = 10^5 \bmod 23 \equiv 19$, which uses the public value $P_A = 10$ together with the private value $S_B = 5$. Thus both sides end up with the same value of $K = 19$. Effectively, both are calculating $K = 5^{3 \times 5} \bmod 23 \equiv 19$.

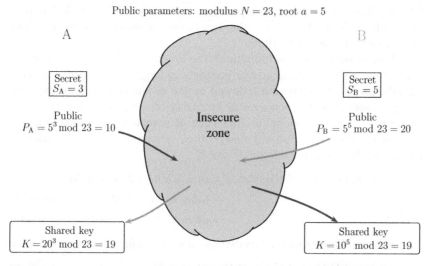

Public parameters: modulus $N = 23$, root $a = 5$

A

B

Secret
$S_A = 3$

Secret
$S_B = 5$

Public
$P_A = 5^3 \bmod 23 = 10$

Insecure
zone

Public
$P_B = 5^5 \bmod 23 = 20$

Shared key
$K = 20^3 \bmod 23 = 19$

Shared key
$K = 10^5 \bmod 23 = 19$

Figure 6.32 Key exchange: a simplified numerical example. Parameters $N = 23$, $a = 5$ are fixed. A places the calculated value of $P_A = 10$ in the public domain, and likewise B finds $P_B = 20$ and reveals it. From the separate pieces, both A and B can compute the secret key $K = 19$. Since $S_A = 3$ and $S_B = 5$ are kept secret, no other party could easily determine K.

It appears that $5^{3\times5} \approx 3 \times 10^{10}$ is a very large number, and so there would be a lot of possibilities for an unknown eavesdropper to try an exhaustive search. However, the modulo operator reduces this space substantially, down to at most $N - 1$. Thus a large value of N is imperative.

The space of all possibilities is quite large. If the modulus N and individual secrets S_A, S_B were hundreds of digits long, it would effectively become impossible to calculate the shared key by iterative solution. Of course, we say "effectively impossible" because it depends on the available computing resources. A secure system designed at any given time may become insecure if greater computing power is harnessed for the task of "cracking" the secret key. Also, weaknesses in the algorithm and/or implementation in practice could provide a point at which it could be exploited.

Since large calculation results are required, it is very likely that numerical values will overflow standard calculation registers. The number of bits required to store the previous result can be estimated as $\lceil \log_2 5^{15} \rceil = 35$. However, the values given are impractically small. If, say, $a = 37$ and the secret values were only 6 and 8, more than 250 bits would be required to store the result. The large number of bits required reveals a significant problem if a direct implementation were to be attempted. It would appear that the result may be determined by directly calculating $37^{6\times8} \approx 1.87 \times 10^{75}$. This is, however, misleading – because floating-point formats have a fixed precision for very large numbers, typically

52 bits. Since an exact result is imperative, any rounding off due to lack of precision affects the final result. In other words, the calculated keys will likely not agree, due to numerical overflow.

So how to overcome this limitation? The fundamental problem is that the computations involve raising large numbers to large powers. If this is done directly, using exponentiation followed by the modulus operation, then the integer limits of any processor will invariably be exceeded. The trick is to perform the modulo operation after each of several smaller operations. If we take $((a \bmod N)(b \bmod N)) \bmod N$, then we can apply the definition of modulo arithmetic using integers k_1 and k_2,

$$((a \bmod N)(b \bmod N)) \bmod N = (a + k_1 N)(b + k_2 N) \bmod N$$
$$= (ab + k_2 aN + k_1 bN + k_1 k_2 N^2) \bmod N$$
$$= ab \bmod N \qquad (6.45)$$

The last line follows from the fact that $kN \bmod N = 0$ for any integer k. Hence

$$((a \bmod N)(b \bmod N)) \bmod N = ab \bmod N \qquad (6.46)$$

If $b = a$, then

$$a^2 \bmod N = ((a \bmod N)(a \bmod N)) \bmod N \qquad (6.47)$$

That is, instead of squaring a number and subsequently taking the modulo remainder, it is possible to perform the computations in a different order. First, the modulo of each number is taken, the multiplication performed, and finally the modulo remainder taken once again. This guarantees the same numerical answer, without the danger of overflow. For example,

$$13^2 \bmod 9 = 169 \bmod 9$$
$$= 7$$

since $169 = 18 \times 9 + 7$. Rearranging the computation order with the modulus operation at each stage results in

$$13^2 \bmod 9 = ((13 \bmod 9)(13 \bmod 9)) \bmod 9$$
$$= (4 \times 4) \bmod 9$$
$$= 16 \bmod 9$$
$$= 7$$

Thus, it is possible to just take the modulo after each multiplication. Calculating $5^{3 \times 5} \bmod 23$ directly results in

```
mod(5^(3*5), 23)
ans =
    19
```

The intermediate result is $5^{3\times5} = 3.0518 \times 10^{10}$, which requires 35 bits to represent exactly. Larger quantities for the intermediate calculation will result in overflow, and thus give an incorrect result. Using the above modulo theory for powers, we can rewrite the exponent and modulo as a loop, as follows.

```
% exponentiation modulo N
% rv = a^ev mod N
% ev may be a scalar or a vector
function [rv] = expmod(a, ev, N)

% loop caters for exponent ev being a vector
rv = zeros(length(ev), 1);
for k = 1:length(ev)

    e = ev(k);

    % exponentiation with modulo at each stage- does not
    % overflow
    r = 1;
    for nn = 1:e
        r = mod(r*a, N);
    end
    r = mod(r, N);

    % save in case vector result
    rv(k) = r;
end
```

The same computation could thus be performed using this function:

```
expmod(5, 3*5, 23)
ans =
    19
```

Suppose the calculation required was 11^{15}, which would fit in 52 bits. With a modulus of $N = 23$, both mod(11^15, 23) and expmod(11, 15, 23) yield the same result of 10. However, if the calculation is altered to 13^{15}, the results differ due to overflow.

The choice of N as a prime number was stated as a requirement earlier. Additionally, a should not be an arbitrary choice either. To see why this is so, consider an exhaustive search by an attacker attempting to guess the secret key. Trying all exponents of a from 1 to $N - 1$ results in a series of guesses for K. Suppose $a = 5$ and $N = 23$. The guesses for $a^k \bmod N$ would then be computed as

```
a = 5;
N = 23;

sort(mod(a.^[1:N−1], N))
ans =
     1 2 3 4 5 6 7 8 9 10 11 12 13 14 15 16 17 18 19 20 21 22
```

It may be observed that the result spans the full range from 1 to $N - 1$. For larger a and/or N, it is desirable to use the modulo exponentiation developed above:

```
a = 5;
N = 23;

sort(expmod(a, [1:N−1], N))'
ans =
     1 2 3 4 5 6 7 8 9 10 11 12 13 14 15 16 17 18 19 20 21 22
```

But suppose we tried at random $a = 6$ and $N = 22$. The iterative trials of an attacker would then result in

```
a = 6;
N = 22;

sort(mod(a.^[1:N−1], N))
ans =
     2 2 4 4 6 6 6 8 8 10 10 12 12 14 14 16 16 18 18 20 20
```

This clearly has a number of repetitions, making the search space somewhat smaller for iterative guesses. In this case, N is clearly not a prime number. However, N being prime is a necessary, but not sufficient condition. Consider again the prime case of $N = 23$, together with $a = 6$. This yields

```
a = 6;
N = 23;

sort(mod(a.^[1:N−1], N))
ans =
     1 1 2 2 3 3 4 4 6 6 8 8 9 9 12 12 13 13 16 16 18 18
```

Once again, there is considerable repetition. The first case ($a = 5, N = 23$) occurs when a is a *primitive root* of N, and there are no repetitions.

6.6.4 Digital Signatures and Hash Functions

Related to the above discussion on secrecy is the need to ensure the integrity of a message and also to authenticate the originator. This may be accomplished by a so-called *digital signature*. This is a unique representative pattern, generated from the message, that indicates that the message has not been tampered with. Extending this idea, the identity of the sender may be incorporated so as to generate an authenticating signature.

Of course, we could use an approach such as that described earlier for computing a checksum for the message. Recall that a checksum is generated in order to check the contents of a message for errors in transmission. If we view the notion of errors as deliberately introduced (rather than random), it is effectively the same problem: computing a unique bit pattern that matches one, and only one, message. When used in the context of securing a communication, such a signature is also termed a *message digest*. A simple checksum is not really suitable as a message digest, though, since simple changes to the data, with knowledge of the checksum algorithm, could be used to create the same checksum for a message that has been altered.

Perhaps the best-known message digest is the MD5 algorithm, as documented in RFC1321 (Rivest, 1992). MD5 produces a 128-bit *message digest* or *fingerprint* of that message. The primary concern in generating a message digest is the uniqueness of the resulting value. This class of algorithms – termed *hashing functions* – is used to produce a digest or *hash*, which is a concept well known in other areas such as database information retrieval. Another hash or digest algorithm is *Secure Hash Algorithm* (SHA), documented in the Secure Hash Standard (NIST, 2015).

Signature or hash functions are also used in challenge-response systems. Consider the problem of transmitting a password over a possibly insecure link. Certainly one would wish to encrypt the password, but that may be vulnerable to brute-force or dictionary-guessing attacks. Even if encrypted, replaying of the same encrypted password would result in a compromised system – without the password itself being decrypted. Such *replay attacks* need to be thwarted somehow.

One approach to address these problems is – surprisingly – to dispense with the requirement to transmit the password at all. In this scenario, when a client wishes to access a resource, the server issues a "challenge" string, which is usually a bitstream based on a pseudorandom number generation. This challenge is sent to the client, who must produce a response based on a hash function whose input is the challenge string together with the known authentication credentials (such as a password or PIN). Since the challenge is random, it is not reused. Importantly, the password or credentials themselves are never actually transmitted; only the computed hash is ever sent. A challenge-response handshake can be employed in HTTP web requests using the *digest authorization* mechanism so that passwords are never sent as cleartext (Fielding et al., 1999; Franks et al., 1999).

6.6.5 Public-key Encryption

If encryption was a one-way function, such that the message hash could not be decrypted, then no one would be able to decrypt the message – including the intended recipient. This is hardly a useful scenario but serves to highlight the fundamental insight of public-key encryption: that only the intended recipient can decode the message. In public-key encryption, there is not one key for encryption and decryption, but two keys: one for encryption and one for decryption. The encryption key is known publicly, but the decryption key is known only to the receiver. Thus, not even the sender can decrypt their own message. Clearly, the encryption and decryption must be a related pair in some way. Less obvious is the fact that there should be no way to deduce the decryption key given the encryption key and an arbitrary amount of ciphertext and/or corresponding cleartext.

The RSA algorithm solved this seemingly impossible puzzle (Rivest et al., 1978). Named after its inventors (Ron Rivest, Adi Shamir, and Leonard Adleman), this public-key algorithm derives security from the difficulty inherent in factoring large prime numbers (very large – hundreds of digits or longer).

In public-key encryption, the entire message to be sent is split up into smaller blocks M of more manageable proportions – of course, any arbitrary length message can be encrypted by splitting it into smaller chunks. The size of the block M must be smaller than some integer N, where N has certain properties that will be described shortly. The resulting ciphertext block C may then be sent on an insecure channel.

To set up RSA encryption, an encryption key e and matching decryption key d are required. Since the intention is to have a public key for encryption, e is made available to all who would wish to send a message to a certain recipient. The recipient must generate e according to certain rules, keeping the decryption key d secret. The roles of the sender, receiver, and availability of the public and private keys are illustrated in Figure 6.33. The cloud indicates an untrusted channel – that is, we assume that anyone can access data in this domain. Although such interception may be difficult, good security implies that we must assume this worst-case scenario.

To ascertain d and e, it is necessary to choose two large random prime numbers p and q, of about the same length. Recall that a prime number has only itself and unity as factors, and no others. Then, defining $N = pq$, the recipient can choose a public encryption key e such that e and $(p-1)(q-1)$ are *relatively prime*. Defining two numbers as being relatively prime means that they have no factors in common. This condition requires the greatest common divisor (gcd) to satisfy

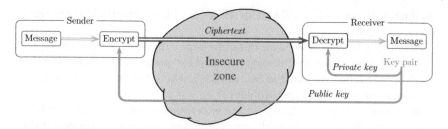

Figure 6.33 Public-key encryption. The "cloud" indicates an insecure channel. This could be a wireless network at the local level, or indeed, the entire Internet.

$$\gcd(e, (p - 1)(q - 1)) = 1 \tag{6.48}$$

The private decryption key d is computed such that

$$ed \bmod (p - 1)(q - 1) = 1 \tag{6.49}$$

Once the encryption and decryption are set up in this way, the sender can proceed to encrypt blocks M of a message into ciphertext C using modulo-exponentiation, similar to that demonstrated earlier for key exchange:

$$C = M^e \bmod N \tag{6.50}$$

Because of the way the encryption parameters were chosen, decryption according to

$$M' = C^d \bmod N \tag{6.51}$$

gives the original message text back. That is, M', the recovered message block, will equal the original message block M. The entire message may then be assembled from all the decrypted blocks.

This may sound complicated, and so a simplified numerical example to illustrate is given. In the setup phase, the following steps are required:

1) Choose $p = 47$ and $q = 79$.
2) Compute $n = pq = 3713$
3) Compute $(p - 1)(q - 1) = 3588$.
4) Choose $e = 37$, so that e and 3588 are relatively prime.
5) Compute d such that $37d \bmod 3588 = 1$. The value $d = 97$ satisfies this condition.

To encrypt a message block M, it is only necessary to compute

$$C = M^{37} \bmod 3713 \tag{6.52}$$

If the first block is $M = 58$, then the corresponding ciphertext is $C = 58^{37} \bmod 3713 = 1671$. To decrypt the message, it is necessary to compute

$$M' = C^{97} \bmod 3713 \tag{6.53}$$

Evaluating this with $C = 1671$ yields $M' = 58$, which is identical to the original message block M.

Given that large numbers are raised to large exponents in both encryption and decryption, there is significant potential for numerical overflow. The issue is identical to that discussed in Section 6.6.3, where numerical solutions were presented to circumvent this problem.

In summary, the steps to set up the encryption–decryption process are as follows: first determine the public encryption key e and private decryption key d using the following steps:

1) Choose large random prime numbers p and q, and find the modulo value $N = pq$.
2) Choose e such that e and $(p − 1)(q − 1)$ are *relatively prime*.
3) Compute d such that $ed \bmod (p − 1)(q − 1) = 1$.

Then, encipherment of an arbitrary message requires splitting it into blocks M of size less than N and finding $C = M^e \bmod N$. Deciphering of the received ciphertext C requires the calculation at the receiver of $M' = C^d \bmod N$, with the result that $M' = M$.

6.6.6 Public-key Authentication

In addition to the abovementioned procedure for *encryption* of a message, the RSA paper showed how essentially the same procedure could be used for *authentication* (Rivest et al., 1978). This is illustrated in Figure 6.34. Comparing to the encryption case, the fundamental change is the location of the key generation. Whereas for encryption the keys are generated by the recipient, for authentication the keys are generated by the sender. The idea is that if a message can be decoded by a public key, then that message must have been encrypted using a matching private key.

Note well that this does *not* address the encryption issue, since the public key is, by definition, public and can be decrypted by anyone. To ensure both confidentiality (via encryption) and guarantee authentication (via the ability to decrypt), then *both* the preceding algorithms must be applied, essentially as independent steps.

6.6.7 Mathematics Underpinning Public-key Encryption

This section aims to provide further insight into the underlying mathematics of the public-key approach described above. This leads to the derivation of the public-key algorithm, an understanding of how and why it works, and in the process touches on some of the issues such as choice of encryption parameters and computational feasibility. For further details and examples, the original paper (Rivest et al., 1978) and references therein is recommended.

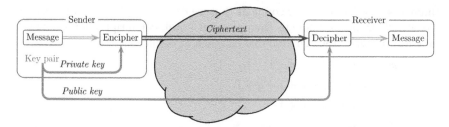

Figure 6.34 Public-key authentication. If the message can be decrypted, only the sender possessing the matching private key could have produced the message. Secrecy is not provided by this stage alone.

One tool needed in public-key cryptography is the ability to determine how many numbers less than some arbitrary number N are prime. We could of course step through all numbers less than N and test each one, in turn, to see if it has any factors. To test each number, say k, we could check all numbers below it – that is, from 2 to $k - 1$. Of course, the highest factor of k would at most be $\sqrt{k}$, so this alone substantially reduces the number of tests required. But remember that *very* large numbers are required for public-key encryption and key exchange, so as to thwart exhaustive search attacks.

Considering the problem from the opposite perspective, we could assume that all numbers are prime, and starting from 2, we could mark off all the multiples of each number. This algorithm dates to antiquity and is termed the Sieve of Eratosthenes. This approach rapidly eliminates a large pool of numbers and does so by multiplication (which is generally simpler) rather than division (which is more complex).

A key result required in the derivation of the encryption–decryption property is the determination of how many numbers less than N are relatively prime to N. This is formally called Euler's totient function $\varphi(N)$. If N is prime, then $\varphi(N) = N - 1$. However, this is a special case. Consider the case where N is the product of two prime numbers. If we select, say, $p = 3$ and $q = 7$, then $N = pq = 21$, then the multiples of 7 are crossed off as

1 2 3 4 5 6 **7** 8 9 10 11 12 13 **14** 15 16 17 18 19 20 $\boxed{21}$

Similarly, the multiples of 3 are crossed off as

1 2 **3** 4 5 **6** 7 8 **9** 10 11 **12** 13 14 **15** 16 17 **18** 19 20 $\boxed{21}$

so we have left

1 2 **3** 4 5 **6** **7** 8 **9** 10 11 **12** 13 **14** **15** 16 17 **18** 19 20 $\boxed{21}$

We immediately see that we have 6 ($= q - 1$) multiples of 3 and 2 ($= p - 1$) multiples of 7. So we conclude that the number of numbers less than 21 that are relatively prime to 21 is found as follows. First, start with $N - 1 = 21 - 1 = 20$

numbers, because we don't count 21 (but we do count 1, because that is how "relatively prime" is defined). Then cross off 7 and its multiples, followed by 3 and its multiples. The equation that can then be deduced is

$$\begin{aligned}
\varphi(N) &= (N-1) - (p-1) - (q-1) \\
&= pq - 1 - p + 1 - q + 1 \\
&= pq - p - q + 1 \\
&= (p-1)(q-1)
\end{aligned} \tag{6.54}$$

So in this case, there are $(3-1)(7-1) = 12$ numbers less than 21 that are not boxed.

Both DH key exchange and RSA public-key encryption require the raising of a number to a power using modulo arithmetic. Suppose, for example, $N = 12$, then numbers less than 12 that are relatively prime to 12 are $1, 5, 7, 11$, and thus $\varphi(12) = 4$. Given a number $a = 5$, which is relatively prime to 12, then multiplying each relatively prime number in the set by a and taking the result modulo 12 gives the set of results

$$\begin{aligned}
1a &\quad \text{mod } 12 = 5 \\
5a &\quad \text{mod } 12 = 1 \\
7a &\quad \text{mod } 12 = 11 \\
11a &\quad \text{mod } 12 = 7
\end{aligned}$$

Note that the resulting remainders on the right-hand side after the modulo operation range through all the same remainders. Multiplying the left-hand side out, and the right-hand side out, and equating,

$$\begin{aligned}
1a \cdot 5a \cdot 7a \cdot 11a &\quad \text{mod } 12 = 5 \cdot 1 \cdot 11 \cdot 7 \text{ mod } 12 \\
a^4 &\quad \text{mod } 12 = 1
\end{aligned}$$

The exponent 4 is actually $\varphi(N)$. The critical result here is that the right-hand side is always unity. This in turn is important in proving that decryption will always succeed.

To generalize this result, consider a number a multiplied by one of the remainders, which will be an integer k times the modulo N, plus a different remainder. That is,

$$ar_i = kN + r_j \tag{6.55}$$

or, using modulo notation,

$$ar_i \mod N = r_j \tag{6.56}$$

As with the numerical example, we take the product of all the left-hand terms and equate to the product of all the right-hand terms:

$$\prod_{i \,\in\, \text{relative primes}} ar_i \mod N = \prod_{j \,\in\, \text{relative primes}} r_j \mod N \tag{6.57}$$

$$\therefore \quad a^{\varphi(N)} \prod r_i \mod N = \prod r_j \mod N \qquad (6.58)$$

$$a^{\varphi(N)} \mod N = 1 \qquad (6.59)$$

This is a key result in the working of the RSA public-key algorithm: any number a raised to the power of the totient function, modulo N, will always be one.

Now we return to the public-key encryption and prove that decryption always yields the original message. To recap, for encryption we had

$$C = M^e \mod N \qquad (6.60)$$

and to decrypt

$$M' = C^d \mod N \qquad (6.61)$$

To set up the system, we choose two large random prime numbers p and q and set

$$N = pq \qquad (6.62)$$

The number of numbers less than N that are relatively prime to N (Euler's totient function) is

$$\varphi(N) = (p-1)(q-1) \qquad (6.63)$$

The public encryption key e is chosen such that e and $\varphi(n)$ are *relatively prime*. The private decryption key d is computed such that

$$ed \mod \varphi(N) = 1 \qquad (6.64)$$

Substituting the ciphertext C into the decryption equation gives

$$M' = C^d \mod N$$
$$= M^{ed} \mod N \qquad (6.65)$$

Because of the choice of ed, that

$$ed \mod \varphi(N) = 1 \qquad (6.66)$$

then modulo-arithmetic says that

$$k\,\varphi(n) + 1 = ed \qquad (6.67)$$

where k is an integer. So the decrypted message M' is

$$M' = M^{k\varphi(N)+1} \mod N \qquad (6.68)$$

Using Euler's theorem (totient function), for any a which is relatively prime to N,

$$a^{\varphi(N)} \mod N = 1 \qquad (6.69)$$

So the recovered message is

$$M' = M^{k\varphi(N)+1} \bmod N$$
$$= M^{k\varphi(N)} \cdot M^1 \bmod N$$
$$= (M^{\varphi(N)})^k \cdot M \bmod N$$
$$= 1^k \cdot M \bmod N$$
$$= M \bmod N$$
$$= M \tag{6.70}$$

As a result, the decrypted message M' will always equal the original message M.

6.7 Chapter Summary

The following are the key elements covered in this chapter:

- Error detection, using checksums and CRCs.
- Error correction, using the Hamming code and convolutional codes.
- Secret-key encryption methods.
- Key-exchange methods.
- Public-key encryption methods.

Problems

6.1 Explain the difference between *error detection* and *error correction*. In relation to these concepts:
   a) Define the terms *error distance* and *Hamming distance*.
   b) Explain how two-dimensional parity could be used to correct errors.
   c) To detect d errors, what Hamming distance is required in the codewords? Explain how this comes about.
   d) To correct d errors, what Hamming distance is required in the codewords? Explain how this comes about.

6.2 Hamming codes are able to detect, and in many cases correct, errors.
   a) Derive the Boolean check bit generating equations for a Hamming (7,4) error-correcting code.
   b) Calculate the check bits for the 4-bit message block 1110.
   c) Calculate the syndrome bits if the message is correctly received.
   d) Calculate the syndrome bits if the message is erroneously received as 1111. Does the syndrome correctly identify the error?
   e) Calculate the syndrome bits if the message is erroneously received as 1101. Does the syndrome correctly identify the error?

6.3 Create a table similar to Table 6.4 and show that a Hamming $(15, 11)$ code is feasible.

6.4 With reference to error-detecting codes:
   a) Explain what is meant by a checksum. What practical systems use a checksum for error checking?
   b) Explain what is meant by a CRC. What practical systems use a CRC for error checking?

6.5 Using the CRC generator 1011 and message 1010 1110,
   a) Calculate the CRC remainder if there are no errors. Check your CRC by replacing it and performing the CRC division process a second time to show that the remainder is all zero.
   b) If there is a 3-bit error burst 111 starting at the third bit transmitted, show that the bit sequence becomes 1001 0110 101. Then show that the error is detected.
   c) If there is an error burst that happens to be identical to the generator, starting at bit position 4, show that the bit sequence becomes 1011 1000 101. Then show that this error is not detected.

6.6 Figure 6.35 shows a simple trellis with starting node A and ending H. Each intermediate node is labeled and is assumed that only the two middle blocks $t_1 \rightarrow t_2$ and $t_2 \rightarrow t_3$ are fully interconnected. With this simple topology:
   a) By tracing all possible paths, determine the number of possible paths and the cost of each.
   b) If another intermediate stage was added, how many possible paths would there be? If two intermediate stages were added, how many possible paths? Does this justify the statement that the number of possible paths goes up exponentially?
   c) Verify that, at each stage, the cumulative path costs $\begin{pmatrix} C_u \\ C_l \end{pmatrix}$ shown as calculated according to the Viterbi algorithm are correct.

6.7 For the trellis and associated parameters described in Section 6.5.4, use the MATLAB code given to verify that the correct state sequence and bit sequence is produced. Note that the trellis shown in Figure 6.27 uses 0 as the starting state index, whereas the MATLAB code assumes states start at 1, with 0 reserved for "unknown state."
   What happens if an incorrect codeword of 111 is received at time step t_1?

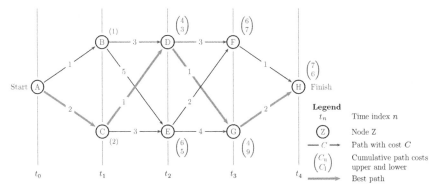

Figure 6.35 Tracing the best path through a trellis, as required for convolutional codes.

6.8 With reference to data security, explain the following terms:
 a) Confidentiality
 b) Integrity
 c) Authentication
 d) Brute-force attack
 e) Encryption key
 f) Public-key encryption
 g) A hash function
 h) A digital signature

6.9 Key exchange using the Diffie–Hellman–Merkle approach, as well as public-key encryption employing the Rivest–Shamir–Adleman algorithm, require exponents of large numbers to be computed. This creates numerical overflow problems if straightforward calculation approaches are implemented.
 a) Use MATLAB to calculate 11^{15}, and by taking the logarithm of this result to base 2, estimate how many bits are required for precise numerical representation. Repeat for the quantity 13^{15}.
 b) Use MATLAB's inbuilt mod function to calculate 11^{15} mod 23. Then use the expmod function developed in this chapter.
 c) Use MATLAB's inbuilt mod function to calculate 13^{15} mod 23. Then use the expmod function developed in this chapter.
 d) Explain the discrepancies observed in each of these cases.

6.10 With reference to the storage and transmission of passwords:
 a) Why is it advantageous to store only encrypted passwords, and how could this be achieved in practice?
 b) Explain the steps involved in the challenge-handshake authentication protocol. What problem does it solve?

6.11 With reference to encryption schemes:
   a) How could a one-stage XOR function perform encryption, and what would be the shortcomings of such a method?
   b) Prove mathematically that the XOR function can be used to decrypt a message that has been encrypted with a simple XOR operation.
   c) How could public-key encryption improve upon this? Draw a block diagram of such a system, defining where the keys are generated, what is transmitted, what is made public, and what is kept secret.

References

Acharya, T. and Ray, A.K. (2005). *Image Processing – Principles and Applications*. New York: Wiley.

Ahmed, N., Natarajan, T., and Rao, K.R. (1974). Discrete cosine transform. *IEEE Transactions on Computers* C-23 (1): 90–93. doi: 10.1109/T-C.1974.223784.

Aho, A.V., Hopcroft, J.E., and Ullman, J.D. (1987). Chapter 6.3. The single source shortest paths problem. In: *Data Structures and Algorithms*. Reading, MA: Addison-Wesley.

Allman, M., Paxson, V., and Blanton, E. (2009). TCP congestion control. https://www.rfc-editor.org/info/rfc5681. DOI 10.17487/RFC5681 (accessed 29 March 2018).

Allman, M., Paxson, V., and Stevens, W. (1999). TCP congestion control. https://www.rfc-editor.org/info/rfc2581. DOI 10.17487/RFC2581 (accessed 29 March 2018).

Armstrong, E.H. (1921). A new system of short wave amplification. *Proceedings of the Institute of Radio Engineers* 9 (1): 3–11.

Armstrong, E.H. (1936). A method of reducing disturbances in radio signaling by a system of frequency modulation. *Proceedings of the Institute of Radio Engineers* 24 (5): 689–740. doi: 10.1109/JRPROC.1936.227383.

Baldwin, R. and Rivest, R. (1996). The RC5, RC5-CBC, RC5-CBC-Pad, and RC5-CTS Algorithms. https://www.rfc-editor.org/info/rfc2040. DOI 10.17487/RFC2040 (accessed 29 March 2018).

Barclay, L.W. (1995). Radiowave propagation – the basis of radiocommunication. In: *Proceedings of the 1995 International Conference on 100 Years of Radio*, 89–94. London: Institution of Electrical Engineers. doi: 10.1049/cp:19950796.

Barclay, L. (2003). *Propagation of Radiowaves*, 2e. London: Institution of Electrical Engineers.

Barnoski, M.K. and Jensen, S.M. (1976). Fiber waveguides: a novel technique for investigating attenuation characteristics. *Applied Optics* 15 (9): 2112–2115. doi: 10.1364/AO.15.002112.

Barnoski, M.K., Rourke, M.D., Jensen, S.M., and Melville, R.T. (1977). Optical time domain reflectometer. *Applied Optics* 16 (9): 2375–2379. doi: 10.1364/AO.16.002375.

Belshe, M., Peon, R., and Thomson, M. eds. (2015). Hypertext transfer protocol – HTTP/2. https://www. rfc-editor.org/info/rfc7540. DOI 10.17487/RFC7540 (accessed 29 March 2018).

Bennett, W.R. (1984). Secret telephony as a historical example of spread-spectrum communications. *IEEE Transactions on Communications* COM-31 (1): 98–104. doi: 10.1109/TCOM.1983.1095724.

Bennett, J., Partridge, C., and Shectman, N. (1999). Packet reordering is not pathological network behavior. *IEEE/ACM Transactions on Networking* 7 (6): 789–798. doi: 10.1109/90.811445.

Berners-Lee, T., Fielding, R., and Frystyk, H. (1996). Hypertext transfer protocol – HTTP/1.0. https://www.rfc-editor.org/info/rfc1945. DOI 10.17487/RFC1945 (accessed 29 March 2018).

Braden, R., Borman, D., and Partridge, C. (1988). Computing the Internet checksum. https://www.rfc-editor.org/info/rfc1071. DOI 10.17487/RFC1071 (accessed 29 March 2018).

Brandenburg, K. (1999). MP3 and AAC explained. http://www.aes.org/e-lib/ browse.cfm?elib=8079 (accessed 29 March 2018).

Burrows, M. and Wheeler, D.J. (1994). A block-sorting lossless data compression algorithm. HP Labs Technical Reports. http://www.hpl.hp.com/techreports/ Compaq-DEC/SRC-RR-124.html (accessed 29 March 2018).

Carrel, R. (1961). The design of log-periodic dipole antennas. *1958 IRE International Convention Record* 9: 61–75. doi: 10.1109/IRECON.1961. 1151016.

Carson, J.R. (1922). Notes on the theory of modulation. *Proceedings of the Institute of Radio Engineers* 10 (1): 57–64. doi: 10.1109/JRPROC.1922.219793.

Chao, H.J. (2002). Next generation routers. *Proceedings of the IEEE* 90 (9): 1518–1558. doi: 10.1109/JPROC.2002.802001.

Cooley, J.W. and Tukey, J.W. (1965). An algorithm for the machine calculation of complex Fourier series. *Mathematics of Computation* 19: 297–301.

Cooper, G.R. and Nettleton, R.W. (1978). A spread-spectrum technique for high-capacity mobile communications. *IEEE Transactions on Vehicular Technology* 27 (4): 264–275. doi: 10.1109/TVT.1978.23758.

Costas, J.P. (1956). Synchronous communications. *Proceedings of the IRE* 44 (12): 1713–1718. doi: 10.1109/JRPROC.1956.275063.

Cotton, M., Eggert, L., Touch, J. et al. (2011). Internet Assigned Numbers Authority (IANA) procedures for the management of the service name and transport protocol port number registry. https://www.rfc-editor.org/info/ rfc6335. DOI 10.17487/RFC6335 (accessed 29 March 2018).

Crypto Museum (30 October 2016). SIGSALY. http://www.cryptomuseum.com/ crypto/usa/sigsaly/index.htm (accessed 29 March 2018).

Delp, E. and Mitchell, O. (1979). Image compression using block truncation coding. *IEEE Transactions on Communications* 27 (9): 1335–1342. doi: 10.1109/TCOM.1979.1094560.

Diffie, W. and Hellman, M.E. (1976). New directions in cryptography. *IEEE Transactions on Information Theory* IT-22 (7): 644–654. doi: 10.1109/TIT.1976.1055638.

DuHamel, R. and Isbell, D. (1957). Broadband logarithmically periodic antenna structures. In: *1958 IRE International Convention Record*, Vol. 5, 119–128. New York: IEEE. DOI 0.1109/IRECON.1957.1150566.

Elias, P. (1954). Error-free coding. *Transactions of the IRE Professional Group on Information Theory* 4 (4): 29–37. doi: 10.1109/TIT.1954.1057464.

Farrell, D., Oakley, A., and Lyons, R. (2005). Discrete-time quadrature FM detection. *IEEE Signal Processing Magazine* 22 (5): 145–149. doi: 10.1109/MSP.2005.1511836.

Fielding, R., Gettys, J., Mogul, J. et al. (1999). Hypertext transfer protocol – HTTP/1.1. https://www.rfc-editor.org/info/rfc2616. DOI 10.17487/RFC2616 (accessed 29 March 2018).

Fletcher, J.G. (1982). An arithmetic checksum for serial transmissions. *IEEE Transactions on Communications* COM-30 (1): 247–252. doi: 10.1109/TCOM.1982.1095369.

Floyd, S. and Jacobson, V. (1994). The synchronization of periodic routing messages. *IEEE/ACM Transactions on Networking* 2 (2): 122–136. doi: 10.1109/90.298431.

Forster, R. (2000). Manchester encoding: opposing definitions resolved. *Bell System Technical Journal* 9 (6): 278–280. doi: 10.1049/esej:20000609.

Franks, J., Hallam-Baker, P., Hostetler, J. et al. (1999). HTTP authentication: basic and digest access authentication. https://www.rfc-editor.org/info/rfc2617. DOI 10.17487/RFC2617 (accessed 29 March 2018).

Fredkin, E. (1960). Trie memory. *Communications of the ACM* 3 (9): 490–499. doi: 10.1145/367390.367400.

Frerking, M.E. (2003). *Digital Signal Processing in Communication Systems*, 9e. New York: Springer.

Friis, H.T. (1944). Noise figures of radio receivers. *Proceedings of the IRE* 32 (7): 419–422. doi: 10.1109/JRPROC.1944.232049.

Fuller, V. and Li, T. (2006). Classless inter-domain routing (CIDR): the internet address assignment and aggregation plan. https://www.rfc-editor.org/info/rfc4632. DOI 10.17487/RFC4632 (accessed 29 March 2018).

Gallager, R.G. (1978). Variations on a theme by huffman. *IEEE Transactions on Information Theory* IT-24 (6): 668–674. doi: 10.1109/TIT.1978.1055959.

Gast, M.S. (2002). *802.11 Wireless Networks – The Definitive Guide*. Sebastopol, CA: O'Reilly.

Giancoli, D.C. (1984). *General Physics*. Englewood Cliffs, NJ: Prentice Hall.

Gupta, P. (2000). Algorithms for routing lookup and packet classification. PhD thesis. Stanford University.

Guru, B.S. and Hiziroğlu, H.R. (1998). *Electromagnetic Field Theory*. Boston: PWS.

Hall, E.A. (2000). *Internet Core Protocols*. Sebastopol, CA: O'Reilly.

Hamming, R.W. (1950). Error detecting and error correcting codes. *The Bell System Technical Journal* 29 (2): 147–160. doi: 10.1002/j.1538-7305. 1950.tb00463.x.

Hartley, R.V.L. (1923). Relations of carrier and side-bands in radio transmission. *Proceedings of the IRE* 1 (1): 34–56. doi: 10.1109/JRPROC.1923.219862.

Hartley, R.V.L. (1928). Transmission of information. *Bell System Technical Journal* 7 (3): 535–563. doi: 10.1002/j.1538-7305.1928.tb1236.

Haykin, S. and Moher, M. (2009). *Communications Systems*, 5e. Hoboken, NJ: Wiley.

Hecht, J. (2004). *City of Light – The Story of Fiber Optics*. Oxford, UK: Oxford University Press.

Hecht, J. (2010). *Beam – The Race to Make the Laser*. Oxford, UK: Oxford University Press.

Hecht, J. (n.d.). Fiber optic history. http://www.jeffhecht.com/history.html (accessed 29 March 2018).

Hedrick, C. (1988). Routing information protocol. https://www.rfc-editor.org/info/rfc1058. DOI 10.17487/RFC1058 (accessed 29 March 2018).

Henry, P. (1985). Introduction to lightwave transmission. *IEEE Communications Magazine* 23 (5): 12–16. doi: 10.1109/MCOM.1985.1092575.

Huffman, D.A. (1952). A method for the construction of minimum-redundancy codes. *Proceedings of the IRE* 40 (9): 1098–1101. doi: 10.1109/JRPROC. 1952.273898.

IANA (2002). Special-use IPv4 addresses. https://www.rfc-editor.org/info/rfc3330. DOI 10.17487/RFC3330 (accessed 29 March 2018).

IANA (n.d.). Internet assigned numbers authority. http://www.iana.org/ (accessed 29 March 2018).

IEC (2014). *Safety of Laser Products, Standard IEC 60825*. International Electrotechnical Commission (IEC) https://webstore.iec.ch/home (accessed 29 March 2018.

IEEE (1997a). *IEEE Standard Definitions of Terms for Radio Wave Propagation, Standard IEEE Std 211-1997*. Piscataway, NJ: Institution of Electrical and Electronics Engineers.

IEEE (1997b). *IEEE Standard Letter Designations for Radar-Frequency Bands, Standard IEEE Std 521-2002*. Piscataway, NJ: Institution of Electrical and Electronics Engineers.

IEEE (2012). *IEEE Standard for Information Technology – Telecommunications and Information Exchange between Systems Local and Metropolitan Area Networks – Specific requirements Part 11: Wireless LAN Medium Access Control (MAC) and Physical Layer (PHY) Specifications, Standard 802.11*. Piscataway,

NJ: Institution of Electrical and Electronic Engineers http://standards.ieee.org/ about/get/ (accessed 29 March 2018.

IEEE (2013). *IEEE Standard for Definitions of Terms for Antennas, Standard IEEE Std 145-2013*. Piscataway, NJ: Institution of Electrical and Electronics Engineers.

Isbell, D. (1960). Log periodic dipole arrays. *IRE Transactions on Antennas and Propagation* 8 (3): 260–267. doi: 10.1109/TAP.1960.1144848.

ISO (2009). *Optics and Photonics – Spectral Bands, Standard ISO 20473*. International Organization for Standardization http://www.iso.org/iso/ catalogue_detail.htm?csnumber=39482 (accessed 29 March 2018.

ITU (n.d.). *Radiowave Propagation, Standard ITU P Series*. International Telecommunication Union https://www.itu.int/rec/R-REC-P (accessed 29 March 2018.

ITU-R (n.d.). ITU radiocommunication sector. http://www.itu.int/ITU-R/index .asp (accessed 29 March 2018).

Jacobson, V. (1988). Congestion avoidance and control. *ACM SIGCOMM Computer Communication Review* 18 (4): 314–329. doi: 10.1145/52325.52356.

Jayant, N.S. and Noll, P. (1990). *Digital Coding of Waveforms: Principles and Applications to Speech and Video*. Englewood Cliffs, NJ: Prentice Hall Professional Technical Reference.

Johnson, J.B. (1928). Thermal agitation of electricity in conductors. *Physical Review* 32: 97–109. doi: 10.1103/PhysRev.32.97.

Kahn, D. (1984). Cryptology and the origins of spread spectrum. *IEEE Spectrum* 21 (9): 70–80. doi: 10.1109/MSPEC.1984.6370466.

Kao, K.C. and Hockham, G.A. (1966). Dielectric-fibre surface waveguides for optical frequencies. *Proceedings of the Institution of Electrical Engineers* 113 (7): 1151–1158. doi: 10.1049/piee.1966.0189.

Karn, P. and Partridge, C. (1987). Improving round-trip time estimates in reliable transport protocols. *ACM SIGCOMM Computer Communication Review* 17 (4): doi: 10.1145/55483.55484.

Kozierok, C. (2005). *The TCP/IP Guide*. No Starch Press www.tcpipguide.com (accessed 29 March 2018.

Kraus, J.D. (1992). *Electromagnetics*. New York: McGraw-Hill.

LaSorte, N., Barnes, W.J., and Refai, H.H. (2008). The history of orthogonal frequency division multiplexing. IEEE 2009 Global Communications Conference, Honolulu, Hawaii, 1–5.

Lim, H., Kim, H.G., and Yim, C. (2009). IP address lookup for internet routers using balanced binary search with prefix vector. *IEEE Transactions on Communications* 57 (3): 618–621. doi: 10.1109/TCOMM.2009.03.070146.

Lyons, R.G. (2011). *Understanding Digital Signal Processing*, 3e. Upper Saddle River, NJ: Prentice-Hall.

Magill, D.T., Natali, F.D., and Edwards, G.P. (1994). Spread-spectrum technology for commercial applications. *Proceedings of the IEEE* 82 (4): 572–584. doi: 10.1109/5.282243.

Mallory, T. and Kullberg, A. (1990). Incremental updating of the internet checksum. https://www.rfc-editor.org/info/rfc1141. DOI 10.17487/RFC1141 (accessed 29 March 2018).

Malvar, H.S. (1990). Lapped transforms for efficient transform/subband coding. *IEEE Transactions on Acoustics, Speech, and Signal Processing* 38 (6): 969–978. doi: 10.1109/29.56057.

Mathis, M., Mahdavi, J., Floyd, S., and Romanow, A. (1996). TCP selective acknowledgment options. https://www.rfc-editor.org/info/rfc2018. DOI 10.17487/RFC2018 (accessed 29 March 2018).

Maxino, T.C. and Koopman, P.J. (2009). The effectiveness of checksums for embedded control networks. *IEEE Transactions on Dependable and Secure Computing* 6 (1): 59–72. doi: 10.1109/TDSC.2007.70216.

Morrison, D.R. (1968). PATRICIA – practical algorithm to retrieve information coded in alphanumeric. *Journal of the ACM* 15 (4): 514–534. doi: 10.1145/321479.321481.

Moy, J. (1998). OSPF Version 2. https://www.rfc-editor.org/info/rfc2328. DOI 10.17487/RFC2328 (accessed 29 March 2018).

Narashima, M.J. and Peterson, A.M. (1978). On the computation of the discrete cosine transform. *IEEE Signal Processing Magazine* COM-26 (6): 934–936. doi: 10.1109/TCOM.1978.1094144.

Narten, T., Huston, G., and Roberts, L. (2011). IPv6 address assignment to end sites. https://www.rfc-editor.org/info/rfc6177. DOI 10.17487/RFC6177 (accessed 29 March 2018).

NASA (n.d.). *What Wavelength Goes With a Color?* National Aeronautics and Space Administration https://web.archive.org/web/20110720105431/http://science-edu.larc.nasa.gov/EDDOCS/Wavelengths_for_Colors.html (accessed 29 March 2018.

NIST (1999). *Data Encryption Standard (DES), Federal Information Processing Standards (withdrawn) FIPS 46-3.* United States National Institute of Science and Technology https://beta.csrc.nist.gov/publications.

NIST (2001). *Advanced Encryption Standard (AES), Standard FIPS 197.* United States National Institute of Science and Technology https://beta.csrc.nist.gov/publications (accessed 29 March 2018.

NIST (2015). *Secure Hash Standard (shs), Standard FIPS 180-4.* United States National Institute of Science and Technology https://beta.csrc.nist.gov/publications (accessed 29 March 2018.

NIST (2017). *NIST Special Publication 800-63b Digital Identity Guidelines, Draft Standard Special Publication 800-63B.* United States National Institute of Science and Technology https://beta.csrc.nist.gov/publications (accessed 29 March 2018.

Nyquist, H. (1924a). Certain factors affecting telegraph speed. *Journal of the American Institute of Electrical Engineers* 43 (2): 124–130. doi: 10.1002/j.1538-7305.1924.tb1361.x.

Nyquist, H. (1924b). Certain topics in telegraph transmission theory. *Bell System Technical Journal* 3 (2): 324–346. doi: 10.1109/T-AIEE.1928.5055024.

Nyquist, H. (1928). Thermal agitation of electric charge in conductors. *Physical Review* 32: 110–113. doi: 10.1103/PhysRev.32.110.

Oberhumer, M.F. (n.d.). LZO. www.oberhumer.com (accessed 29 March 2018).

Painter, T. and Spanias, A. (1999). A review of algorithms for perceptual coding of digital audio signals, DSP97. DOI 10.1109/ICDSP.1997.628010.

Painter, T. and Spanias, A. (2000). Perceptual coding of digital audio. *Proceedings of the IEEE* 88 (4): 451–515. doi: 10.1109/5.842996.

Paschotta, R. (2008). *The Encyclopedia of Laser Physics and Technology*. RP Photonics Consulting GmbH https://www.rp-photonics.com/encyclopedia .html (accessed 29 March 2018.

Paxson, V., Allman, M., Chu, J., and Sargent, M. (2011). Computing TCP's retransmission timer. https://www.rfc-editor.org/info/rfc6298. DOI 10.17487/RFC6298 (accessed 29 March 2018).

Personick, S.D. (1977). Photon probe – an optical-fiber time-domain reflectometer. *The Bell System Technical Journal* 56 (3): 355–366. doi: 10.1002/j.1538-7305.1977.tb00513.x.

Postel, J. ed. (1981). Transmission control protocol. https://www.rfc-editor.org/info/rfc793. DOI 10.17487/RFC0793 (accessed 29 March 2018).

Postel, J. ed. (1991). Internet protocol. https://www.rfc-editor.org/info/rfc791. DOI 10.17487/RFC0791 (accessed 29 March 2018).

Pozar, D.M. (1997). Beam transmission of ultra short waves: an introduction to the classic paper by H. Yagi. *Proceedings of the IEEE* 85 (11): 1857–1863. doi: 10.1109/JPROC.1997.649661.

Price, R. (1983). Further notes and anecdotes on spread-spectrum origins. *IEEE Transactions on Communications* 31 (1): 85–97. doi: 10.1109/TCOM.1983.1095725.

Frana, P.L. and Misa, T.J. (2010). An interview with Edsger W. Dijkstra. *Communications of the ACM* 53 (8): 41–47. doi: 10.1145/1787234.1787249.

Princen, J.P., Johnson, A.W., and Bradley, A.B. (1987). Subband/transform coding using filter bank designs based on time domain aliasing cancellation. *International Conference on Acoustics, Speech, and Signal Processing (ICASSP)*, Dallas, TX, 2161–2164, New York: IEEE. DOI 10.1109/ICASSP.1987.1169405.

Razavi, B. (1998). *RF Microelectronics*. Upper Saddle River, NJ: Prentice Hall.

Rivest, R. (1992). The MD5 message-digest algorithm. https://www.rfc-editor.org/info/rfc1321. DOI 10.17487/RFC1321 (accessed 29 March 2018).

Rivest, R.L. (1994). The RC5 encryption algorithm. *Proceedings of the 1994 Leuven Workshop on Fast Software Encryption*, Leuven, Belgium (December 1994),

86–96. Berlin: Springer. http://people.csail.mit.edu/rivest/pubs.html (accessed 29 March 2018).

Rivest, R.L., Shamir, A., and Adleman, L. (1978). A method for obtaining digital signatures and public-key cryptosystems. *Communications of the ACM* 21 (2): 120–126. doi: 10.1145/359340.359342.

Scholtz, R.A. (1982). The origins of spread-spectrum communications. *IEEE Transactions on Communications* COM-30 (5): 822–854. doi: 10.1109/TCOM.1982.1095547.

Schroeder, M.R., Atal, B.S., and Hall, J.L. (1979). Optimizing digital speech coders by exploiting masking properties of the human ear. *The Journal of the Acoustical Society of America* 66 (6): 1647–1652. doi: 10.1121/1.383662.

Sedgewick, R. (1990). Chapter 17. Radix searching. In: *Algorithms in C*. Reading, MA: Addison-Wesley.

Shannon, C.E. (1948). A mathematical theory of communication. *Bell System Technical Journal* 27 (3): 379–423. doi: 10.1002/j.1538-7305.1948.tb1338.x.

Sklower, K. (1993). *A Tree-Based Routing Table for Berkeley Unix*. Technical Report. Berkeley: University of California.

Srisuresh, P. and Holdrege, M. (1999). IP network address translator (NAT) terminology and considerations. https://www.rfc-editor.org/info/rfc2663. DOI 10.17487/RFC2663 (accessed 29 March 2018).

Stevens, W.R. (1994). *TCP/IP Illustrated, Volume 1. The Protocols*. Boston, MA: Addison-Wesley.

The Fiber Optic Association (n.d.). Guide to fiber optics & premises cabling. http://www.thefoa.org/tech/ref/basic/fiber.html (accessed 29 March 2018).

Tierney, J., Rader, C., and Gold, B. (1971). A digital frequency synthesizer. *IEEE Transactions on Audio and Electroacoustics* 19 (1): 48–57. doi: 10.1109/TAU.1971.1162151.

Ueno, Y. and Shimizu, M. (1976). Optical fiber fault location method. *Applied Optics* 15 (6): 1385–1388. doi: 10.1364/AO.15.001385.

van der Pol, B. (1946). The fundamental principles of frequency modulation. *Electrical Engineers – Part III: Radio and Communication Engineering, Journal of the Institution of* 93 (23): 153–158. doi: 10.1049/ji-3-2.1946.0024.

van Etten, W.C. (2006). *Appendix F: The Q(_) and erfc(_) Functions*, 243–244. Hoboken, NJ: Wiley. doi: 10.1002/0470024135.app6.

Viterbi, A. (1967). Error bounds for convolutional codes and an asymptotically optimum decoding algorithm. *IEEE Transactions on Information Theory* 13 (2): 260–269. doi: 10.1109/TIT.1967.1054010.

Waldvogel, M., Varghese, G., Turner, J., and Plattner, B. (1997). Scalable high speed. IP routing lookups. *Proceedings of the ACM SIGCOMM'97 Conference on Applications, Technologies, Architectures, and Protocols for Computer Communication, Cannes, France (September 1997)* 27 (4): 25–36. doi: 10.1145/263105.263136.

Weaver, D. Jr., (1956). A third method of generation and detection of single-sideband signals. *Proceedings of the IRE* 44 (12): 1703–1705. doi: 10.1109/JRPROC.1956.275061.

Weinstein, S.B. (2009). The history of orthogonal frequency-division multiplexing. *IEEE Communications Magazine* 47 (11): 26–35. doi: 10.1109/MCOM.2009.5307460.

Weinstein, S.B. and Ebert, P. (1971). Data transmission by frequency-division multiplexing using the discrete Fourier transform. *IEEE Transactions on Communications Technology* 19 (5): 628–634. doi: 10.1109/TCOM.1971. 1090705.

Weisstein, E.W. (2004). Sequence a00797. On-Line Encyclopedia of Integer Sequences. https://oeis.org/A091704 (accessed 29 March 2018).

Welch, T. (1984). A technique for high-performance data compression. *IEEE Computer* 17 (6): 8–19. doi: 10.1109/MC.1984.1659158.

Woodward, P.M. and Davies, I.L. (1952). Information theory and inverse probability in telecommunication. *Proceedings of the IEE* 99 (58): 37–43. doi: 10.1049/jiee-2.1952.0023.

Wright, G.R. and Stevens, W.R. (1995a). *TCP/IP Illustrated, Volume 2. The Implementation*. Boston, MA: Addison-Wesley.

Wright, G.R. and Stevens, W.R. (1995b). Chapter 18. Radix tree routing tables. In: *TCP/IP Illustrated, Volume 2. The Implementation*. Boston, MA: Addison-Wesley.

Yagi, H. (1928). Beam transmission of ultra short waves. *Proceedings of the Institute of Radio Engineers* 16 (6): 715–740. doi: 10.1109/ JRPROC.1928.221464.

Ziv, J. and Lempel, A. (1977). A universal algorithm for sequential data compression. *IEEE Transactions on Information Theory* IT-23 (3): 337–343. doi: 10.1109/TIT.1977.1055714.

Ziv, J. and Lempel, A. (1978). Compression of individual sequences via variable-rate coding. *IEEE Transactions on Information Theory* IT-24 (5): 530–536. doi: 10.1109/TIT.1978.1055934.

Index

a

adaptive prediction   417
address
  classful   294
  Ethernet   283
  hardware   283
  IP   294
  IPv4   286
  IPv6   288
  loopback   294
  MAC   283
  NAT   298
  physical   283
  private   295
  subnetwork   296
Address Resolution Protocol (ARP)
      301
Advanced Encryption Standard (AES)
      512
AM
  frequency analysis   167
  modulation index   165
  power analysis   170
Amplitude Modulation (AM)   164
Amplitude Shift Keying
      (ASK)   226
analysis by synthesis   434
antenna
  array   117
  definitions   105
  dipole   106
  elemental   108
  log periodic   114
  parabolic   115
  Yagi   112
application layer   279
arctangent
  Costas loop   209
  range   222
  unwrapping   222
arp command   301
audio coding   442
  modified DCT   444

b

baseband   50
basis vectors   428
Bellman-Ford optimality   353
Bessel function for FM analysis
  proof   195
  using   189
binary digit   272
binary tree   331
bit   272
bit error rate (BER)   456
block coders   405
block truncation coding   422

c

Carrier Sense Multiple Access with
      Collision Avoidance
      (CSMA/CA)   285

Communication Systems Principles Using MATLAB®, First Edition. John W. Leis.
© 2018 John Wiley & Sons, Inc. Published 2018 by John Wiley & Sons, Inc.
Companion website: www.wiley.com/go/Leis/communications-principles-using-matlab

Carrier Sense Multiple Access with
Collision Detection
(CSMA/CD)   283
challenge-handshake authentication
519
challenge-response protocol   519
channel capacity   369
characteristic impedance
  calculations   75
  definition   73
  reflection coefficient   77
  RG coax   52
  short and open circuit   74
checksum   478
  calculating   292, 481
  Fletcher   481
  IP   479
  IP header   290
cipher algorithm   508
  substitution   510
  transposition   510
ciphertext   508
circuit switching   270
Classless Inter-Domain Routing
(CIDR)   326
cleartext   508
coaxial cable   51
Code Division Multiple Access
(CDMA)   255
color   431
common-mode signal   51
companding   379
companding, μ law/A law   380
complex numbers
  multiplication of   160
  polar and rectangular   159
  use of   159
constellation   239
convolutional encoding   489
  trellis   493
correlate-integrate   228
Costas loop   205
crosstalk   51

cutoff frequency   8
Cyclic Redundancy Check (CRC)   482
  examples   483
  failure to detect error   486
cyclic redundancy check (CRC)
  generator polynomial   483

d
Data Encryption Standard (DES)   512
data packets   271
datagram
  IP   286
  TCP   279
  UDP   279
decibel
  common values   25
  dBm   23
  dBW   23
  definition   23
  SNR   29
  system gain   24
demodulation
  AM using IQ signals   216
  diode   171
  FM using IQ signals   222
  IQ methods   215
  PM using IQ signals   219
  synchronous   173
difference equation   366
Differential PCM (DPCM)   409
differential signal   50
differentiation   11
Diffie-Hellman-Merkle algorithm   513
diffraction   99
digital signature   519
Dijkstra's algorithm   349
diode laser   133
direct conversion receiver   127
Direct Digital Synthesis (DDS)   17
Discrete Cosine Transform (DCT)
425
Domain Name System (DNS)   302
Double Sideband (DSB)   173

downconversion   120
  heterodyne   121
  superheterodyne   122
  TRF   120
Dynamic Host Configuration Protocol
  (DHCP)   302

e

end-around carry   479
entropy   390
entropy of Gaussian   371
ephemeral port   305
Ethernet
  about   281
  address   283
  MAC address   283
  wired   282
  wireless   285
Euclid's algorithm   460
Euler's totient function   524
exclusive-OR (XOR)   470

f

Fast Fourier Transform (FFT)
  example   170
  OFDM signal generation   247
fiber loss calculations   145
filter types
  bandpass   8
  bandstop   8
  highpass   8
  lowpass   8
filterbank   443
Fletcher checksum   481
Fourier series
  coefficients   39
  expansion   39
Fourier transform
  coding   43
  definition   42
  window   42
Frame Check Sequence (FCS)   482
frequency and phase   181

frequency components   38
Frequency Division Multiple Access
  (FDMA)   254
Frequency Division Multiplexing
  (FDM)   243
Frequency Modulation (FM)   180
  Bessel functions   189
  generation   185
  modulation index   188
  spectrum   186
Frequency Shift Keying (FSK)   227

g

gain of a system   4
greatest common factor   460

h

Hamming code   472
  block interleaving   477
  construction   475
  error correction performance   474
  error detection performance   474
  error probability   475
Hamming distance   473
Hartley image rejection   126
Hartley modulator   175
hash   519
heterodyne receiver   121
histogram   364
Homodyne receiver   127
Huffman code   392
HyperText Transfer Protocol (HTTP)
  definition   308
  HTTP 2.0   308

i

image coding   421
image frequency   125
image rejection   126
impulse response   231
instantaneous frequency   181
integration   10
Inter-Symbol Interference (ISI)   58

Intermediate Frequency (IF)   119
  increasing   126
  multiple stages   126
intermodulation distortion   128
Internet Protocol (IP)
  about   286
  addressing   294
  checksum   290
  IPv4 datagram   287
  IPv6 datagram   288
  MTU   286
  probing MTU   359
  version 4   286
  version 6   286
inverse DCT   428
IQ signal definition   215

k
key exchange   512

l
laser diode   133
laser linewidth   136
Lempel-Ziv coding   406
line code   62
linear prediction
  analysis by synthesis   439
  block derivation   434
  code   417
  definition   412
Linear Predictive Coding (LPC)
  435
linear vs nonlinear   5
link layer   279
Local Area Network (LAN)   281
local oscillator   163
lossless coding   390

m
Manchester encoding   63
matched filter   228
MATLAB®
  obtaining   xix

MATLAB
  class   273
    constructor   274
    methods   274
    properties   273
  handle   275
  object   273
    by reference   275
    by value   275
  pass-by-reference   276
  pass-by-value   275
  struct   273
Maximum Segment Size (MSS)   307
Maximum Transmission Unit (MTU)
  286
MD5 algorithm   519
message digest   519
Modified DCT (MDCT)
  code example   446
  definition   444
modulation
  definition   156
modulo arithmetic   458, 512
MP3 coding   443

n
Network Address Translation (NAT)
  298
Network Allocation Vector (NAV)
  285
network layer   279
NIC address   283
noise factor   30
noise figure   30
noise ratio   30
numerical aperture   144
Numerically Controlled Oscillator
  (NCO)   209
Nyquist Law   369

o
optical fiber
  multi-mode   140
  single-mode   140

optical time-domain reflectometry (OTDR) 148
Orthogonal Frequency Division Multiplexing (OFDM) 242
orthogonality
  definition 237
  proof 238

p

packet switching 270
parity 471
patricia trie data structure 332
phase and frequency 181
Phase Locked Loop (PLL) 204
Phase Modulation (PM)
  generation 185
Phase Modulation (PM) 181
Phase Shift Keying (PSK) 226
physical layer 280
plaintext 508
point-to-point link 283
port 305
prediction 410
  adaptive 418
predictive encoding 410
prime numbers 459
private key encryption 509
probability density function 364
protocol
  encapsulation 281
  stack 279
public key authentication 522
public key encryption 520
  decipherment 521
  encipherment 521
  example 521
pulse detection
  correlate-integrate 228
  matched filter 228

q

quadrature
  definition 238

Quadrature Amplitude Modulation (QAM) 239
Quadrature Phase Shift Keying (QPSK) 239
quadtree encoding 431
quantization
  adaptive 383
  definition 364
  Lloyd-Max optimal 382
  noise 378
  nonuniform step size 382
  scalar 373
  vector 385
quantizer
  optimization 378
  signal to noise 379
quantizer types 374

r

raised cosine filter 59
RC5 512
reflection coefficient 77
relatively prime 459, 521
replay attack 519
Request for Comments (RFC) 270
Request to Send/Clear to Send (RTS/CTS) 285
route
  command 323
  default 327
route loop 330
routing
  CIDR 326
  count to infinity problem 346
  definition 322
  distance vector 343
  example 323
  hold-down interval 348
  hop 323
  link state 344
  longest matching prefix 331
  lookup speed 330
  poison-reverse 347

routing (*cont'd*)
  route loop 329
  split-horizon 346
  Time-To-Live (TTL) 324
  triggered update 346
  update 345
Routing Information Protocol
      (RIP) 344

s

scalar quantization 373
scramblers 66
secret key encryption 508
security requirements 507
segment 279
self-synchronizing scrambler 71
Shannon bound
    derivation 371
Shannon-Hartley Law 371
signal-to-noise ratio
        (SNR) 29, 370
Snell's Law 141
socket 299, 306
source coding 389
Spectrum Analyzer 44
speech coding
  linear prediction 434
  noise shaping 440
  noise weighting 440
spread spectrum 254
SSB Demodulation 177
standing wave 84
subnet 296
  address checking 324
  masking 324
subnet mask 296
subnetwork 296
sum and difference frequency 163

t

TCP/IP 309
Time Division Multiple Access
      (TDMA) 254

topology
  bus 282
  star 283
total internal reflection 143
totient function 523
tracert command 323
Transmission Control Protocol (TCP)
  acknowledgment field 315
  acknowledgments 309
  congestion avoidance 317
  congestion control 314
  congestion window 315
  CWND 315
  details 303
  duplicate acknowledgment 317
  fast retransmit 318
  flags 311
  Karn's algorithm 321
  layer in protocol stack 278
  Maximum Segment Size (MSS)
      307
  port 304
  retransmission timer 317
  RFC 309
  round-trip time (RTT) 319
  RTT variance 320
  segment layout 305
  segment within MTU 310
  self-clocked 313
  slow-start 316
  steady-state flow 313
  timeout calculation 319
transport layer 279
traveling wave 87
trellis diagram 493
trie data structure 332
trigonometry formulas 157
Tuned Radio Frequency (TRF) 120
twisted pair 51

u

Unshielded Twisted Pair (UTP) 63
unwrap phase 221

upconversion   120
User Datagram Protocol (UDP)
    details   303
    layer in protocol stack   279
    port   304
    datagram layout   305

V

Vector Quantization (VQ)
    definition   385
    distortion criteria   387
    exhaustive search   387
    training   388

W

Weaver modulator   175
Wide Area Network (WAN)   278
wireless Ethernet   284

X

XOR function   470

Z

z transform   366
zero-IF   128

Printed and bound by CPI Group (UK) Ltd, Croydon, CR0 4YY

16/04/2025

14658417-0003